# Springer Complexity

Springer Complexity is an interdisciplinary program publishing the best research and academic-level teaching on both fundamental and applied aspects of complex systems-cutting across all traditional disciplines of the natural and life sciences, engineering, economics, medicine, neuroscience, social and computer science.

Complex Systems are systems that comprise many interacting parts with the ability to generate a new quality of macroscopic collective behavior the manifestations of which are the spontaneous formation of distinctive temporal, spatial or functional structures. Models of such systems can be successfully mapped onto quite diverse "real-life" situations like the climate, the coherent emission of light from lasers, chemical reaction-diffusion systems, biological cellular networks, the dynamics of stock markets and of the internet, earthquake statistics and prediction, freeway traffic, the human brain, or the formation of opinions in social systems, to name just some of the popular applications.

Although their scope and methodologies overlap somewhat, one can distinguish the following main concepts and tools: self-organization, nonlinear dynamics, synergetics, turbulence, dynamical systems, catastrophes, instabilities, stochastic processes, chaos, graphs and networks, cellular automata, adaptive systems, genetic algorithms and computational intelligence.

The three major book publication platforms of the Springer Complexity program are the monograph series "Understanding Complex Systems" focusing on the various applications of complexity, the "Springer Series in Synergetics", which is devoted to the quantitative theoretical and methodological foundations, and the "Springer Briefs in Complexity" which are concise and topical working reports, case-studies, surveys, essays and lecture notes of relevance to the field. In addition to the books in these two core series, the program also incorporates individual titles ranging from textbooks to major reference works.

# Understanding Complex Systems

**Founding Editor: J.A. Scott Kelso**

Future scientific and technological developments in many fields will necessarily depend upon coming to grips with complex systems. Such systems are complex in both their composition – typically many different kinds of components interacting simultaneously and nonlinearly with each other and their environments on multiple levels – and in the rich diversity of behavior of which they are capable.

The Springer Series in Understanding Complex Systems series (UCS) promotes new strategies and paradigms for understanding and realizing applications of complex systems research in a wide variety of fields and endeavors. UCS is explicitly transdisciplinary. It has three main goals: First, to elaborate the concepts, methods and tools of complex systems at all levels of description and in all scientific fields, especially newly emerging areas within the life, social, behavioral, economic, neuro- and cognitive sciences (and derivatives thereof); second, to encourage novel applications of these ideas in various fields of engineering and computation such as robotics, nano-technology and informatics; third, to provide a single forum within which commonalities and differences in the workings of complex systems may be discerned, hence leading to deeper insight and understanding. UCS will publish monographs, lecture notes and selected edited contributions aimed at communicating new findings to a large multidisciplinary audience.

NECSI

## New England Complex Systems Institute Book Series

### Series Editor

Ali A. Minai

210 Broadway, Suite 101
Cambridge, MA 02139, USA

**New England Complex Systems Institute Book Series**

The world around is full of the wonderful interplay of relationships and emergent behaviors. The beautiful and mysterious way that atoms form biological and social systems inspires us to new efforts in science. As our society becomes more concerned with how people are connected to each other than how they work independently, so science has become interested in the nature of relationships and relatedness. Through relationships elements act together to become systems, and systems achieve function and purpose. The study of complex systems is remarkable in the closeness of basic ideas and practical implications. Advances in our understanding of complex systems give new opportunities for insight in science and improvement of society. This is manifest in the relevance to engineering, medicine, management and education. We devote this book series to the communication of recent advances and reviews of revolutionary ideas and their application to practical concerns.

More information about this series at http://www.springer.com/series/5394

Philip Vos Fellman • Yaneer Bar-Yam
Ali A. Minai
Editors

# Conflict and Complexity

## Countering Terrorism, Insurgency, Ethnic and Regional Violence

Springer

NECSI

*Editors*
Philip Vos Fellman
American Military University
Charles Town, WV, USA

Yaneer Bar-Yam
New England Complex Systems Institute
Cambridge, MA, USA

Ali A. Minai
University of Cincinnati
Cincinnati, OH, USA

This volume is part of the NECSI Studies on Complexity collection.

ISSN 1860-0832 ISSN 1860-0840 (electronic)
ISBN 978-1-4939-1704-4 ISBN 978-1-4939-1705-1 (eBook)
DOI 10.1007/978-1-4939-1705-1
Springer New York Heidelberg Dordrecht London

Library of Congress Control Number: 2014954129



Printed on acid-free paper

Springer is part of Springer Science+Business Media (www.springer.com)

# Foreword

This volume represents a considerable investment of time, energy, and effort (not to mention patience) on the part of the many authors involved, but above all by Yaneer Bar-Yam and Ali Minai, with whom I began discussing the project at the sixth international conference on complex systems in 2006. While it took us a bit of time to actually move the project to completion (approximately 7 years), that time has been used to create a volume that is quite literally the first and only work of its kind. There have been other books which contain pieces of the puzzle of approaching terrorism with the tools of complex adaptive systems research. However, nowhere has there been such a comprehensive volume in which authors from such a wide variety of disciplines (political science, economics, history, sociology, geography, psychology, mathematics, computer science, and physics, to name just the most prominent disciplines drawn upon here) bring their skills to bear on the problems of terrorism, insurgency, battlefield warfare, and ethnic and regional violence.

Above and beyond this diversity, there is a unity of participation across companies, countries, and institutions which really marks this book as a first. Just a few of the participating organizations in this volume, which was organized under the umbrella of the New England Complex Systems Institute (NECSI), include the Institute for Operations Research and Management Science (INFORMS) and its Military Applications Society (MAS), the Virginia Modeling and Simulation Center (VMASC), American Military University (AMU), and its parent organization the American Public University System (APUS) as well as the Metropolitan College of Boston University. Leaders in mathematical modeling, computational design, defense logistics, and military operations research have all come together to create a novel book combining, for the first time, both the "hard" and "soft" approaches of complexity science to take what I can only characterize as a giant step in bringing the full array of complex systems tools to bear on the problems of terrorism, insurgency, irregular warfare, cyber-warfare, and ethnic and regional violence.

I have for many years been an admirer of Dr. Bar-Yam's previous volumes, The Dynamics of Complex Systems, and Making Things Work. I can only hope that this volume, to which Dr. Bar-Yam has made a very substantial contribution, not only in terms of his five chapters which comprise Part III of the book, but also in his unwavering support for the entire project across the 7 years which it has taken to bring all the participants together, will stand equally tall with those volumes in our effort to make things work in combating terrorism, insurgency, irregular conflict, and ethnic and regional violence.

Finally, I would like to thank all the authors for their insightful contributions to this volume, and Michelle Lim for helping immensely with the editing process.

Manchester, NH, USA Philip Vos Fellman

# Preface

This volume presents a complex systems approach to analyzing, modeling, understanding, and combating terrorism and conflict, and is a unique and timely contribution to a topic of critical importance. Much of the effort in this area has used—and continues to use—classical methods based on intelligence, statistical and game-theoretic modeling, and military operations. The need for other methods has become increasingly clear. Recognizing and conflict, at least in part, as a social phenomenon suggests that methods that have succeeded in analyzing other social systems may also work well in this case. This has led to the application of network modeling and analysis to terrorism and conflict. Other complex systems concepts such as chaotic dynamics, self-organization, emergent patterns, and fractals have also been applied, generating important insights. This book reviews and discusses these efforts.

Part I of the book, comprising the first six chapters, focuses on theoretical ideas. Chapter 1 by Fellman is a comprehensive overview of the issues and methods in the study of terrorism, and of complex systems approaches that have been used. It can, therefore, be seen as an extended introduction to the volume. Chapter 2 by Mesjasz discusses the complex systems methods that are relevant to the study of terrorism and provides several specific ways in which they can be applied. Chapter 3 by Mastors focuses on the psychology of terrorism, discussing the motivations and characteristics of terrorists, and the processes that lead to their recruitment into the activity. Chapter 4 by Upal describes an agent-based approach to modeling social identity dynamics, which is crucial to both understanding terrorists' actions and countering their recruitment methods. Chapter 5 by Hartley describes a comprehensive model called DIME/PMESII for analyzing possible counter-terrorism strategies using a state defined in terms of political, military, economic, social, information and infrastructure (PMESII) variables, operated on by diplomatic, information, military and economic (DIME) means. Finally, in Chap. 6, Ray describes how a distributed, complex systems-based logistics approach may thwart terrorists more effectively than a traditional centralized one.

The chapters in Part II of the book describe applications and case studies based on complex systems approaches. Chapter 7 by Passman presents the extension of a traditional model of combat attrition to one based on fractals, which better encompasses all the dimensions of modern warfare. Chapter 8 by Fellman, Clemens, Wright, Post, and Dadmun shows how the decision dynamics of terrorist networks can be understood using a fitness landscape approach. Chapter 9 by Ezell and Parnell provides a detailed comparative analysis of six specific methods used for risk analysis, including logic trees and game-theoretic methods. Chapter 10 by Fellman, Frutos, Thanakijsombat, Teekasap, and Schear uses a quantitative approach to explore whether there is a "learning curve" in the activity of maritime piracy. Chapter 11 by Jacobs, Chitkushev, and Zlateva looks at identity and anonymity in cyber-warfare.

Finally, the five chapters in Part III go well beyond the narrow confines of terrorism to look at conflict in general—often with terrorism as a subtext. Such conflict may arise through direct combat, but also increasingly as a result of socioeconomic and environmental factors. Climate change, water scarcity, overpopulation, ethnic rivalries, religious friction, and political turmoil are all producing conflicts in many parts of the world—notably in Africa and the Middle East. Chapter 12 by Rutherford et al. discusses a theory to predict where ethnic violence may occur, using the former Yugoslavia, India, and Switzerland as examples. Chapter 13 by Lagi, Bertrand, and Bar-Yam considers the role of food prices in producing political instability, identifying a price threshold at which instability may be triggered. Chapter 14 by Bar-Yam, Lagi, and Bar-Yam continues this theme, looking at the role of food prices in triggering unrest in South Africa and considering what it might portend in other places around the world. In Chapter 15, Gros, Gard-Murray, and Bar-Yam look at rioting triggered by food prices in Yemen, a country with an active terrorist presence. Finally, Chapter 16 by Gard-Murray and Bar-Yam looks at the case of the Arab Spring from a complex systems and evolutionary perspective, hypothesizing that there is significant danger of reversion to autocracy—a prediction being borne out by events already.

As a whole, this volume provides the reader with a broad and diverse view of how complex systems thinking is helping in the analysis and mitigation of terrorism, conflict, and upheaval. While this book cannot claim to have included all significant efforts in this area, these chapters and their bibliographies will guide the reader to almost everything of significance.

Cincinnati, OH, USA — Ali A. Minai

# Contents

# Part I
# The Theoretical Background

# Chapter 1
# Modeling Terrorist Networks: The Second Decade

**Philip Vos Fellman**

> *The wrath of the terrorist is rarely uncontrolled. Contrary to both popular belief and media depiction, most terrorism is neither crazed nor capricious. Rather, terrorist attacks are generally as carefully planned as they are premeditated.*
>
> Bruce Hoffman, RAND Corporation

> *The best method to control something is to understand how it works.*
>
> J. Doyne Farmer, Santa Fe Institute

## Introduction: The View from a Decade Farther Forward

The original version of "Modeling Terrorist Networks" was prepared for a NATO conference in 2003.[1] There have been subsequent re-publications, most notably in The Intelligencer,[2] and on the London School of Economics website.[3] In those original versions of the paper, we sought to elucidate how the techniques of nonlinear dynamical systems modeling, combined with first principles of counter-intelligence, could be brought to bear on various problems regarding the structure of terrorist networks and the appropriate methods to counter those groups. Because we worked from first principles, many of the insights presented in that original paper remain true today. However, as we began to develop our approach, we noticed almost immediately that there were several constraints on our method, some simply challenging and others just plain awkward.

These ranged from the habits and prejudices of the audiences to whom we presented, which were themselves quite varied in makeup (on one occasion, military officers, on another academics, on a third intelligence professionals and on other occasions, a highly educated and interested audience, but one without military or

---

[1] Fellman et al. [1].

[2] Fellman and Wright [2].

[3] Fellman and Wright [3].

P.V. Fellman (✉)
American Military University and InterPort Police, Charles Town, WV, USA
e-mail: shirogitsune99@yahoo.com

 
P.V. Fellman et al. (eds.), *Conflict and Complexity*, Understanding Complex Systems,
DOI 10.1007/978-1-4939-1705-1_1

intelligence community experience) to the resource constraints of those whom we were advising to a general lack of hard data outside units specifically tasked with combating terrorism.

By prejudices, we mean the organizational orientation and needs of the group in some cases, and in other cases a predisposition on the part of either individuals or groups to view terrorist organizations in a particular fashion, which might make the presentation of new ideas especially difficult, particularly if the audience were inclined to see the nature of terrorist groups as fundamentally static, fundamentally random or any one of a dozen other "prejudices" (in this sense, audience prejudice is really a reference to cognitive or motivated bias and not to any imagined audience hostility).

As the war in Iraq heated up and we confronted a hostile insurgency, a different dimension of terrorist attacks, particularly the presence of IED and IIED distribution networks among the insurgents made approaching the scientific problem of terrorism more difficult, probably by a full order of magnitude.[4] At the same time, military intelligence professionals were frequently confronted by inadequate intelligence in the field, whether tactical intelligence at the company level or strategic intelligence at the division level.[5]

To go back to our work of a decade ago, while many principles remain the same, the evolution of computational systems and computational power has given us the ability to model many aspects of terrorism with an accuracy and scope impossible 10 years ago. Further, when we started writing about modeling terrorist networks, only Kathleen Carley, her students, and her colleagues were truly using dynamical models (i.e., DNA or dynamic network analysis) even though many of us were engaged in applying the insights of dynamical systems modeling to terrorism.

Over the years, the tools have changed, the techniques have spread and with additional input from a number of gifted authors, but especially from Derek Jones of the U.S. Special Forces Command[6] as well as the Military Applications Society of the Institute for Operations Research and Management Science (among the contributors whose names come to mind, Dean Hartley III, Darryl Ahner, and Gregory Parnell particularly stand out), modelers have been able to develop a new, more useful perspective on the relationship between modeling terrorism and field applications.

Additionally, the modeling techniques themselves have changed, been refined and new techniques have been developed. Kathleen Carley remains, perhaps, the pre-eminent leader in this field, and in my conversations with military personnel and academics (including those teaching military officers) she is now well known for her presentations at DARPA, ONI, ONR, USMA, and other government facilities. Moreover, her techniques have gained a widespread acceptance over the past decade, so that she no longer represents an isolated group in the ivory tower work-

[4] Fellman [4].

[5] Flynn et al. [5].

[6] Jones [6].

ing on arcane solutions to extremely complex problems, but is rather a well-known leader providing inspiration and information (as well as some extremely useful software tools, such as ORA) to the rest of us. Indeed, as Derek Jones indicates, while an open-source solution to terrorist network dynamics may not capture all the relevant dimensions of the problem, it may well be impossible to accurately capture the dynamics of a terrorist network without using the techniques pioneered by Carley et al.

At the same time, the entire field of quantitative modeling of terrorist activities has grown and evolved. In the same fashion as Doyne Farmer has done for the application of statistical physics techniques to modeling economic phenomena, other physicists, perhaps most notably my co-editor Yaneer Bar-Yam, and our colleague, Serge Galam have discovered novel and significant applications of physical systems theory to modeling and countering terrorism, insurgency, regional and ethnic violence. This makes the landscape of knowledge about terrorism (as well as insurgency and regional and ethnic violence) much richer and far more diverse than what it was just a decade ago. While we still draw upon the same first principles of intelligence and statecraft when analyzing terrorism, the techniques which we apply to these problems have evolved considerably. Thus, as I look back on how we approached this problem a decade ago, those "first principles" take on a new life in the context of the more advanced modeling and simulation tools available today. No doubt, some author a decade from now will find much of what we do today quite simplistic or limited when compared to the then current repertoire of techniques, although I would argue that the nature of the first principles which guide our research provides a common thread across the decades, going back in some cases to Machiavelli or even Aristotle and Plato.

## Theoretical Versus Practical Considerations

In the original paper, quite a bit before Derek Jones established a typology for the current generation of modelers, we had to think quite a bit about even the basic organization of even a moderately comprehensive piece on modeling terrorist networks. Among the difficulties we had to confront was where to draw the metaphorical line in the sand between theory and practice (as the differences between theorists and practitioners prior to 9/11 were much deeper and the two areas of inquiry were often entirely or nearly completely disjunct).

My original instincts led me back to the work of an old mentor, Paul Bracken, of Yale University. I recalled two statements of his which had powerfully impressed me as a graduate student. This was back in 1984–1985, when Bracken had just recently left the Hudson Institute, (where he had worked a number of years for Hermann Kahn, the "father of modern nuclear strategy") and had just finished his landmark book "The Command and Control of Nuclear Forces," which was focused directly upon the central issues of national security policy at that time. Early on in our studies, Bracken was quick to distinguish two key factors:

(1) National security policy is characterized by *irreducible levels of ambiguity and complexity*; and
(2) That for the previous decade and a half (dating roughly to the beginning of the Kennedy administration and Robert McNamara's tenure as Secretary of Defense) *academics commenting on security issues had been largely theorizing in a vacuum.*

Given the changes in the research landscape, (2) is less of a problem than it was a decade ago, and as a result, we have a great deal of more knowledge and many more tools for dealing with (1). However, from the academic side, much of the new knowledge is vested in research institutes, such as the Santa Fe Institute, the New England Complex Systems Institute, or in academic departments such as Computing Organization and Society at Carnegie Mellon University which maintains its disciplinary currency through its connection to Dr. Carley's CASOS. The remainder of the professionals working in academia who are not in operations research departments are scattered among various business and social science departments, all of which have either a largely qualitative approach to their discipline, or operate on a paradigm which either does not address the issues of terrorism, insurgency, regional and ethnic violence, or does so in the same traditional manner that Bracken decried nearly 30 years ago.

In at least one sense, academics live on either one side or the other of what I think of as "The Great Computational Divide." There are those who work with concepts (even here the computer screen has replaced pen and paper) and then gather and transform data (largely through simple statistical procedures, such as linear and multiple regression) in order to advocate a particular position and those who are able to apply the techniques and insights of physical science to modeling the problems (broadly speaking) of social science. In this regard, the emergence of complex adaptive systems research may be the end point of what Edward O. Wilson described in Sociobiology as the difference between the "advocacy" method and the hypothetical-deductive scientific method.[7] In a practical sense, it is difficult to explain the methods of disrupting terrorists networks based on their positioning on a dynamic fitness landscape[8] to an audience which has never heard of that particular representation of evolutionary dynamics.[9] So, curiously enough, while a decade later we have emerged with a new and powerful toolbox, those tools themselves have introduced a new kind of problem which was not really present a decade ago.

To return to our original paper, we began by arguing in a very Bracken-like fashion:

> The mathematical revolution of chaos theory and complexity studies has given us powerful new modeling tools that were unthinkable just a generation ago. The pace of technological change has matched the emergence of new sciences in ways which likewise would have been difficult to conceive of in even the relatively recent past. For example, one of the primary purposes of the 1979 U.S. Export Administration Act was to prevent the migration of dual-use technologies like the 32 bit architecture of the Intel 80486 Microprocessor,

[7] Wilson [7].
[8] Fellman et al. [8].
[9] Kauffman [9].

> which could be used to MIRV ICBM's. Today, far more sophisticated architectures are freely available on the market. A camcorder has far more technological and computing power than the guidance system of an MX missile. There's more science and technology involved in controlling the scaling dynamics of internet traffic packet delay than there is in designing the navigational system for mid-course correction on MIRV re-entry vehicles.[10] When these developments are combined with the new power of autonomous non-state actors and various persistent vulnerabilities of complex, self-organizing systems ("Avoiding Complexity Catastrophe", McKelvey, 1999)[11] the challenges of national security in the 21[st] century truly take on an entirely different character and require, tools, techniques, resources, models and knowledge which are fundamentally different from their 20[th] century predecessors.

In this regard, I believe that both academics and practitioners who are able to employ novel scientific means in the service of first principles are still properly focused on the central issues of modeling terrorism, just as we were a decade ago. The greatest difference is that we now have far more information available through the internet, and for all their damning of Western values, even Al Qaeda relies on the internet to sustain their organization and operations. Given the foregoing, one of the novel contributions of our earlier work was to pinpoint where and how the modeling community could make its strongest contribution to fighting terrorism. We elucidated this in the version presented at the London School of economics[12]:

> In the context of this newly emerging dynamics, a proper approach to modeling terrorist networks and disrupting the flow of information, money, material needs to be found in such a way that at the ***mid-range***, various government agencies can efficiently share information, spread and reduce risk (particularly risk to sensitive infrastructure) as well as pro-actively ***target*** and ***dismantle*** the specific components of embedded terrorist networks while deconstructing their dynamics.
>
> In addition to the twin problems of ambiguity and irreducible complexity, I'd like to add a third problem, one which in military jargon means we've got to add but another c, for ***counterintelligence*** to the traditional **C³I** (Command, Control, Communications and Intelligence) or **C⁴I** (Command, Control, Communications, Computing and Intelligence) structures. If we are going to address terrorist threats in any serious way, then we have to establish clear command authorities, mission priorities, funding procedures and operational capabilities to bring **C⁵I** up to a functional standard. While I'm primarily here today to explain how recent advances in science, collectively referred to as ***chaos theory*** and ***complexity science*** can improve the identification, tracking and dismantling of terrorist groups, to ignore the very real administrative and political obstacles to implementing these techniques in a well-functioning organization would throw us right back onto the bone-yard of dead ideas. To engage in a dialogue about systems which cannot, whether for physical, financial or political reasons, be built, would involve us in what Russian philosopher G.I. Gurdjieff so aptly characterized as "pouring from one empty vessel into another".[13]

---

[10] Li and Mills [10].

[11] Kauffman [9].

[12] Fellman and Wright [11].

[13] Richard N. Hass explains this problem rather nicely in terms of the first principles governing the relationship between intelligence, foreign policy, and policy makers, noting that "to be more than the

It is interesting to look back a decade later and see how some of these organizational dynamics, or in the language of Harvard's Graham Allison, "bureaucratic micropolitics" actually played out.[14] In at least one sense, a number of military actors at the theater or sub-theater level were able to substantially advance the discipline beyond all expectation. Dr. Greg Parlier, the current president of the Military Applications Society of INFORMS (Institute for Operations Research and Management Science) who came out of retirement to deploy with J-4 Task Force Troy, has composed an unclassified (however, still official use only) memoir of his tour of duty explaining the application of advanced operations research techniques to many of the problems on the battlefield in Iraq. New definitions of battlespace, asymmetric warfare, and Irregular Warfare (which replaced the previous, short-lived $C^5I$ concept) have also emerged in a fashion which has made military practitioners and planners unusually receptive to the new science of nonlinear dynamical systems modeling and to complex adaptive systems research. Dr. Passman's chapter on "A Fractal Concept of Warfare" developed in part while he was working for the British Ministry of Defence (again, this research is entirely open-source) is a bold illustration of the changing conceptualization of armed conflict. Where the new science has met resistance, there are now both protagonists and antagonists, as exemplified by Dr. Gregory Parnell's work heading up the National Academies (National Academy of Science, National Academy of Engineering and National Institute of Health) review of DHS' Bioterrorism Threat Risk Assessment of 2006.[15] Here, we can see distinguished scientists pushing back at the administration, refusing to accept "theorizing in a vacuum" when there is a real scientific basis for an alternative approach.

However, after two wars and two presidential elections, the policy landscape of national security affairs, as well as the "Global War on Terrorism" has changed quite a bit. During the Bush administration an artificial divide between substantive experts and the White House was created on largely ideological grounds, especially

accumulation of responses to separate crises, a successful foreign policy depends upon bridging the intellectual gap between the imperatives of the present and the potential of the future. In turn, this often depends upon bridging the gap between policymakers and the Intelligence Community. After all, as Robert Bowie—a predecessor of mine as Director of the Policy Planning Staff who later served as a deputy director of the CIA—insightfully defines it, "intelligence" is "knowledge and analysis designed to assist action." *Information and insights that do not "assist action" remain lifeless.* Successful intelligence, therefore, requires a mutual understanding between policymakers and the Intelligence Community that is all too often lacking. Policymakers need to ensure that the Community is not working in a vacuum, that analysts know what is on our minds and what questions we need answered. *At the same time, members of the Intelligence Community have a responsibility to seek out policymakers, understand their concerns, and tell them what they should be paying attention to. It is important to tell policymakers what they need to hear, not what they want to hear."*

[14] Allison and Zelikow [12].

[15] Department of Homeland Security Bioterrorism Risk Assessment: A Call For Change Committee on Methodological Improvements to the Department of Homeland Security's Biological Agent Risk Analysis, Board on Mathematical Sciences and Their Applications, Division on Engineering and Physical Sciences, Board on Life Sciences, Division on Earth and Life Studies, National Research Council Of The National Academies, The National Academies Press, Washington, DC, [13].

with respect to the Iraqi insurgency.[16] This was not so much a methodological divide, as discussed above, as a basic policy divide, as well as a basic difference over a fundamental interpretation of the facts (for example, the misconception that Saddam Hussein was responsible for the events of 9/11). Even using a very simple regression methodology to cover the quantitative aspects of the conflict, Anthony Cordesman of the Center for Strategic and International Studies compiled an overwhelming amount of material documenting the causes, nature, and implications of this divide.[17] Applying complex adaptive systems tools to Dr. Cordesman's research led several experts to conclude that through a misdirected approach, we may have actually enabled a self-organizing insurgency in Iraq.

While the nature of insurgent recruitment might only be a small element of developing a more effective counter-IED methodology, it does go to show in a very Hobbesian fashion that even a small amount of poor policy will stop the best scientific approach dead in the water. In that sense, the global war on terror has introduced another dimension to the application of science to countering terrorism and that is a failure to recognize the nature and value of new scientific advances at the policy level. Thus, the modern debate is less about what academics believe as opposed to what IC professionals or military professionals believe and more about what a particular political leadership believes, supports, and regards as an imperative.[18] We may have the tools to achieve new and surprising victories, but we have to be allowed to use them in order to produce results. Finally, and this is jumping ahead a bit, one of the things we found in our subsequent research applying complex adaptive systems tools to countering terrorism is that opportunity cost plays a much higher role than previously imagined.[19] The implication of our work in this area is that disrupting terrorist networks is more a function of a targeted high-level approach and far less the result of a massive, broad based effort than previously believed. We used a common-sense management term to summarize our results by arguing that in the global war on terror, "sometimes less is more."[20]

## Terrorism Is Not Random

In our earlier paper, we devoted a considerable amount of attention to explaining why terrorism is not random, and focused particularly on the work of Bruce Hoffman. Dr. Hoffman has spent a lifetime in the study of terrorism and despite some apparent lack of "political correctness" there could hardly be a better or more respected source of "first principles" for approaching the study of terrorism. In the

[16] See Packer [14]. See also Robb, "The Bazaar of Violence in Iraq" http://globalguerrillas.typepad.com/globalguerrillas/2004/07/the_bazaar_of_v.html.

[17] See, most recently, Cordesman et al. [15].

[18] Ibid., No. 13 and No. 16.

[19] Ibid. No. 8.

[20] Fellman [16].

earlier paper, we drew an analogy between Dr. Hoffman's work at the RAND corporation and Dr. Farmer's work at the Santa Fe Institute. I have replicated that section of the paper in its entirety below.

Take a look at our opening quotations. The important work on terrorism done by Dr. Hoffman, Graham Fuller,[21] and their colleagues at the RAND corporation shares a strong connection with the more abstract mathematical modeling which has been going on at the Santa Fe Institute for the past decade and a half. J. Doyne Farmer captures the opposing viewpoints highlighted by Dr. Hoffman in an *Edge* (http://www.edge.com) question when he notes that.[22]

> Randomness and determinism are the poles that define the extremes in any assignment of causality. Of course reality is usually somewhere in between. Following Poincare', we say that something is random if the cause seems to have little to do with the effect. Even though there is nothing more deterministic than celestial mechanics, if someone gets hit in the head by a meteor, we say this is bad luck, a random event, because their head and the meteor had little to do with each other. Nobody threw the meteor, and it could just as well have hit someone else. The corresponding point of view here is that bin Laden and his associates are an anomaly, and the fact that they are picking on us is just bad luck. We haven't done anything wrong and there is no reason to change our behavior; if we can just get rid of them, the problem will disappear. This is the view that we would all rather believe because the remedy is much easier.

Those of us who have had to deal directly with the problems of terrorism know just how deeply this kind of wishful thinking runs counter to the real-world terrorist threat. There are a number of institutions who calculate the total number of injuries and fatalities arising from terrorist attacks on a daily or weekly basis. Much as Bruce Russett and Paul Kennedy at Yale University have shown in the case of war casualties rising exponentially by century,[23] what we see in those calculations is an emerging pattern in which the overall number of casualties resulting from terrorism is also growing at exponential rate, and that rate is clearly accelerating. Obviously then, this problem is not going to go away on its own, and even with considerable attention to reducing terrorist threats it is not going to go away either very easily or very rapidly. Moreover, I think one clear implication from the work by Dr. Hoffman[24] and his colleagues is that while there is a kind of dampening or dissipative effect at the edges of the terrorist *ideological* distribution when mapped within a Western perspective (i.e., from the far left-wing to the far right), one of the consequences of *globalization* is that terrorism from Non-Western extremist groups is increasing. Generally, these are groups who, not only do not fit into the traditional political calculus, but also who, for various reasons, often oppose the very foundations of modernity. In consequence, they oppose institutions or even nations, such as the United States, whom they perceive as using the globalization process to enforce

[21] Fuller [17].

[22] Farmer [18].

[23] (a) Russett et al. [19]; (b) Kennedy [20].

[24] Hoffman and Carr [21].

values upon their culture or their constituency which undermine the terrorists' own independence, prestige, and power.[25]

A good example of this is provided by the Taliban. Again contrary to popular belief, the Taliban, at least at a leadership level, was not composed of ignorant, primitive militants. As Dr. Hoffman recounts when he describes the structure of terrorist organizations, their skill sets, their planning, and their evolution as learning organizations, the Taliban had a sophisticated executive or managerial element (here we are speaking about those running the operational and strategic programs, not the visible figureheads like Bin Laden or Mullah Omar). Recruiting cadres at the rank and file level certainly did take place in poor, unmodernized areas (like the Saudi Arabian province of Asir). But a perusal of the recruitment literature reveals a sophisticated knowledge of psychology, and an operational knowledge which is typical of professional military and intelligence officers. Using the techniques of the very society which they eschew, the Taliban built an elaborate psychological campaign around the *Salafist* concept that "jihad" was a central duty of all true Muslims. Now clearly, an approach like this is not going to fly in a modernized or secular Muslim state. It's an approach tailored to a constituency which has benefited little, not at all or has even been punished by the changes accompanying modernization, and its appeal is, at least in part, one of redress of grievances, combined with a pattern as old as Hassan Ibn Al Sabah, of offering great future rewards to those who take up the fight today. The terrorism that arises from this type of powerful, deeply held social and psychological belief, in many ways, lies outside our traditional terrorist typologies, and in the absence of countervailing influences and positive efforts, is only going to increase in the decades to come.

Yet, getting at the root causes of terrorism is one of those things that falls into the category of irreducible complexity and ambiguity. It is, in fact, the very difficulty of the enterprise which leads us towards looking at solutions at the mid-range rather than proposing some system, set of techniques, or methodology which would render terrorist acts either highly predictable (and hence, theoretically avoidable) and which would allow us to dismantle terrorist organizations as soon as they form. In terms of formal properties of the system, terrorist behavior falls somewhere between the purely chaotic and the fully deterministic realms. For those of you with at least a passing familiarity with chaos theory, you'll recognize what we're talking about as a nonlinear dynamical system, characterized by a low-order strange attractor. As a pattern of behaviors, it can be modeled in the same way as other phenomena which exhibit regularity but not periodicity (i.e., locally random, but globally defined).[26] Farmer, for example describes the two principal approaches to dealing with prediction in a "chaotic" system. The first is a formal predictive methodology. Relating terror-

[25] For a full description of this process, see Hudson et al. [22] available at: http://www.loc.gov/rr/frd/Sociology-Psychology%20of%20Terrorism.htm.

[26] At a basic mathematical level, this kind of phenomenon is explained very clearly by Peters [23]. A more rigorous treatment can be found in "Statistical Mechanics of Complex Networks" by Reka Albert and Laszlo Barbási, arXiv:cond-mat/0106096v1 6Jun2001, http://www.nd.edu/~networks/Papers/review.pdf.

ism to "simple" systems[27] such as roulette wheels, turbulent fluids and stock markets, he explains[28]:

> To predict the trajectory of something, you have to understand all the details and keep track of every little thing. This is like solving terrorism by surveillance and security. Put a system in place that will detect and track every terrorist and prevent them from acting. ***This is a tempting solution, because it is easy to build a political consensus for it***, and ***it involves technology, which is something we are good at***. But if there is one thing I have learned in my twenty five years of trying to predict chaotic systems, it is this: ***It is really hard, and it is fundamentally impossible to do it well***. This is particularly so when it involves a large number of independent actors, each of which is difficult to predict. We should think carefully about similar situations, such as the drug war: As long as people are willing to pay a lot of money for drugs, no matter how hard we try to stop them, drugs will be produced, and smugglers and dealers will figure out how to avoid interception. We have been fighting the drug war for more than thirty years, and have made essentially no progress. If we take the same approach against terrorism we are sure to fail, for the same reasons.

While Farmer could reasonably be described as the world heavyweight champion of scientific modeling, his idea that fundamentally changing U.S. foreign policy behavior would substantially defuse or deflect terrorist activities is, unfortunately, mostly wishful thinking.[29] The alternative of altering the behavior of actors at the state level takes us out of the realm of solutions at the mid-range and into an area which is closer to academic speculation. Even the speculation is weak if one simply follows the empirical typology which Dr. Hoffman has developed. From an empirical perspective, there is no clear relationship of cause and effect which would lead one to believe that altering behavior at the state level is likely to result in any significant drop-off in the number or intensity of terrorist acts[30].

---

[27] Here, Farmer is having a bit of a laugh at the expense of his audience. By "simple" systems he means complex, nonlinear systems whose strange attractor is one of sufficiently low dimension that there is an observable phenomenon of closely packed state space and mapping of the phase space with Lyapunov exponents is relatively tractable. A "complex" system in this context would be one with a sparsely populated state space, with bifurcations taking place so frequently that even if there is a strange attractor, the "curse of dimensionality" makes it computationally intractable. The definitive work on the subject is (a) Farmer's paper [24]. A good representative demonstration of the techniques involved is (b) Shampine and Thompson's "Solving Delay Differential Equations with dde23" available on the world wide web at http://www.cs.runet.edu/~thompson/webddes/tutorial.html#CITEjdf . (c) Stuart Kauffman also draws on Farmer's treatment in "The Structure of Rugged Fitness Landscapes", Chapter 2 of The Origins of Order, Oxford University Press, 1993, pp. 33–67.

[28] Ibid., No. 3.

[29] See Lewis [25]; See also Sageman [26]; and Mastors and Deffenbaugh [27].

[30] Hoffman [28]. In this paper Dr. Hoffman explains a number of factors driving terrorist behavior, from the psychological to the operational. In the psychological sense, he argues, (and we agree with the assessment) that "All terrorists have one trait in common: they live in the future: that distant—yet imperceptibly close—point in time when they will assuredly triumph over their enemies and attain the ultimate realization of their political destiny." (p. 6) As psychological motivation moves into operational capabilities and objectives, terrorist behavior suddenly takes on characteristics which are essentially the same as those which underlie the planning and execution of any covert military, paramilitary, or intelligence operation. In terms of target selection, Hoffman points out that 'the terrorists' ability to attract—and moreover, to continue to attract—attention, however is most often predicated on the success of their attacks." Given their desire for continued success, terrorists of necessity must develop an operational tradecraft which closely resembles that of

For this reason, along with the reasons cited in our references, we do acknowledge the impossibility of total predictivity. However, we also still believe that over and above any action which might be taken at the state level, the greatest room for improving the performance of those organizations tasked with preventing or combating terrorism is at the *mid-range*. That is, we think the application of the most recent advances in science is most likely to bear fruit in the fight against terrorism not at the level of state leadership, and not at the level of mapping and predicting the behavior of each individual terrorist, but rather at an intermediate or organizational level, which, following the late Theda Skocpol, we characterize as "action at the mid-range." In this context we will be drawing on several species of network analysis and interpreting them in the context of three fundamental principles of counter-intelligence: Compartmentation, Coverage, and Penetration.

## Terrorism is Still Not Random a Decade Later

A decade later, we have a great deal of additional research describing in detail the dynamics of terrorism, including some formal predictors for risk, damage, and prevention in the area of bioterrorism. As Dr. Parnell and I argued recently, following the methodology of the National Academies review of BTRA 2006, Bioterrorism,

their opponents. They generally have good intelligence, even exceptionally good intelligence, careful planning, well rehearsed, ergonomic execution, and as Hoffman explains "success for the terrorist is dependent not only on their ability to keep one step ahead of the authorities, but of the counter-terrorist technology curve as well. (p. 11). At this point we are looking at terrorism not at the long-range, socio-cultural level, or even at the mid-range, which is the central focus of our paper, but at the operational level which is about as far removed from Farmer's social reform model as you can get. Just one example of this kind of operational thinking should help clarify how little likely long-range planning is to be in dealing with terrorist problems. In an investigation involving an unidentified group detonating explosives and destroying commercial property, the investigation was largely handled by a police element of the Ministry of Internal Affairs. One of the dangerous vulnerabilities that had to be corrected in a very rapid and powerful fashion was the fact that the investigators had not considered that the group undertaking the bombings might be monitoring their investigations, including signals traffic, much of which was in clear, unencrypted form. Nobody had thought that if they continued to leave their communications channels open to interception, that exactly in the spirit of staying both one step ahead of the law and one step ahead of the technology, should the investigation get close to providing a solution to the identity of the bombers that their own headquarters might very likely be the next target for a bomb. Add in the exponential increases in computing power, commercial and black market descrambling technologies and this kind of threat falls right into Hoffman's category of "staying one step ahead of the technology." Not being social theorists, we cannot professionally comment on the policy aspects of US or Russian governmental behavior with respect to moderating terrorist behavior against their citizens. However, at the operational level, the organization which better manages its compartmentation, which is more adept at either penetrating its adversary's compartments or forcing its adversaries into a degree of over-compartmentalization that impedes the ability of signals to flow through the network (another form of command decapitation) and which better monitors and interprets its adversary's communications (coverage) is the one which is vastly more likely to emerge as the winner.

in particular, is not random.[31] The number and kind of biological agents can be calculated and assigned probabilities. Immunization and treatment programs can be assessed and valued and most importantly, following Dr. Hoffman's arguments about the evolutionary nature of terrorist organizations, we can develop a predictive methodology parameterizing the behavior of an intelligent adversary rather than simply assigning random probabilities to bioterrorist attacks. In that sense, our new scientific tools allow us to develop a full-fledged quantitative approach with a rich internal structure and dynamics where the old approach crudely assigns random variables following a profoundly flawed policy first principle. I have reproduced some key sections from our recent work below. As one can see, particularly from the graphic material, we are now able to show very precisely how and why terrorism is not random. Now, what remains is to get this message out to a broad audience (practitioners, scholars, and most of all, to draw upon that great scholar of politics, Bill Murray, registered voters!)

## Excerpts from Bioterrorism Threat Risk Assessment and Biowar

Assessing the risk of terrorism, and terrorist threats is a difficult and complex undertaking.[32] As we have argued elsewhere, official government estimates produced by national boards have a tendency to use probabilistic language in a rather loose fashion, placing excessive emphasis on often ill-defined or incomplete analytical models.[33] In this paper, we review some of the methodological difficulties highlighted by the National Research Council Report "Department of Homeland Security Bioterrorism Risk Assessment: A Call for Change" which analyzed the approach of the Department of Homeland Security's 2006 Bioterrorism Threat Risk Assessment program.[34]

[31] Fellman et al. [29].

[32] Fellman [16].

[33] Fellman [30].

[34] In 2004, the President issued a homeland security directive that, along with the National Strategy for Homeland Security published in 2002, mandated assessments of the biological weapons threat to the nation and assigned responsibility for those assessments to the Department of Homeland Security (DHS). The first such assessment—the Biological Threat Risk Assessment (BTRA) of 2006—is a computer-based tool to assess the risk associated with the release of each of 28 biological threat agents. To assist in its preparation of this version of BTRA as well as the 2008 version, DHS asked the NRC to carry out a study of the methodology used by the agency to prepare BTRA of 2006. This NRC report presents an introduction to the challenge; an analysis of the critical contribution of risk analysis to risk management; a description of the method used to produce the BTRA of 2006, which is the foundation for later assessments; a discussion of risk assessment for unknown and engineered bio-threats; and ways to improve bioterrorism consequence assessment and the BTRA methodology (from "[31]: A Call for Change, Committee on Methodological Improvements to the Department of Homeland Security's Biological Agent Risk Analysis, National Research Council, ISBN: 0-309-12029-2, 92008) http://www.nap.edu/catalog/12206.html.

Following this review, we explore two alternative approaches for more carefully and usefully modeling bioterrorism threats, Intelligent Adversary Risk Analysis, as developed by Parnell, Smith and Moxley,[35] and Merrick and Parnell[36] as well as the Biowar model developed by Carley.[37]

## The Intent of BTRA 2006

The model used by DHS for BTRA 2006 was a computer-based tool designed to assess the relative likelihood and consequences of terrorists' employing each of the 28 specific pathogens identified by CDC as possible terrorist threats. This methodology relied upon largely static probabilities and treated the probabilistic occurrence of an attack as being essentially similar to modeling the risk of an uncertain hazard rather than modeling the behavior of an intelligent adversary. A constructive methodology for intelligent adversary modeling is presented in section three of this paper, where we discuss the Parnell, Smith, and Moxley model.

Additional description of the intent of BTRA 2006 is provided by the executive summary of the NRC report:

> DHS intended that the BTRA of 2006 be an "end-to-end risk assessment of the bioterrorism threat" with potential catastrophic consequences to human health and the national economy and that it "assist and guide biodefense strategic planning" (DHS, 2006, Ch. 1, p. 1) in response to the HSPD-10 directive to "conduct biennial assessments of biological threats." Guided by DHS's customers for information from the assessment, the BTRA of 2006 was designed to produce assessments in the form of risk-prioritized groups of biological threat agents. These prioritized lists could then be used to identify gaps or vulnerabilities in the U.S. biodefense posture and make recommendations for rebalancing and refining investments in the overall U.S. biodefense policy. DHS has assembled a confederation of researchers and subject-matter experts and is collaborating with national laboratories that can contribute to expanding the knowledge base of bioterrorism.

While BTRA 2006 was designed as a comprehensive treatment of bioterrorism, in practice it fell far short of the mark. Following the basic critique summarized above, the NRC report actually recommended that BTRA 2006 not be used as the methodology for dealing with bioterrorism (p. 2):

> The committee met on August 28–29, 2006, with representatives of DHS in response to a DHS request for guidance on its near-term BTRA development efforts. In November 2006, in response to that request and based on the information it had received at the 2-day meeting with DHS, the committee electronically issued its Interim Report (reproduced as Appendix J in this final report). Subsequently the committee received the full DHS (2006) report documenting the analysis in the BTRA of 2006. While DHS agreed with the recommendations of the Interim Report and planned to address them, the committee did not learn of any

[35] Parnell et al. [32].

[36] Merrick and Parnell [33].

[37] Carley et al. [34].

progress up to the conclusion of its deliberations in May 2007 that would obviate those recommendations, which require sustained work.

However, the content of the DHS (2006) report and information gained at additional meetings with DHS and national experts have significantly changed the committee's overall assessment of the BTRA of 2006. The committee identified errors in mathematics, risk assessment modeling, computing, presentation, and other weaknesses in the BTRA of 2006. It recommends against using this current BTRA for bioterrorism risk assessment as presented in the BTRA of 2006 or proposed for 2008. Instead, the committee offers improvements that can significantly simplify and improve future risk assessments. The improved BTRA should be used for risk management as well as risk assessment, as intended by HSPD-10.

Without going into further details from our paper on the existing weaknesses of the BTRA 2006 model, I have excerpted our treatment of the Parnell-Smith-Moxley "Intelligent Adversary" model to illustrate a non-random modeling approach to bioterrorism.

Terrorist attacks are not random, but are purposive, require large organizational commitments (the postulated anthrax attack is more than an order of magnitude larger than the 9/11 attack and would require considerable intelligence and operational resources, if not outright state sponsorship), and are carried out by intelligent adversaries acting and reacting to a dynamic landscape as well as to the counterterrorism strategies of their opponents.[38] While probabilistic risk analysis models uncertain hazards using probability distributions for threats, vulnerabilities, and consequences based on a statistical analysis of past events, the risk analysis of terrorist attacks is fundamentally different than that of uncertain natural disasters and requires a methodology which incorporates the response of an intelligent adversary to changing conditions as shown in Appendix I. In comparing the intelligent adversary approach, Parnell, Smith, and Moxley demonstrate how event trees underestimate intelligent adversary risk by assigning random probabilities to events which are actually decision nodes and which should be modeled as a decision trees rather than event trees. In particular, they develop a canonical intelligent adversary risk model for homeland security which incorporates sequential attacker–defender decisions and outcomes (Appendix II).

The canonical intelligent adversary risk model has six components, the initial actions of the defender to acquire defensive capabilities, the attacker's uncertain acquisition of the implements of attack (e.g., agents A, B, and C), the attacker's target selection and method of attack(s) given implement of attack acquisition, the defender's risk mitigation actions given attack detection, the uncertain consequences, and the cost of the defender's actions. The model consists of three material elements—a decision analysis whether to increase the levels of vaccine, whether to add a city to the BioWatch program, and how to calculate the effects of a pathogen not detected by BioWatch:

In our defender–attacker–defender decision analysis model, we have the two defender decisions (buy vaccine, add a Bio Watch city), the agent acquisition for the attacker is uncertain, the agent selection and target of attack is another decision, the consequences (fatalities and

[38] Hoffman [35].

> economic) are uncertain, the defender decision after attack to mitigate the maximum possible casualties, and the costs of defender decisions are known. The defender risk is defined as the probability of adverse consequences and is modeled using a multiobjective additive model similar to multiobjective value models. We have assumed that the defender minimizes the risk and the attacker maximizes the risk. We implemented this model as a decision tree (Fig. 3) and an influence diagram (Fig. 4) using DPL…

(Figures 3 and 4 are identical to appendices I and II in the current paper—shown below following the body of the text.) The mathematical formulation of the model is contained in appendix four. The model uses COTS software to quantitatively evaluate the potential risk reductions associated with different options and likewise uses COTS software to make cost-benefit decisions. The model then provides outputs with respect to both budget vs. risk as well as the cumulative distribution (Appendix III). Among the conclusions which the model demonstrates are that:

> …spending US$ 0 or US$ 10 million gives the defender a 10% chance of zero risk, whereas spending US$ 20 or US$ 30 million gives the defender an almost 50% chance of having zero risk. The best risk management result would be that option 4 deterministically or stochastically dominates (SD) option 3, option 3 SD option 2, and option 2 SD option 1. The first observation we note from Fig. 6 is that options 2, 3, and 4 stochastically dominate 1 because option 1 has a higher probability for each risk outcome. A second observation is that while option 4 SD option 3, option 4 does not SD option 2 because option 4 has a larger probability of yielding a risk level of 0.4. Along the x-axis, one can see the expected risk (ER) of each alternative. This expected risk corresponds to the expected value of risk from the budget versus risk rainbow diagram. This example illustrates a possibly important relationship necessary for understanding and communicating how the budget might affect the defender's risk and choice of options.
>
> Risk managers can run a value of control or value of correlation diagram to see which nodes most directly affect the outcomes and which are correlated…Because we only have two uncertainty nodes in our canonical model, the results are not surprising.
>
> The graphs show that the ability to acquire the agent is positively correlated with the defender risk. As the probability of acquiring the agent increases, so does defender risk. In addition, the value of control shows the amount of risk that could be reduced given perfect control over each probabilistic node, and that it is clear that acquiring the agent would be the most important variable for risk managers to control.

Admittedly, this is a basic example, but with a more complex model, analysts could determine which nodes are positively or negatively correlated with risk and which uncertainties are most important. In a probabilistic model of this type, which measures intent, and incorporates feedbacks, the interesting feature of the model is the decision driven (i.e., strategic) step function as shown in appendices III and IV. Because the core of the model is based on a Min/Max formulation, the stochastically dominant step function is a natural, if slightly counter-intuitive outcome. More detailed models can be developed with more extensive data about attacker intentions, but the gist of this model is that unlike the BTRA 2006 model, it provides concrete guidance and allows risk managers to peg a given decision to a given cost and expected value.

| | Uncertain hazards | Intelligent adversaries |
|---|---|---|
| Historical data | *Some historical data:*<br>A record exists of extreme events that have already occurred | *Very limited historical data:*<br>Events of September 11, 2001 were the first foreign terrorist attacks worldwide with such a huge concentration of victims and insured damages |
| Risk of occurrence | *Risk reasonably defined:*<br>Well-developed models exist for estimating risks based on historical data and expert's estimates | *Considerable ambiguity of risk:*<br>Adversaries can purposefully adapt their strategy (target, weapons, time) depending on their information on vulnerabilities. Attribution may be difficult (e.g., anthrax attacks) |
| Geographic risk | *Specific areas art risk:*<br>Some geographical areas are well known for being at risk (e.g., California for earthquakes or Florida for hurricanes) | *All areas risk:*<br>Some cities may be considered riskier than others (e.g., New York City, Washington), but terrorists may attack anywhere, any time |
| Information | *Information sharing:*<br>New scientific knowledge on natural hazards can be shared with all the stakeholders | *Asymmetry of information:*<br>Governments sometimes keep secret new information on terrorism for national security reasons |
| Event type | *Natural event:*<br>To date, no one can influence the occurrence of an extreme natural event (e.g., an earthquake) | *Intelligent adversary events:*<br>Government may be able to influence terrorism (e.g., foreign policy; international cooperation; national and homeland security measures) |
| Preparedness and prevention | Government and insureds can invest in well-known mitigation measures | Attack methodologies and weapon types are numerous. Local agencies have limited resources to protect potentially numerous targets. Federal agencies may be in a better position to develop better offensive, defensive, and response strategies |

## *Canonical Intelligent Adversary Risk Management Model*[32]

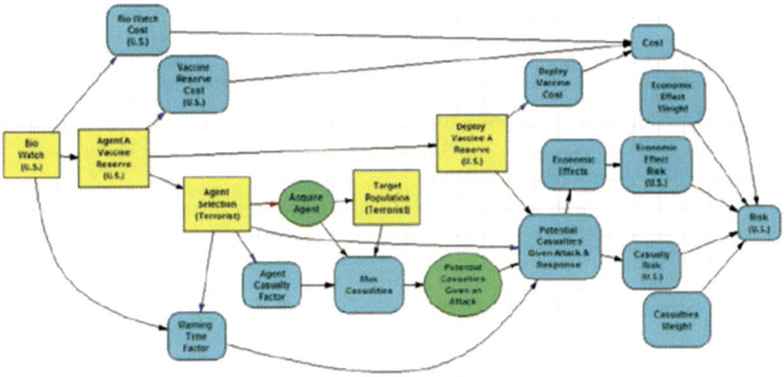

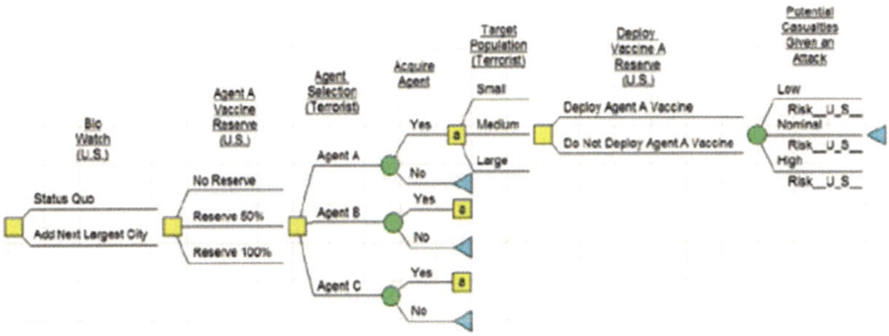

## *Model Probability Distributions*[32]

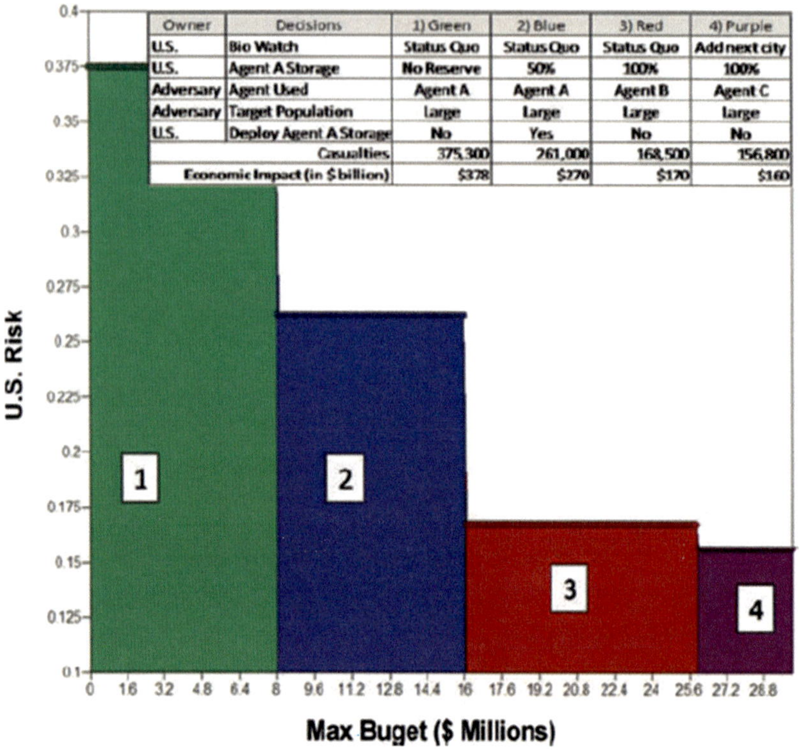


In addition to being able to provide a statistically non-random model of terrorist behavior and the costs of that behavior as well as the costs and risks of countermeasures, we are today, able to quantify the overall number of terrorist attacks, as well as lethal attacks associated with particular organizations with substantial precision. In that regard, we no longer have to simply say that terrorism is increasing with globalization (although that is a good argument about the mechanism of the increase) but as in our chapter on maritime piracy, we can draw upon the results of the Clauset and Gleditsch study at the Santa Fe Institute for data on the cumulative damage caused by terrorist attacks conducted by land-based organizations[39] or our own work based on 10 years of International Maritime Organization data, on the incidence of maritime piracy acts.[40]

---

[39] Clauset and Gleditsch [36].

[40] Fellman et al. [37].

## *Politics and Policy*

Much of our earlier paper then went on to assess the structure of counter-intelligence and counter-terrorism in the US, in NATO, and in the C.I.S. states. A significant amount of that discussion is either now outdated (in the US as a result of the Intelligence Reorganization and Counterterrorism Act of 2004, and the establishment of the Office of the National Counter-Intelligence Executive in 2009 and in NATO and the former Soviet Union by virtue of similar if less salutary changes) and will not be recapitulated here. And while there remains a certain degree of acrimonious politics over national security policy (most of which is really directed at theater warfare rather than counter-terrorism) we have, in the United States, at least to some degree, actually managed post 9/11 to re-instill a substantial degree of the bi-partisan spirit which dominated foreign policy during much of the third quarter of the twentieth century.

The kind of policy debate which interests us at the present time is largely centered around the allocation of resources and requires a more sophisticated understanding of the enemy than in previous eras. The question now is not whether to fight terrorism, but in developing a protocol for prevention and rapid action following a bioterrorism act, should we base our planning on the kind of random characterization used for natural disasters, or should we recognize that we are dealing with an evolving intelligent adversary and that this will change the dynamics of the phenomenon we are dealing with to the extent that if everyone in the target population is immunized against a particular agent, we can then assume that agent will never, in fact be used, whether or not there is a successful bioterrorism act.

Similarly, as I have argued elsewhere [20], there is a significant area of policy which needs to be addressed with respect to opportunity cost. The current state of the US economy is, at the very least, reflective of two long, drawn out wars in Iraq and Afghanistan. If it is true that we can accomplish significantly more by better targeting HUMINT operations, it is quite important from an overall policy perspective to forego the expense of low-yield or poorly performing types of operations so that resources can be allocated to where they are most needed.

Moreover, even at a "mid-range" level, many results of dynamic network analysis and nonlinear dynamical systems modeling turn out to be counter-intuitive. In this regard, one indication provided by such results is that we may need to adjust both policies and institutions to fit with empirical rather than imagined reality or conventional wisdom.

The remainder of the original paper was essentially a primer on Social Network Analysis. While we now understand that Valdis Krebs was only working with the public portion of the 9/11 hijacker network, this is still a good example for demonstrating the fundamentals of dynamic network analysis and how they apply to modeling terrorist networks. Obviously, all of us working in this area have undertaken substantial newer and more detailed and wide ranging work over the past decade. A good, fairly recent summary of some of this work can be had in my article "The Complexity of Terrorist Networks."[41] However, for the purposes of explaining the

[41] Ibid. No. 20.

foundations of the modeling process, I have chosen to present our original exposition of the 9/11 hijacker network and the modeling of the 9/11 network characteristics below.

## September 11 and Network Analysis Models: Simple Distributional Properties

The first difficulty which the analyst must face in constructing a network analysis of terrorist organizations is the difficulty of building an accurate map. Valdis Krebs, who has used network analysis to provide an extensive analysis of the 9/11 Hijackers network, explains three problems he encountered very early on. Drawing on the work of Malcolm Sparrow, he notes that three problems are likely to plague the social network analyst regardless of context. These are[42]:

1. Incompleteness—the inevitability of missing nodes and links that the investigators will not uncover.
2. Fuzzy boundaries—the difficulty in deciding who to include and who not to include.
3. Dynamic—these networks are not static, they are always changing.

In addition, there is rather a bit of a paradox in that even using a more sophisticated methodology, such as measuring the strength of ties in terrorist networks (i.e., vector vs. scalar values), may still not yield a more useful map. One reason for this is that many of the factors which determine the strength of terrorist ties are prior connections, which are not easily measured and which, over the short run, may leave the analyst with an impractically sparse network.[43]

### *Compartmentation, Coverage, and Penetration*

The sparseness of terrorist networks, however, is also a bit of a two-edged sword. In Krebs initial mapping (Fig. 1.1), he found that the 19 member network had an average path length of 4.75 steps, with some of the hijackers separated by more than six steps, and some of the associates over the observable event horizon. Krebs describes this as trading efficiency for secrecy. Another way of describing this process is extreme compartmentation or even “over-compartmentation.” In US defense circles, over-compartmentation has primarily been viewed as an internal planning, or tactical obstacle. However, in the context of terrorist networks, the identification of an organization with large degrees of separation between cells provides the opportunity to degrade the performance of the organization by forcing an already highly compartmented organization in the direction of even greater compartmentation in

[42] Krebs [38].

[43] Ibid.

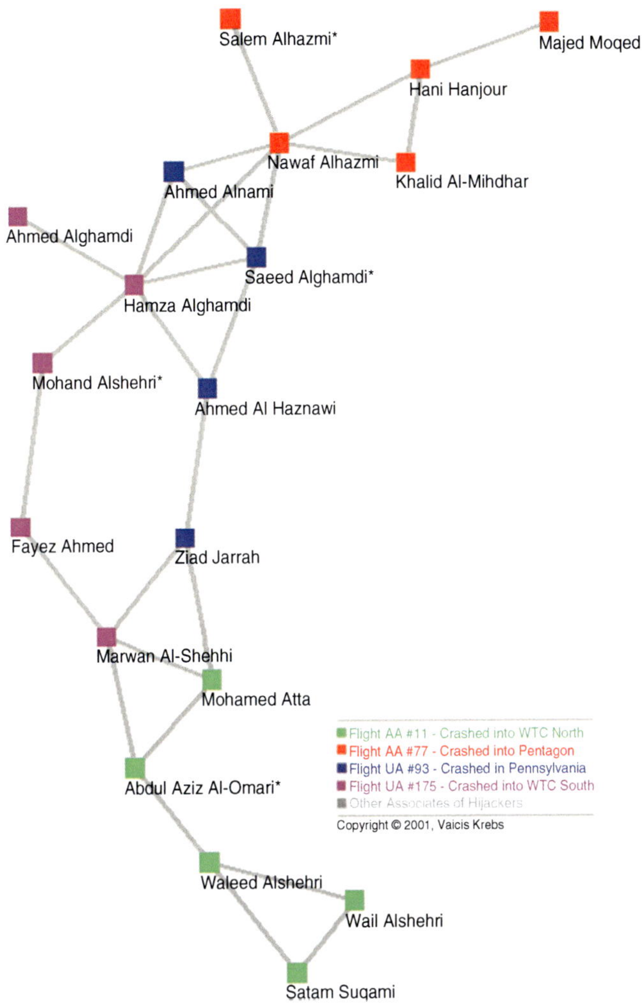


**Fig. 1.1** Valdis Krebs' initial mapping of the 9/11 hijackers' network

order to avoid detection. Inevitably this will lead to coordination problems. In 1993, a Joint Special Operations Publication described this process (for our purposes we would add "compartmentation of terrorist cells") as follows[44]:

> Examples of self-imposed complexity include over-compartmentation of information, ambiguous chains of command, complicated and time-sensitive link-up plans, and movements of multiple elements over multiple routes.

If a terrorist network can be identified as sparsely distributed, the obvious first strategy is to force them into over-compartmentation. Coverage, akin to surveillance

[44] [39], Operational Procedures, Joint Pub 3-053, Joint Doctrine Division, J-7, Joint Staff, Pentagon, Washington, DC (1993).

and identification is a key factor in monitoring and mapping such groups. The actual operations of dismantling such groups more properly belongs to the category of penetration. That there is confusion about both the terminology and methodology of counter-terrorism is evidenced by the approach taken in a number of US government policy studies as well as private consulting agencies. For example, the U.S. Department of State in examining the fund-raising activities of terrorist groups, set as its FY 2001 goal "Increased sensitivity by fund-raising organizations to danger of terrorist penetration." While that is a worthwhile goal, it is a defensive strategy at best and really belongs in the "coverage" or surveillance category. Penetration would be better defined as "covertly entering a terrorist network, and then identifying, diverting, and disrupting its operations." However, in a practical sense, that kind of activity must be assigned to groups whose institutional history and memory is well adapted to serving that purpose, something which the State Department is not.

In the wake of the 9/11 disaster we have seen a lot of scrambling for resources, positioning in information sharing, and acquisition of manpower and technology. What we have not seen is a more positive coordination to enhance the key values of (1) compartmentation (controlling our own and disrupting the enemy's); (2) coverage (increased HUMINT, surveillance and an inflow of personnel to operations who possess the language and cultural knowledge to work autonomously in regions where terrorist recruitment and training is prevalent), (3) penetration, in this case civilian activities comparable to the military activities in Afghanistan and Iraq which have proven efficient and successful, if one disregards the backlash that they have generated throughout the rest of the world (and we're not quite sure the such discounting is appropriate, but that too is beyond the scope of this paper).

Complex systems tools, especially network analysis, offer some very strong insights about the behavior of terrorist networks, but in the flood of data which has hit the market since 9/11 it is often difficult or impossible to tell who has chosen an appropriate methodology and who has not. Equally difficult is discerning whether a complex model truly possesses the ability to model terrorism in newer, more accurate, more powerful ways. Some models may be mathematically complex, but yield little in the way of practical results, simply because their method is static and terrorist cells are a dynamic phenomenon.[45]

[45] A classic mistake in this area is frequently made when the authors apply neoclassical microeconomic "rational actor" assumptions to modeling terrorism which creates a static, homogenous treatment of opponents and sows more confusion than it resolves. While the rational actor methodology was once extremely popular in economics, it has been generally dismissed by "hard" science and is slowly being replaced by complexity science's heterogeneous agent-based modeling. Typically, an agent-based model presumes heterogeneous agent composition, preferences, and behaviors and uses the *stochastic microagent assumption* to replace the rational actor model. For an explanation of agent-based modeling, see Farmer [40].

For applications of agent-based models to terrorism, see Michael Johns and Barry Silverman, "How Emotions and Personality Affect the Utility of Alternative Decisions: A Terrorist Target Selection Case Study" http://www.seas.upenn.edu/~barryg/emotion.pdf, or Ronald A. Woodman, "Agent Based Simulation of Military Operations Other Than War: Small Unit Combat, Thesis, Naval Postgraduate School, Monterey, CA, September, [41] http://diana.gl.nps.navy.mil/~ahbuss/StudentTheses/WoodamanThesis.pdf.

## *Complex Networks: Beyond Simple Heterogeneous Preferences*

The previously unstructured concept of terrorist networks takes on a new character in the treatment advanced by Krebs.[46] Krebs' network mapping draws on the application of social network analysis and develops a software system based methodology which he uses to map knowledge networks within and across the boundaries of an organization in order to uncover the *dynamics of learning and adaptation*. This kind of organizational network analysis combines social network analysis and organizational behavior with chaos theory and complex adaptive systems. The network mapping goes far beyond the formal organizational structure by exposing the real knowledge-sharing dynamics within the functional structures. Krebs describes these *communities of practice* as *emergent groups* in which knowledge is concentrated around common problems and interests, and the core competencies of an organization are shared and developed (his organizational mapping) of terrorist networks derives in large part from his earlier work mapping corporations and studying the dynamics of organizational learning.

The measurement of these "complex human structures" focuses on individual *network centrality*, which reveals key individuals in the information flow and knowledge exchange. *High-centrality scores demonstrate extensive access to* "**hidden assets**" within the organization of an entity with high capacity to "get things done." Network centrality relates the performance of the network as shown below.

What is interesting about this second plot is that it illuminates the ways in which terrorist compartmentation dictates the operational parameters of a terrorist attack. Such mappings may also yield previously hidden information about the command structure of terrorist organizations. In "Six Degrees of Mohammed Atta," Thomas Stewart describes several important features of Krebs' network, and points out that[47]:

> It is not a complete picture; among other problems, it shows only those links that have been publicly disclosed. Still, it's possible to make some interesting inferences. First, the greatest number of lines lead to Atta, who scores highest on all three measures, with Al-Shehhi, who is second in both activity and closeness, close behind. However, Nawaf Alhazmi, one of the American Flight 77 hijackers, is an interesting figure. In Krebs's number crunching, Alhazmi comes in second in betweenness, suggesting that he exercised a lot of control, but fourth in activity and only seventh in closeness. But if you eliminate the thinnest links (which also tend to be the most recent – phone calls and other connections made just before Sept. 11), ***Alhazmi becomes the most powerful node in the net. He is first in both control and access, and second only to Atta in activity. It would be worth exploring the hypothesis that Alhazmi played a large role in planning the attacks, and Atta came to the fore when it was time to carry them out.***

To return to measures of centrality, and the dynamic operational advantage which high-centrality incurs, one must first understand how such a system can take advantage of wide degrees of separation between cells. Operational benefits arise from the

[46] Krebs (a) [42] (b) [43], (c) [44].

[47] Stewart [45].

pattern of connections surrounding a node that allows for wide network reach with minimal direct ties. "Structural holes" at the intersection of flows across knowledge communities position unique and superior nodes. It is the individuals spanning these "internal holes of opportunity" that impact the network's functioning and performance. The implicit corollary of this is that if a small number of these critical nodes can be identified and "clipped" from the network, then command signals will not be able to propagate through the system.

In Krebs' mapping, the main centrality measures are degrees (number of direct connections that a node has), betweenness (the ability of an individual to link to important constituencies), and closeness (a position's ability to monitor the information flow and to "see" what is happening in the network). The knowledge flow is facilitated and influenced by boundary spanners with access to information flowing in other clusters, as well as peripheral players that bring fresh information into the network. A network with a low centralization score is more resilient in that it has no single highly central points of failure. These networks "fail gracefully" as the damage of a node does not lead to a breakdown in information flows and coordination links.

## *Social Network Theory*

What social network analysis contributes to counter-terrorism is the ability to map the invisible dynamics inside a terrorist community. The methodology draws upon graphical representation in exploring and presenting the patterns displayed by structural data. In the case of terrorist networks, surveillance of the daily activities and contacts of suspects reveals the network around them and thus adds more nodes and links of intentional contacts to the map. Once the direct links are identified, and the "connections of the connections" are included, the key individuals begin to stand out. In 2000, the Central Intelligence Agency identified al-Qaeda suspects Nawaf Alhazmi and Khalid Almihdhar attending a meeting in Malaysia. The mapping of the links between the terrorists involved in the WTC attacks shows that all 19 hijackers were within two degrees from these original suspects, while they also had multiple ties back into the network.

Based on publicly released information from the investigation of the 9/11 terrorists Krebs mapped and evaluated the links which tied the network together and analyzed its resilience. Each link's strength was evaluated based upon the amount of time members spent together. Interactions were rank ordered so that those living together or attending the same training were assigned the strongest ties, and the terrorists traveling or participating in common meetings were given ties of moderate strength, and finally, those ties which reflected only occasional relations were characterized as weak links. The thickness of the lines in Fig. 1.2 corresponds to the strength of the ties between the terrorists. This mapping exhibits a dispersed but well-defined structure, although as mentioned earlier, the connections between members are more than usually distant.

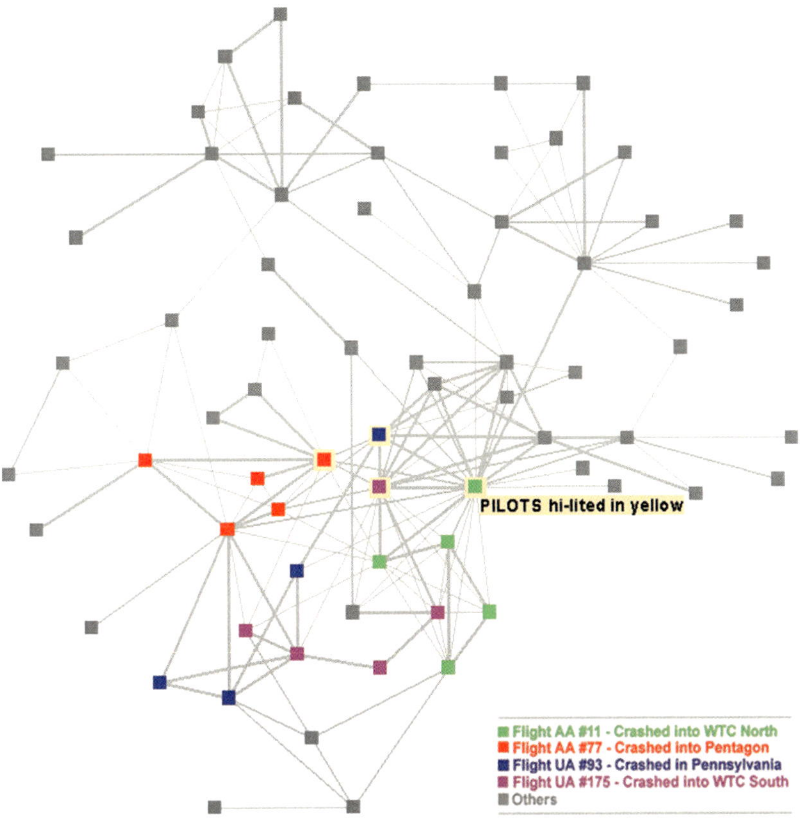


**Fig. 1.2** Krebs' extended model 9–11 hijacker model with measures of centrality

As the positioning of Alhazmi discussed above suggests, strong ties may have been inactive and hidden for relatively long intervals, whereas a minimum of weak ties ensured secrecy. This configuration reveals a network which was consciously constructed on the principle of minimizing damage to the organization as a whole in the event that a link is compromised.

This type of network can only achieve its goals by the use of transitory shortcuts that temporarily balance the need for covertness with the need for intense information flow and coordination in active times.[48] Sources of public information show that the dense and resilient ties forged in the past were "invisible" during the hijackers' stay in the US. This "massive redundancy through trusted prior contacts" is considered one of the major hidden strengths of this network. Such a finding once again highlights the need for human collection, particularly in remote locations. All the

[48] Watts [46].

tasking in the world by Homeland Security or any other security service won't matter a whit if the ability to identify strong connective linkages between terrorists is not developed.

After the 9/11 disaster, there was a trend in Washington to talk about the hijackings as a massive intelligence failure. If we take network analysis seriously, then the failure leading to 9/11 was the result of not having built up human intelligence resources capable of recognizing and responding to the evolution of Al-Qaeda and its field operatives.[49] Michael Porter has been arguing for over two decades in competitive strategy that when firms compete with their buyers and suppliers, the more concentrated group wins, claiming the bulk of the profits for themselves.[50] If I were the head of any nation's intelligence service, I would be asking myself very seriously why terrorist organizations are able to achieve a higher level of coordination and robustness than my counter-terrorist division.

As compared to static models, Krebs' work analyzes the dynamics of the network and also recognizes the centrality measure's sensitivity to changes in nodes and links. In terms of utility as a counter-intelligence tool, the mapping exposes a concentration of links around the pilots, an organizational weakness which could have been used against the hijackers had the mapping been available prior to, rather than after the disaster.[51]

Because nodes with high centrality are potential points of failure they need to be mapped and monitored and whenever possible, removed. If enough nodes with high centrality are removed at the same time, this will cause the network to fragment into unconnected sub-networks. Naturally, the more compartmented the organization is, the fewer nodes of centrality need be removed in order to cause the network to implode. One trigger for moving from monitoring to disabling is a sudden increase in the flow of either money or information flow between links, and rapidly forming connections. In this regard, SIGINT can prove a powerful supplement to HUMINT. If an initial map of members and connections can be assembled, then SIGINT can indicate critical time periods. If we take Krebs' comment about the 9/11 network possessing self-organizing system properties, then the optimal time to intervene and remove high-centrality nodes is just at the beginning of the organizational equivalent of a "phase transition."[52]

---

[49] See Gerecht [47].

[50] See Porter [48,49].

[51] See Klerks [50].

[52] For more detail on this subject see Lissack [51].

## *Social Cohesion and Adhesion: Further Measures of Organizational Structure*

Moody and White provide an expansion of the social solidarity concept and the understanding of linkages between members of a community, the changing interconnections, and the impact on node connectivity in "Social Cohesion and Embeddedness."[53] They argue that the defining characteristic of a strongly cohesive group is that "it has a status beyond any individual group member." The authors define structural cohesion as "the minimum number of actors who, if removed from a group, would disconnect the group," leading to hierarchically nested groups, where highly cohesive groups are embedded within less cohesive groups. Thus, cohesions is an emergent property of the relational pattern that holds a group together.

As the dynamical process of group development unfolds, typically a weak form of structural cohesion begins to emerge as collections of unrelated individuals begin connecting through a single path which reflects new relationships. As additional relations form among previously connected pairs of individuals, multiple paths through the group develop, increasing the community's ability to "hold together."

In situations where relations revolve around a leader, the group is often described as "notoriously fragile," illustrating the fact that *increasing relational volume thru a single individual does not necessarily promote cohesiveness*. Nevertheless, groups with an all-in-one relational organization such as *terrorist networks may be stable and robust to disruptions if "extraordinary efforts" are put into maintaining their weak relational structure*. The spoke-and-hub configuration of these networks thrives on the lack of knowledge that each particular node has about the organization as a whole, and a captured or destroyed link in the network does not put the organization at risk. The stability of such groups depends on the ability to keep the hub hidden, because *the hub then becomes the entire group's fundamental structural weakness*.

Weakly cohesive organizations also promote segmentation into structures that are only minimally connected to the rest of the group, leading to schisms and factions. These organizations are also easily disrupted by individuals leaving the group. *Usually, individuals whose removal would disconnect the group are those in control of the flow of resources in the network*.

On the contrary, *collectivities that do not depend on individual actors are less easily segmented*. These highly cohesive groups benefit from the existence of multiple paths and sets of alternative linkages, with no individual or minority within the group exercising control over resources. The "multiple connectivity" is thus the essential feature of the strong structurally cohesive organizations.

An interesting characteristic of such highly cohesive networks (*HCN's*) is that they are characterized by a reduction in the power provided by structural holes, such that the ability of any individual to have power within the setting is limited as

[53] Moody and White [52].
http://www.santafe.edu/sfi/publications/Working-Papers/00-08-049.pdf.

connectivity increases.[54] For a structurally cohesive group, the information transmission increases with each additional independent path in the network, which may infer that high connectivity leads to more reliability as information is combined from independent multiple sources. "Local pockets of high connectivity" can act as "amplifying substations" of information and/or resources. Moody and White relate this operationalization to the actors' relative involvement depth in social relations, as defined by the concept of embeddedness.

> If cohesive groups are nested within each other, then each successive group is more deeply embedded within the network. As such, one aspect of embeddedness—the depth of involvement in a relational structure—is captures by the extent to which a group is nested within the relational structure.

In a companion paper to Moody and White's "Social Cohesion and Embeddedness," White and Harary distinguish between the adhesion concept which is related to the attractive or charismatic qualities of leaders (or attractions to their followers) and the cohesion concept which is defined by the many-to-many ties created among individuals as they form into clusters.[55]

The authors reiterate the intuitive aspects of the cohesion's definition: a group is cohesive to the extent that the social relations of its members are resistant to the group being pulled apart, and a group is cohesive to the extent that the multiple social relations of its members pull it together. In revisiting the idea that *minimal cohesion occurs in social networks with a strong group leader or popular figure*, White and Harary introduce the concept of ***"adherents"*** of a social group to specify ***"the many-to-one commitments of individuals to the group itself or to its leadership"***.

> ***What holds the group together where this is the major factor in group solidarity is the strength of adhesion of members to the leader, not the cohesiveness of group members in terms of social ties amongst themselves***. The model of "adhesion" rather than cohesion might apply to the case of a purely vertical bureaucracy where there are no lateral ties.

As a general definition, a group is adhesive to the extent that the social relations of its members are pairwise-resistant to being pulled apart.[56] Another element of group robustness is the redundancy of connections[57]:

> The level of cohesion is higher when members of a group are connected as opposed to disconnected, and further, when the group and its actors are not only connected but also

[54] In terms of strategy, destabilizing this kind of network means pressuring the group to increase its recruitment and raise its connectivity as opposed to the previously discussed strategy of forced over-compartmentation. Induced excess connectivity represents a different kind of complexity overload.

[55] The Cohesiveness of Blocks in Social Networks [53].

[56] The concept of cohesion is formalized through the use of graph theory. The graph is defined that the vertices represent the set of individuals in the network, and the edges are the relations among actors defined as paired sets. The subsets of nodes that link non-adjacent vertices will disconnect actors if removed. Any such set of nodes is called an (i, j) cut-set if every path connecting i and j passes through at least one node of the set. The "cut-set resistance to being pulled apart" criterion and the multiple independent paths "held together" criterion of cohesion are formally equivalent in this formal specification. This kind of graph, if constructed with complete information, also provides a predictive mechanism for exactly which nodes need to be removed in order to remove the possibility of signals propagating through the system.

[57] Ibid. No. 36.

have redundancies in their interconnections. The higher the redundancies of independent connections between pairs of nodes, the higher the cohesion, and the more social circles in which any pair of persons is contained

The important consideration for counter-intelligence here is that the higher the level of redundancy, the more likely the existence of the group is to be revealed and the easier it is to create a map of social network relationships. A major part of the successful exploitation of this group characteristic is another of the three basic counter-intelligence principles—coverage. Good coverage will yield good observations from which good social network maps can be derived. The caution here, as we have already noted, is like many other HUMINT activities, good coverage is not possible to achieve solely by satellite reconnaissance or any other national technical means (NTM). Of course, once target individuals have been identified, dedicated remote sensing technology can, in fact, be a very helpful adjunct to the processes of coverage, compartmentation, and penetration.

Again, just as technology cannot substitute for on the ground coverage, effectively covering the broad range of terrorist groups in the world today is also not possible with case officer staffing at the 15 % level in embassies. It is far less achievable without a highly trained group of multi-lingual, multi-talented individuals who are both willing and able to work successfully for extended periods of time in local environments as local citizens with no official cover or identifying features as "foreign" nationals (NOC's). One thing that mid-range analysis tells any intelligent decision maker is that you can have all the science in the world at your fingertips, but without an effective human organization capable of learning and carrying out the missions which that science prescribes you are paralyzed. Remember, the 9/11 hijackers took those planes with box-cutters and pocket knives, not AK-47's, machine guns, intelligent weapons, or recoilless rifles and TOW launchers.

To return to social network analysis, another logical inference is that *measurable differences in cohesiveness should have predictive consequences* for social groups and their members across many different social contexts. In terms of counter-terrorism, this is a clear, predictive model which can be utilized for the optimization of effort and resources. It's not merely a case of getting the most "bang for the buck," but it is an ideal mid-range solution for empirically validated scientific research to provide a typology which allows groups to be characterized in such a way that when resources are directed at preventing the group from executing terrorist attacks, the resources are used in an operationally efficient fashion.

In our original paper we argued that "to bring it down to the micro level, *it won't do any good to remove the leader of a group which is characterized by a strong measure of adhesion*. Just as *it would do no good to attempt to counter a group with strong cohesion by removing its charismatic leader*." While I don't wish to debate the wisdom of killing Osama Bin Laden (although personally, I do see it as a desirable moral victory for the US, extremist backlash notwithstanding) it is really just a symbolic victory (and possibly a learning curve victory because we learned not to trust Pakistan and Pakistani intelligence). The reason it is only a symbolic victory is because (as Elena Mastors has argued in considerable detail elsewhere) Bin Laden was not at the helm of Al Qaeda planning and strategy and his removal is little likely to change either their tactics or strategy. On the other hand, the degree to which we have been able to identify and seize Al Qaeda's financial assets will definitely impact Al Qaeda's operations and the degree to which we are able to degrade their communications and logistical networks will also likewise have significant impacts. Among the current subgroups of Al Qaeda, *Al Qaeda in Arabia* is likely the most virulent Al Qaeda group (following on our comments nearly a decade ago about Al Qaeda recruiting in Asir province in Saudi Arabia) and this is really where the application of nonlinear dynamical systems modeling as well as social network analysis is most likely to yield fruitful results.

## References

1. Fellman, P. V., Sawyer, D., & Wright, R. (2003). Modeling terrorist networks – complex systems and first principles of counter-intelligence. In *Proceedings of the NATO conference on central Asia: Enlargement, civil – military relations, and security*, Kazakh American University/North Atlantic Treaty Organization (NATO).
2. Fellman, P. V., & Wright, R. (2004). Modeling terrorist networks. *The Intelligencer U.S. Journal of Intelligence Studies*, Vol. 14, No. 2, pp. 59–68.
3. Fellman, P. V., & Wright, R. (2003). Modeling terrorist networks – complex systems at the mid-range. In *Paper presented at the joint complexity conference*, London School of Economics.
4. Fellman, P. V. (2011). Understanding the complexity of IIED terrorist networks. In H. Sayama, A. Minai, D. Braha, & Y. Bar-Yam (Eds.), *Unifying themes in complex systems* (Vol. 3). Cambridge: NECSI Knowledge Press.
5. Flynn, M. T., Pottinger, M., & Batchelor, P. D. (2010). Fixing Intel: A blueprint for making intelligence relevant in Afghanistan. *Marine Corps Gazette.* http://www.mca-marines.org/gazette/article/fixing-intel.
6. Jones, D. (2006). *Understanding the form, function, and logic of clandestine cellular networks: the first step in effective counternetwork operations.* Fort Leavenworth, KS: School of Advanced Military Studies, United States Army Command and General Staff College.
7. Wilson, E. O. (2000). *Sociobiology: The new synthesis.* Cambridge: Belknap Press of Harvard University Press. Twenty-Fifth Anniversary Edition.
8. Fellman, P. V., Clemens, J. P., Wright, R., Post, J. V., & Dadmun, M. (2006). Disrupting terrorist networks, a dynamic fitness landscape approach. In *Proceedings of the 2006 annual meeting of the North American association for computation in the social and organizational sciences.*
9. Kauffman, S. A. (1993). *The origins of order: Self organization and selection in evolution.* New York: Oxford University Press.
10. Li, Q., & Mills, D. Investigating the Scaling Behavior, Crossover and Antipersistence, Global Telecommunications Conference, Institute of Electrical and Electronic Engineers, Rio de Janeiro, Dec. 5-9, 1999.
11. Fellman, P. V., & Wright, R. (2003). *Modeling terrorist networks: Complex systems at the mid-range, the joint complexity conference*, London School of Economics. http://www.psych.lse.ac.uk/complexity/conference/FellmanWright.pdf.
12. Allison, G. T., & Zelikow, P. (1999). *Essence of decision: Explaining the Cuban missile crisis* (2nd ed.). New York: Pearson.
13. Department of Homeland Security Bioterrorism Risk Assessment (2008). *A call for change committee on methodological improvements to the department of homeland security's biological agent risk analysis.* Washington, DC: Board on Mathematical Sciences and Their Applications, Division on Engineering and Physical Sciences, Board on Life Sciences, Division on Earth and Life Studies, National Research Council Of The National Academies, The National Academies Press.

14. Packer, G. (2006). *The assassins gate: America in Iraq, Farrar, Strauss and Giroux*. See also John Robb, *The bazaar of violence in Iraq* http://globalguerrillas.typepad.com/globalguerrillas/2004/07/the_bazaar_of_v.html.
15. Cordesman, A. H., Mausner, A., Khazai, S., Alsis, P., & Loi, C. (2013). *The real outcome of the Iraq war: US and Iranian strategic competition in Iraq*. Center for Strategic and International Studies. http://csis.org/publication/real-outcome-iraq-war-us-and-iranian-strategic-competition-iraq.
16. Fellman, P. V. (2011). The complexity of terrorist networks. *Open Source Intelligence and web Mining, 1*(2), 4–14. Special Edition of the International Journal of Networking and Virtual Organisations.
17. Fuller, E. to Fuller, G. (1997). Redrawing the world's borders. World Policy Journal;Spring97, Vol. 14 Issue 1, pp. 11-21.
18. Farmer, J. D. (2001). "What Now?" http://www.edge.org/documents/whatnow/whatnow_farmer.html.
19. Russett, B., Starr, H., & Kinsella, D. (1999). *World politics: The menu for choice* (6th ed.). Belmont: Wadsworth.
20. Kennedy, P. M. (1989). *The rise and fall of the great powers: Economic change and military conflict from 1500 to 2000*. New York: Vintage Books.
21. Hoffman, B., & Carr, C. (1997). Terrorism: Who is fighting whom? World Policy Journal; Spring 97, Vol. 14 Issue 1, pp. 97-104.
22. Hudson, R. A., Majeska, M., Savada, A., & Metz, H. C. (1999). *The sociology and psychology of terrorism: Who becomes a terrorist and why?* Washington, DC: Federal Research Division, Library of Congress. Available at: http://www.loc.gov/rr/frd/Sociology-Psychology%20of%20Terrorism.htm.
23. Peters, E. (1992). At a basic mathematical level, this kind of phenomenon is explained very clearly. In *Chaos and order in the capital markets*. New York: John Wiley and Sons. A more rigorous treatment can be found in *Statistical mechanics of complex networks* by Reka Albert and Laszlo Barbási, arXiv:cond-mat/0106096v1 6 Jun 2001. http://www.nd.edu/~networks/Papers/review.pdf.
24. The definitive work on the subject of "simple" systems is (a) Farmer's paper (1982). Chaotic attractors of an infinite-dimensional dynamical system. *Physica D*, 4, 366–393. A good representative demonstration of the techniques involved is (b) Shampine and Thompson's "Solving Delay Differential Equations with dde23" available on the world wide web at http://www.cs.runet.edu/~thompson/webddes/tutorial.html#CITEjdf. (c) Stuart Kauffman also draws on Farmer's treatment in "The Structure of Rugged Fitness Landscapes", Chapter Two of The Origins of Order, Oxford University Press, 1993, pp. 33–67.
25. Lewis, B. (2001). *The multiple identities of the Middle East*. New York: Schocken.
26. Sageman, M. (2004). *Understanding terror networks*. Philadelphia: University of Pennsylvania Press.
27. Mastors, E., & Deffenbaugh, A. (2007). *The lesser Jihad: Recruits and the al-Qaida network*. Lanham: Rowman and Littlefield.
28. Hoffman, B. (1997). *The modern terrorist mindset: Tactics, targets and technologies*. Scotland: Centre for the Study of Terrorism and Political Violence, St. Andrews University.
29. Fellman, P. V., Carley, K., & Parnell, G. (2011). Biowar and bioterrorism threat risk assessment. In H. Sayama, A. Minai, D. Braha, & Y. Bar-Yam (Eds.), *Unifying themes in complex systems* (Vol. 8). New York: NECSI Knowledge Press.
30. Fellman, P. V. (2011). The complexity of intelligence estimates. In H. Sayama, A. Minai, D. Braha, & Y. Bar-Yam (Eds.), *Unifying Themes in Complex Systems* (Vol. 8). New York: NECSI Knowledge Press.
31. Department of Homeland Security Bioterrorism Risk Assessment: A Call for Change, Committee on Methodological Improvements to the Department of Homeland Security's Biological Agent Risk Analysis, National Research Council, ISBN: 0-309-12029-2, 92008. http://www.nap.edu/catalog/12206.html.

32. Parnell, G. S., Smith, C. M., & Moxley, F.I. (2010). Intelligent adversary risk analysis: A bioterrorism risk management model. *Risk Analysis, 30*, 32–48. doi: 10.1111/j.1539-6924.2009.01319.x.
33. Merrick, J., & Parnell, G. S. (2011). A comparative analysis of PRA and intelligent adversary methods for counterterrorism risk management. *Society for Risk Analysis*. doi: 10.1111/j.1539-6924.2011.01590.x.
34. Carley, K. M., Altman, N., Kaminsky, B., Nave, D., & Yahja, A. (2004). *BioWar: A city-scale multi-agent network model of weaponized biological attacks,* CASOS Technical Report, CMU-ISRI-04-101.
35. Hoffman, B. (2006) *Inside terrorism*. New York: Columbia University Press.
36. Clauset, A., & Gleditsch, C. S. (2012). The developmental dynamics of terrorist organizations. Santa Fe Institute Working Paper. http://www.santafe.edu/media/workingpapers/09-07-021.pdf.
37. Fellman, P. V., Frutos, D., Thanakijsombat, N., & Teekasap, P. (2011). The complex landscape of maritime piracy. In H. Sayama, A. Minai, D. Braha, & Y. Bar-Yam (Eds.), *Unifying themes in complex systems* (Vol. 8). New York: NECSI Knowledge Press.
38. Krebs, V. (2001). Uncloaking terrorist networks. *First Monday*. http://pear.accc.uic.edu/ojs/index.php/fm/article/view/941/863.
39. Joint Special Operations, Operational Procedures, Joint Pub 3-053, Joint Doctrine Division, J-7, Joint Staff, Pentagon, Washington, DC. (1993)
40. Farmer, J. D. (1999). Toward agent based models for investment. http://www.santafe.edu/~jdf/aimr.pdf, and Physicists attempt to scale the ivory towers of finance. *Computing in Science and Engineering*. http://www.santafe.edu/sfi/publications/Working-Papers/99-10-073.pdf.
41. Johns, M., & Silverman, B. (2000). *How emotions and personality affect the utility of alternative decisions: A terrorist target selection case study*. http://www.seas.upenn.edu/~barryg/emotion.pdf, or Woodman, R.A. Agent based simulation of military operations other than war: Small unit combat, Thesis. Monterey, CA: Naval Postgraduate School. http://diana.gl.nps.navy.mil/~ahbuss/StudentTheses/WoodamanThesis.pdf.
42. Krebs, V. (2001). Mapping networks of terrorist cells. *Connections, 24*(3), 43–52. http://www.orgnet.com/MappingTerroristNetworks.pdf.
43. Krebs, V. (2002). *Surveillance of terrorist networks*. http://www.orgnet.com/tnet.html.
44. Krebs, V. (2002). *Social network analysis of the 9–11 terrorist network*. http://www.orgnet.com/hijackers.html.
45. Stewart, T. A. (2001). *Six degrees of Mohammed Atta*. Business2.com is an electronic publication now owned by Forbes, Inc. http://www.business2.com/articles/mag/0,1640,35253,FF.html.
46. Watts, D. J. (1999). Networks, dynamics, and the small-world phenomenon. *American Journal of Sociology, 13*(2), 493–527.
47. Gerecht, R. M. (2001). The counterterrorist myth. *The Atlantic Monthly.*
48. Porter, M. (1979). How competitive forces shape strategy. *Harvard Business Review.*
49. Porter, M. (1996). What is strategy. *Harvard Business Review*.
50. Klerks, P. (2001). The network paradigm applied to criminal organisations: Theoretical nitpicking or a relevant doctrine for investigators? Recent developments for the Netherlands. *Connections, 24*(3), 53–65. http://www.sfu.ca/~insna/Connections-Web/Volume24-3/klerks.pdf.
51. Lissack, M. (1996). Chaos and complexity – What does that have to do with knowledge management? In J. F. Schreinemakers (Ed.), *Knowledge management: Organization, competence and methodology* (Vol. 1, pp. 62–81). Wurzburg: Ergon. http://www.lissack.com/writings/knowledge.htm.
52. Moody, J., & White, D. R. (2000). *Social cohesion and embeddedness: A hierarchical conception of social groups*, Santa Fe Institute Working Papers, 00-08-049, http://www.santafe.edu/sfi/|publications/Working-Papers/00-08-049.pdf.
53. The Cohesiveness of Blocks in Social Networks (2001). *Node Connectivity and Conditional Density*. Submitted to Sociological Methodology. http://www.santafe.edu/files/workshops/dynamics/sm-wh8a.pdf.

# Chapter 2
# Complex Systems Studies and Terrorism

Czeslaw Mesjasz

## Introduction

Although terrorism has been present in social life for decades, only after the World Trade Center attacks of September 11, 2001, did the topic gain new significance as a sociopolitical phenomenon and as a method of warfare. Terrorism has been analyzed from a variety of viewpoints, ranging from security and military research to cultural studies and anthropological analyses. Scholars in the field of Complexity Science (Complex Adaptive Systems [CAS] Research and Nonlinear Dynamical Systems Modeling) have recently been developing a number of models for the identification and prediction of terrorist activities (Bar-Yam, Carley, Clauset and Gleditsch, Fellman, Galam, etc.). It may even be argued that ideas are drawn from CAS research (the term complexity theory, or complexity science, is purposely avoided in this chapter), which could be applied not only in specialized research on how to deal with terrorism as a form of warfare, but also in a broader sense, for studying terrorism as a sociopolitical phenomenon.

The aim of this chapter, which is part of a broader research project both here and in another forthcoming publication from Springer Verlag, is to present a survey of the applications of ideas drawn from CAS research in a theoretical discourse on terrorism. In so doing, I shall examine the use of these concepts both in policymaking and in more applied settings. Other authors in this book will focus on more specific applications of various tools and models, while in this chapter, I shall endeavor to provide a more general theoretical foundation for the entire volume.

This chapter is based upon three assumptions. First, that contemporary terrorism has become a peculiar facet of modern society, which itself can be viewed as a complex system [1]. Second, that concepts drawn from complex systems research

C. Mesjasz (✉)
Kraków University of Economics, Kraków, Poland
e-mail: mesjaszc@uek.krakow.pl

 
P.V. Fellman et al. (eds.), *Conflict and Complexity*, Understanding Complex Systems,
DOI 10.1007/978-1-4939-1705-1_2

broadly defined can be applied as new instruments both to facilitate the understanding of terrorism as a pervasive phenomenon affecting the security of modern society and to provide new applications for improving counter-terrorism measures. Third, that expectations regarding the ability of complex systems research to predict terrorist activities (including those models developed by theoreticians as well as programs designed for policymakers engaged in counter-terrorism activities) can only be partially fulfilled.

Due to the fuzziness of the term "complexity" and its multitude of uses and abuses, the meaning of the term "the complexity of social systems" will be explained in detail in the present chapter. The concepts associated with complexity studies will also be defined and analyzed. Subsequently, an overview of the interpretations of terrorism in contemporary security theory will be presented along with an analysis of the relationship between the complexity of terrorism and the complexity of modern society.

This survey of the linkages between complexity studies and terrorism is divided into two parts. The first part of the analysis examines the role of systems thinking in the theoretical consideration of terrorism as well as in practical applications and responses to terrorism. This line of reasoning, which originates from the definitions of the complexity of social systems, is an attempt to identify those characteristics of terrorism that are typical of modern complex society. In the second part of the survey, the methods of dealing with terrorism and terrorist activities related to complex systems studies are the primary focus. The scope of the topics related to terrorism is obviously too broad for a single book chapter. Therefore, not all complex models applied to the study of terrorism are included. Rather the selection of models focuses upon those models which are most representative of the ways in which CAS research approaches terrorism. For the same reason, two issues are purposely left for separate consideration—cyber-terrorism and the applications of complex systems-related ideas in combating terrorism.

The conclusions stemming from this study should allow for a better understanding of terrorism as a factor of security/insecurity in contemporary complex society. This chapter is also designed to function as a point of departure for understanding the potential changes and improvements which can be made to various counter-terrorism activities and programs as the result of a better understanding of CAS research and models.

## *Complexity and Social Systems: The Characteristics of Complexity*

Often used as a byword, or even a "buzzword," the term *complexity* has gained a specific role in the language of modern science and social practice. Simultaneously, "complexity scholars," i.e., the authors claiming to study the complexity of nature and society, whether purposely or not, directly or indirectly, stimulate the expectations of policymakers by assigning marketable titles to their works: "Hidden Order" [2],

"Harnessing Complexity" [3], "Order out of Chaos" [4], "Understanding Complex Organizations" (repeated in various contexts), etc.

The demand for actionable knowledge from policymakers, military planners, bankers, financiers, managers, and others as well as the attempts to provide useful responses by the scholarly community is nothing unusual in and of itself. A new element in the discourse between practice and "complexity studies" has resulted from the awareness of the limited possibilities, or even the impossibility, of the analysis and prediction of various social phenomena. This impossibility is often expressed in a declaration of fuzzily defined concepts which refer to the "complexity" of the subject of research.

The question thus arises: how can we understand the complexity of social systems and social phenomena?[1] If limited possibilities for explanation, unpredictability, or low reliability of prediction are key features of complexity, then ideas drawn from complexity studies may be able to help social scientists to better understand complex social systems. This question holds a special significance in policy-oriented, normative sciences dealing with society—economics, finance, management theory, international relations, security theory, peace, and conflict studies, which aim not only at description and explanation, but also at providing guidance for action.

The need or the fashion of studying the complexity of society has brought about a tremendous wave of writings in which the authors, whether they be sociologists, political scientists, economists, mathematicians, physicists, or biologists, attempt to describe, explain and, in particular, to predict phenomena occurring in social life. There are numerous writings in the social sciences, in economics, management, and finance where the authors use such concepts as "systems theory," "complexity," "equilibrium or stability," "non-equilibrium dynamics," "the onset of turbulence," "strange attractors," "catastrophe theory," "emergence," "self-organization," "chaos," "fractals," etc. The nature and character of the applications of those concepts varies tremendously, beginning with very precisely defined mathematical models and ending with analogies and metaphors, with the vast majority of the literature being dominated by the latter approaches.

At the same time, in works which are rooted in broadly defined systems thinking, for example, research using an applied systems approach, or research based on general systems theory, CAS theory, complexity studies, or even more ambitiously, "complexity science," many authors have discovered that the concepts which they have developed when embodied in mathematical models, or even when used as analogies and metaphors, can help to encapsulate various aspects of social reality [5]. Since "complexity" is the key concept in all these discussions, it is necessary to ask the following question: "Is it possible to describe the complexity of social systems knowing that unequivocal definitions are unachievable?"

Numerous problems arise in defining terms associated with "studies of complexity," "complex systems studies," or the like. The author refrains from using

[1] Social system is understood herein as human system. In general sense social systems may also include other actors—animals or artificial agents.

the terms "complexity theory," or "complexity science," although an idea of the "emerging science of complexity" has been already proposed by Waldrop [6], the first attempts to explicitly study complexity and complex systems go back at the very least to the works of Weaver [7] (disorganized complexity and organized complexity), and includes those of Simon [8] (the architecture of complexity), and Ashby [9] (the Law of Requisite Variety). In his search for explaining the meaning of complexity, Lloyd [10] identified 45 definitions of complexity. In other writings, numerous definitions and interpretations of complexity and of its characteristics have been proposed. In particular, the following authors have been recognized as conceptual leaders in this enterprise: Prigogine and Stengers [4], Gleick [11], Gell-Mann [12], Holland [2], Kauffman [13], Bar-Yam [14], Axelrod and Cohen [3], Biggiero [15], Kwapień and Drożdż [16].

The most universal characteristics of complex systems are: large numbers of constituent elements and interactions, non-linearity of the characteristics depicting its behavior, various forms of hierarchical structure, non-decomposability, unpredictability, and self-organization. Complexity can also be characterized by a multitude of other ideas and exemplified by a variety of phenomena such as: artificial life, *autopoiesis,* universal bifurcation, co-evolution, emergent properties, stability at far-from-equilibrium states, fractal dimensionality and scaling behavior, power-law behavior, self-organized criticality, sensitivity to initial conditions ("butterfly effect"), spontaneous self-organization (such as stereo-specific autocatalysis), and other similar phenomena typically observed at the edge of chaos.

In some instances, complexity studies or complexity science is identified solely with CAS, which are treated as a specific case of multi-agent systems (MAS). As of the time of this writing there is no universally accepted interpretation of the term "complex adaptive systems." Following the initial concepts of CAS [2], their most representative properties are the following: non-linearity of interactions (internal and external), emergent properties arising from simple rules of behavior for their constituent elements, self-organization, diversity of internal structure, existence at the edge of chaos, and co-evolution with other complex entities or with the environment. The above list is obviously not complete. CAS are regarded at present as an instrument of modeling of collective phenomena in all disciplines of science. Due to the possibility of creating elements of theoretically unlimited varieties of behavior, they are perceived as the most promising tool of modeling for broadly defined social phenomena and social systems.

The methods applied in complexity studies include: agent-based modeling (less commonly known as generative computer simulation), cellular automata or iterative arrays, catastrophe theory, CAS research, data mining, nonlinear dynamical systems modeling (otherwise known as "chaos theory"), fractal geometry, genetic algorithms, neural networking (otherwise known as distributed artificial intelligence), power-law scaling, scale-free network dynamics, self-organized criticality, and synergetics.

In order to identify the meaning of complexity, one must base the meaning on some particular properties of the relationships between human observers (or the observation of systems in general) and various types of observed systems. These systems may be natural or artificial, and include social systems. Biggiero [15, pp. 3, 6] treats the predictability of the behavior of an entity as the fundamental criterion for distinguishing various kinds of complexity. In their novel solution to the problem of clustered volatility in financial economics, Ilija Zovko of the University of Amsterdam and J. Doyne Farmer of the Santa Fe Institute [17] describe the complex phenomena which they treat as an "observed behavioral regularity" which is characteristic of Farmer's treatment of scientific laws and law-like behavior in complex systems.

To return to Biggiero, the conceptual foundation which he proposes is an interpretation of complexity as a property of objects which are neither deterministically nor stochastically predictable (Gleick describes a similar line of reasoning with respect to Stephen Smale's work and Smale's discovery of the properties of systems which are neither periodic nor random in their behavior). In Biggiero's words, "Complexity refers to objects which are predictable only in a short run and that can be faced only with heuristic and not optimizing strategies" [15, p. 6]. He proposes three characteristics of complexity: (a) objects not deterministically or stochastically predictable at all; (b) objects predictable only with infinite computational capacity; (c) objects predictable only with a transcomputational capacity (beyond the Bremermann's limit) [15, 18].

Edgar Peters [19, 20] has proposed similar definitions, with the additional typology of chaotic systems which exhibit one of two kinds of behavior, globally deterministic but locally random (such as the weather, which at one level of analysis contains global boundaries recognized as seasons, within which virtually any local variation might be observed but which many readers will also recognize as being distinguished by a global strange attractor—the Lorenz Attractor) or objects which exhibit behavior which is locally deterministic over short periods of time, but which is random or unpredictable over extended periods of time and for which no final state of the system can be predicted. Peters discusses this in terms of financial systems such as foreign exchange futures, which are anti-persistent because they have no underlying "fundamentals" (and no well-defined second moment). He has further developed a sophisticated system of autoregressive fractal integrated moving averages, ARFIMA, to more accurately replace the historical methods of ARCH (autoregressive conditional heteroskedasticity), GARCH (generalized autoregressive conditional heteroskedasticity), and I-GARCH (integrated generalized autoregressive conditional heteroskedasticity).

Peters' approach is designed to capture a relatively low dimensional strange attractor present in the ill-behaved time series returns of financial markets [21] and resembles the moving average depth of the order book approach used by Smith, Farmer, Gillemot, and Krishnamurthy [22] in the more fully developed, long-form solution to clustered volatility in financial markets, "Statistical Theory of the Continuous Double Auction." Examples of complex systems defined by strange attractors (closed form global systems) and chaotic attractors (open form global systems) are also discussed in mathematical detail by Ali Bulent Cambel [23]. Finally, the vast majority of the foregoing materials are comprehensively linked together in Bar-Yam's [14] study "The Dynamics of Complex Systems."

## *"Hard" and "Soft" Complexity*

Systems thinking, complex systems studies, etc., can be divided into two basic streams relevant to social science research methodology. The first stream was developed through the use of mathematical modeling and can be called "hard" complexity research by way of analogy to "hard" systems thinking. "Soft" complexity research, a term also coined as the result of an analogy with "soft" systems thinking, includes qualitative concepts of complexity elaborated in other areas such as cybernetics and systems thinking, social science research, and research in psychology [24]. It is necessary to stress that soft complexity initially had two domains—(1) purely verbal considerations about complexity, and then (2) the application of ideas from "hard" complexity in qualitatively defined situations. Subjectivity or qualitative methods are the main aspect of complexity in the "soft" approach. This quality is just a consequence of the fact that complexity is not an intrinsic property of an object but rather depends on the observer.

In the social sciences, and particularly in sociology, special attention is given to the concepts of complexity in social systems proposed by a German sociologist, Niklas Luhmann. First of all, Luhmann is one of only a few authors who has attempted to elaborate a comprehensive definition of a social system based solely on communication and on the concept the *autopoiesis* (self-creation) of biological systems. *Autopoiesis* means "auto (self)-creation" (from the Greek: *auto—αυτό* for self- and *poiesis—ποίησις* for creation or production), and expresses a fundamental dialectic between structure and function. The concept of *autopoiesis* was introduced by Chilean biologists Umberto Maturana and Francisco Varela in the early 1970s. It was originally presented as a system description that was designed to define and explain the nature of living systems [25].

*Autopoiesis* also refers to self-reference and to the role of the observer. It is reflected in the assertion: "everything said is said by an observer" [25, p. xix]. Due to such a self-referential approach, the concept of *autopoiesis* was criticized as a form of solipsistic methodology and radical constructivism.

The concept of *autopoiesis* was used by Luhmann to elaborate an indigenous theory of social systems, and has become one of most popular universal social theories. He defines a social system of conscious units as an *autopoietic* system of meaningful communication. In this case, *Autopoiesis* refers not to the tangible attributes of a system but to communication [26, 27].

The theories of social systems proposed by Luhmann are broadly discussed in social science, especially in Europe. The Luhmann concept of "soft" complexity is likely its most influential interpretation in contemporary social theory. According to Luhmann, a complex system is one in which there are more possibilities than can be actualized. *Complexity of operations* means that the number of possible relations exceeds the capacity of the constituent elements to establish relations. It means that complexity enforces selection. The other concept of complexity is defined as a problem of *observation.* Now, if a system has to select its relations itself, it is difficult to foresee what relations it will select, for even if a particular selection is known, it is not possible to deduce which selections would be made [26, p. 81].

The idea of complexity promulgated by Luhmann has also been applied to defining risk in social systems. The existence of a large number of elements in a given system means that not all elements can relate to all other elements. Complexity implies a need for selectivity, and the need for selectivity means contingency, and contingency means risk [28].

The complexity of social systems developed by Luhmann is strongly linked to self-reference since the irreducibility of complexity is also a property of the system's own self-observation, because no system can possess total self-insight. In "hard" complexity, this approach might be likened to that of Gödel's second incompleteness theorem, which proves that no axiomatic system can demonstrate its own consistency [29]. This phenomenon is representative of the epistemology of post-modern social science, where observation and self-observation, reflexivity and self-reflexivity, and subsequently, self-reference and recursion have been playing increasing theoretical roles. According to this interpretation, social systems are self-observing, self-reflexive entities attempting to solve emergent problems through the processes of adaptation (learning).

## *Social Systems as "Complexities of Complexities"*

Applications of the concept of complexity, with its multitudinous interpretations in the social sciences, are becoming even more difficult to describe and explain due to another obstacle—the vast multitude of meanings attributed to the phrase "social systems." The basic assumption is that social systems are mental constructs of the observers (participants) as interpretations of the behavior of their components and of the entities which make up the system. In this context, the most important distinction in defining social systems lies in defining the role of the participant–observer. If she/he remains outside the system and is not able to interfere with the system's behavior, then a physicalist approach can be applied (obviously without the need to refer to quantum mechanics and its own special observer relations). Such an approach belongs to the tradition of "first order cybernetics" in "hard" systems thinking. If the participant/observer is able to exert an impact on the system, then the consequences of reflexivity and self-reflexivity must be taken into account. Under such circumstances, "second order cybernetics" or "soft" systems thinking become the basic methods of research [30].

The complexity of social systems is more difficult to comprehend since it is always the result of an intersubjective discourse. The "hard" approach allows for the more precisely defined tangible attributes of the system to be described as measurable quantities with a "strong" ratio scale that possesses tangible characteristics. The "soft" approach makes the description much more difficult since inter-subjectivity depends on the transfer of imprecise meanings in the discourse. In both cases it is necessary to consider the limitations stemming from the reification of subjective/intersubjective categories. It may thus be concluded that if studies concentrate upon the "tangible" measurable attributes of social systems, then "hard" complexity methods, mainly mathematical models, including simulations, can be applied. Otherwise, the discussion must also include reflexive ideas taken from "soft" complexity studies. Therefore a mixed approach is necessary—mathematical modeling and/or analogies and metaphors [31].

There is a specific set of factors which allow us to differentiate between traditionally defined systems thinking and complexity research, at least through the mid-1980s. While systems thinking sought holistic ideas and universal patterns in all kinds of systems, complexity research defined its goals in a more specific manner. A common theoretical framework, the vision of underlying unity illuminating nature and humankind, is viewed as an epistemological foundation of complexity studies [6]. This claim for unity results from an assumption that there are simple sets of mathematical rules that, when followed by a computer, give rise to extremely complicated, or rather extremely complex patterns. The world also contains many extremely complex patterns. In consequence, it can be concluded that simple rules underlie many extremely complex phenomena in the world. With the help of powerful computers, scientists can root those rules out. Subsequently, at least some rules of complex systems can be unveiled.

Two important conclusions with respect to studying social systems in particular can be drawn here. Firstly, in all discussions on the complexity of social systems composed of conscious elements, the role of the observer–participant must be taken into account, even when studies concern objectively defined complexity. This postulate does not necessarily mean radical constructivism (in which case the observer invents reality). It should simply be remembered that when the quantitative modeling of social systems is conducted, that no part of the model is absolutely objective. Second, human systems are characterized by the presence of all sources and types of complexity [15]. We might then summarize the discussion by noting that, in a universal sense, all or many collective phenomena may be complex, including, for example, animal or artificial social systems, but human systems made of conscious elements are the "complexities of complexities."

## The Linguistic Approach to the Complexity of Social Systems

All of the aforementioned barriers to the interpretations of social systems' complexity can be analyzed with reference to linguistics. Systems thinking/complex systems studies or whatever name may be used for the subject (viz., our earlier mention of nonlinear dynamical systems modeling, general systems theory, CAS research, etc.) can be used in the social sciences as a great source for analogies and metaphors as well as mathematical models. According to this distinction, the term "(formal) models" refers solely to mathematical structures. Using a deepened approach, attention should be paid to three of Wittgenstein's [32] "language games," including the meaning of three utterances: (1) the meaning of social systems, (2) the meaning of complexity, and (3) the meaning of ideas in which the concepts of social systems and the concept of complexity are together applied.

Mathematical models can be applied in three areas of complexity studies: computationally based experimental mathematics; high precision measurements made across various disciplines and confirming the "universality" of various complex systems properties; and mathematical studies embodying new analytical models, theorems, and results (see, for example, [33–35]).

Models, analogies, and metaphors deriving from systems thinking and complexity studies are gaining a special significance in the social sciences. Mathematical models are associated with "objective" research. Analogies and metaphors taken from complex systems studies are related to ideas drawn from "rational" science. They are treated as "scientific" and may provide additional political influence in the discourse resulting from "sound" normative (precisely prescriptive) legitimacy in any policy-oriented debate.

In the application of complexity-based analogies and metaphors to the social sciences, the following nine approaches can be identified:

1. Descriptive
2. Explanatory
3. Predictive
4. Anticipatory
5. Normative
6. Prescriptive
7. Retrospective
8. Retrodictive (backcasting)
9. Control and Regulation

Following the distinction from traditional cybernetics and control, a regulation approach can be also proposed. In normative social sciences this approach is expressed by the way in which the dominant analogy or metaphor influences the control of a system (i.e., they differ for mechanistic, evolutionary, and learning systems).

Complexity associated with nonlinear dynamics adds some new elements to our knowledge of social dynamics. We are aware that social systems are uncontrollable, but even the desirability of such control has already been put in doubt. Self-organization is regarded as a desired pattern for the dynamics of economics and politics. The value of this proposition is reflected in Hayek's [36] interest in the complexity of social systems as an argument against centrally planned economies.

Another lesson that nonlinear dynamics and complex systems teaches us is that social change, or in a broader sense, evolution, is produced by both deterministic historical factors and chance events that may push social phenomena to new patterns of behavior. Thanks to a better understanding of the confluence of chance and determinism in social systems, we may now have a vastly improved opportunity to learn what kind of actions we have to undertake, or even perhaps, what kind of norms we have to apply in order to reach desired social goals.

It must be also reminded that analogies and metaphors of rather loosely interpreted non-linearity, chaos, complexity, self-organization, etc., in many instances have become the backbone of the post-modernist (post-structuralist) new science.

Reaffirmation of limited predictability has become an epistemological foundation of discourse-based science. Numerous examples could be quoted here, but as an illustration, it is worth recalling the synthesis of the post-modernist ideas of Braudel and Prigogine's concepts on far-from-equilibrium states made by Wallerstein [37, pp. 160–169] in modeling social systems, although solely at a metaphorical level.

The above epistemological links between complexity research and the social sciences are predominantly associated with "hard" complexity. However, the inputs to this area from "soft" complexity research are equally significant. The reflexive complexity of society has become one of the primary foundations of post-modernist social theory [26].

Unfortunately, various abuses and misuses of the theory may occur, particularly when eminent social theoreticians of post-modernism/post-structuralism treat analogies and metaphors drawn from "hard" complexity research carelessly, and to a lesser extent from their use of "soft" complexity research itself. Several examples of such abuses are mimicked in the so-called "Sokal Hoax" and there are other examples which have been described by the originator of that hoax [38].

In summarizing the considerations discussed above, we may conclude that the application of complex systems analogies and metaphors to the social sciences exposes two basic weaknesses of the approach. First, in most of their applications, the authors have failed to explain that these methods are useful primarily, if not entirely as purely descriptive or narrative instruments. The application of such analogies and metaphors for prediction and norm-setting is always limited by their reification. This limitation has brought about two sets of rather undesirable consequences.

First, in theoretical research, a great deal of time, energy, and money has been expended on what are inherently futile efforts to make use of these analogies and metaphors, and the research positions which they support, more "scientific," "objective," or "analytical." Often this produces a non-causal literature which is filled with "objective" terms such as "stability" or "equilibrium" in the attempt to make the exposition sound more "scientific" when what has actually happened is that the author/authors have simply embedded supernumerary terms with hidden normative loading. Secondly, researchers who are limited by inclination, profession, or capacity to employing a non-quantitative analogical or metaphorical approach to complexity may seek to enhance the influence and credibility of their work by finding someone else to add some kind of quantitative data or mathematical treatment, whether that treatment is appropriate to the subject or not and whether or not it brings any additional value to the research. John Baez [39] characterizes this kind of effort as taking place along the lines of "I'm not good at math, but my theory is conceptually right, so all I need is for someone to express it in terms of equations." Likely any researcher who has presented or chaired the presentation of a "hard" complexity application to one or more social science problems at a major research conference has been approached at one time or another to undertake this kind of work for someone who is trying to lend a more scientific flavor to their research. Rather than increasing the legitimacy of such research, the metaphorical and analogical approach, particularly when supplemented by spurious mathematical or quantitative terms drawn from the physical sciences, has frequently served to obscure or to decrease the legitimacy of the research [39–41].

In addition it should also be remembered that contrary to the approach used in physics, where axiomatization is possible, mathematical modeling in the social sciences—with the exception of the axiomatic approaches used in economics—always has its origins in operationalization. However, in such cases, the process of building operationalizable definitions begins from a "central metaphor" or "stylized fact." This core element of the research is a qualitative idea later developed with the use of formal models, e.g. equilibrium, stability, risk, and even complexity! In such cases, the selection of the model is subjective in all possible ways—self-reflexive and self-referential (from the point of view of the modeler).

## Terrorism as Security Threat for Modern Complex Society

### *Terrorism as a Social, Political, and Military Phenomenon*

A plethora of approaches to the study of terrorism have been developed in the literature. They include sociological, psychological, anthropological, political, legal, philosophical, and military approaches to the problem. As a result of this diversity, only a brief overview of the study of terrorism is presented in this chapter. This overview is not a recapitulation of the specialized studies presented in later chapters of the book, but rather is undertaken for the purpose of allowing us to better understand the state of contemporary theories regarding terrorism as a complex social phenomenon.

It is commonly agreed that terrorism is not merely a contemporary phenomenon and that there is no universal definition of terrorism. A study by Schmid and Jongman in 1988 identified 109 definitions of terrorism. The study included a total of 22 different definitional elements. It is not possible to find a consensus in defining terrorism, since in addition to purely conceptual barriers, there are important normative (predominantly ideological), political, and legal obstacles to finding a common definition. Much of this difficulty is derived from the potentially relative character of certain cases of terrorism, which is reflected in the question "terrorists or freedom fighters?"

The needs of society, typically expressed in political, legal, economic, or military terms represent a kind of pluralist demand for dealing with the problems of terrorism [42]. This demand requires researchers at a minimum to attempt to put some order into theoretical discourse in the context of both a domestic and an international framework as a prelude to enlightened or at least more effective policymaking with respect to terrorism than we have seen to date. However, this also exacerbates tensions to the extent that on the one hand, international institutions, especially the UN and its agencies are attempting to elaborate more or less universal interpretations of terrorism, while at the nation-state level efforts are tailored to each individual nation's political climate and institutional demands. Finally, the entire picture is further complicated by the fact that much of what is commonly characterized as terrorism is undertaken by non-state actors (NSAs) who are not easily incorporated into either type of framework (national or international).

A rank ordering of publications and websites is also useful in characterizing existing studies of various aspects of terrorism: Laqueur [43, 44], Schmid [45–47], Schmid and Jongman [48], Defining Terrorism [49], Gadek [50], White [51]. From the vast literature on terrorism, published before, and predominantly, (which is understandable) after 2001, a collection of works and the concepts contained within those works was surveyed in order to provide a background for this chapter's analysis of the linkages between complex systems studies and the study of terrorism.

Schmid as far back as the 1980s proposed an approach basing on the terms applied in definitions of terrorism, which led to the elaboration of an "academic consensus definition," accepted by the UN [46, pp. 39–81; 51, p. 12]: "Terrorism is an anxiety inspiring method of repeated violent action, employed by (semi) clandestine individual, group or state actors, for idiosyncratic, criminal or political reasons, whereby—in contrast to assassination—the direct targets of violence are not the main targets. The immediate human victims of violence are generally chosen randomly (targets of opportunity) or selectively (representative or symbolic targets) from a target population, and serve as message generators. Threat-and-violence-based communication processes between terrorist (organization), (imperiled) victims, and main targets are used to manipulate the main target (audience(s)), turning it into a target of terror, a target of demands, or a target of attention, depending on whether intimidation, coercion, or propaganda is primarily sought."

Schmid has extended this definition [46, pp. 86–87; 47, pp. 158–159], by identifying 12 dimensions of terrorism. This approach may prove helpful in developing a better understanding of terrorism, but at the same time it only reaffirms our inability to elaborate a set of commonly accepted definitions of terrorism. One illustration of the difficulties involved in describing and explaining terrorism is the discrepancies between various definitions created by state institutions in the USA, such as the FBI, the Department of Defense (DoD), the Department of Homeland Security (DHS), the Central Intelligence Agency (CIA), and the US Department of State [50].

Typologies of terrorism and the analysis of terrorist behavior constitute the second unequivocal component of discourse on terrorism. It is relatively easier to identify analytical approaches to terrorism, which can be studied from five different conceptual perspectives [45]: (1) terrorism as/and crime; (2) terrorism as/and politics; (3) terrorism as/and warfare; (4) terrorism as/and communication; and (5) terrorism as/and religious/ideological/political/ fundamentalism. In addition, the sources of terrorism constitute a hierarchy—from global issues to religious fanaticism. Terrorism treated as a method of warfare is an example of asymmetric warfare, or of the net-wars/cyberwar. Although each perspective has its specificity, this survey of the applications of complexity methods for the prediction of terrorism behaviors does not separate the perspectives.

As a consequence of the absence of definitions, the elaboration of typologies of terrorism is, naturally, highly challenging. Delving into the details of each typology and explaining detailed criteria is not necessary at this point in the analysis. Instead, it can be done when discussing specific links between complexity studies and terrorism. As a point of departure from discussing the typologies of terrorism, ten bases of classification can be used to differentiate various terrorist activities [48,

p. 40]: (1) Actor-based, (2) Victim-based, (3) Cause-based, (4) Environment-based, (5) Means-based, (6) Political-orientation-based, (7) Motivation-based, (8) Purpose-based, (9) Demand-based, (10) Target-based.

These bases can be applied in producing a multitude of typologies which can be augmented by other types of groupings. Applying geographical criteria, it is necessary to distinguish between terrorism on a local, national, or transnational level. Institutional criteria define state and non-state terrorism. Economic criteria help to treat terrorism as an economic phenomenon and to consider, for example, funding of terrorism, and its economic consequences [52]. With respect to military considerations, terrorism can be an element of asymmetric or irregular warfare (the "weapon of the weak") and in many instances can be linked to guerilla warfare [53].

There is a very specific type of terrorism, which in addition to the lack of definitional clarity, brings about political and ideological disputes. The role of the state, not only as a defender, but also in part as a supporter of terrorism and/or a performer of terrorist activities is also an issue for both theory and policymaking. Two different roles can be distinguished here: state-sponsored terrorism and state terrorism. While the former concerns the activities by proxies, the latter concerns direct involvement of state institutions. The main difference in the interpretations of these links between terrorism and the state lies between those who claim that the state may commit terrorist acts and those who deny such views referring to those definitions of terrorism acts that are committed only by no-state actors [49].

In the era of the development of Information Technologies, cyber-terrorism, a new form of terrorism has become an important threat to modern society. Taking its popular name by adding "cyber" to any social phenomena, e.g. cyber-punk, cyber-society, cyber-space, cyber-warfare, etc., shows the pervasiveness of the connections between the applications of advanced computer networks, and nearly every aspect of modern society. Cyber-terrorism is viewed as the most recent but at the same time one of most dangerous forms of terrorism. Similarly, as in the case of universal definitions of terrorism, no agreement about defining cyber-terrorism has been achieved. The most quoted definition of cyber-terrorism was proposed by Denning in 2000 [54]: "Cyber-terrorism is the convergence of terrorism and cyberspace. It is generally understood to mean unlawful attacks and threats of attack against computers, networks, and the information stored therein when done to intimidate or coerce a government or its people in furtherance of political or social objectives. Further, to qualify as cyber-terrorism, an attack should result in violence against persons or property, or at least cause enough harm to generate fear. Attacks that lead to death or bodily injury, explosions, plane crashes, water contamination, or severe economic loss would be examples. Serious attacks against critical infrastructures could be acts of cyber-terrorism, depending on their impact."

There is another form of terrorism that is very difficult to identify and to prevent—"lone wolf terrorism." Terrorist acts by individuals not aligned directly to any social group are particularly difficult to identifying and combat, and are even difficult to study. The key factor of the counter-terrorist response concerning locating lone wolf attacks is in knowing not who will carry out an attack (almost an impossibility) but rather in knowing how such attacks are formulated [55, p. 47].

When studying the links between complex systems studies and terrorism, it is also necessary to recall examples from the philosophical discourse on terrorism in which the ontological, epistemological, and axiological aspects of terrorism are considered in reference to individuals and society. As one of the most representative examples, the ideas of a French influential philosopher Jean Baudrillard provide an excellent foundation for study of this aspect of terrorism.

Terrorism as a subject of analysis after September 11, 2001 was present in several works of Baudrillard [56, 57] who is sometimes viewed as disputable and accused of biases and anti-Americanism. The level of analysis of terrorism proposed by Baudrillard reaches the axiological roots of the functioning of modern society referring at the same time to its systemic properties at the global and local levels. His primary ideas about the systemic consequences of terrorism refer to the internal fragility of the modern world. The more the system is globally concentrated toward ultimately constituting a single unified network, the more it becomes vulnerable to single point failure—examples of hacking and September 11, 2001 only confirm that observation [56].

## *Security Theory and Terrorism*

Although terrorism has always been an important aspect of security studies (both domestic and international) it was not present in mainstream discourse until after the 9/11 attacks. In fact, the terrorist attacks of September 11, 2001 have made it one of the main focal points for the theoretical consideration of security. Both prior to and during the Cold War, when classical security theory, which focused primarily on deterrence, was developed terrorism was not analyzed as a principal component of either domestic or international security, but rather as a "specialty" subject, treated apart from the mainstream of the discourse on security theory, e.g. Schmid and Jongman [58], Laqueur [44, 59].

Contemporary discussions of the theoretical aspects of terrorism are generally conducted in the context of one of three major competing International Relations (IR) theories: realism, liberalism, or constructivism. The realist approach depicts international relations as a struggle for power among strategic, self-interested states. International order is based upon power or force projection capabilities. However, realism is not a single theory. There are two cross-cutting, dichotomous versions of realism. The first is classical realism, which came out of World War II and the failure of the 1930s legalist school of international relations, and which is uncompromising in its placing of the national interests of each sovereign nation against those of every other nation. The second, and more popular version of the theory is neorealism, which comes out of mid-1970s regime theory, which is based on the observation of the growing importance of self-organized international agreements (primarily economic, financial, educational, and technological agreements) which transcend the traditional boundaries of national self-interest [60, 61]. One might characterize these two theories as an offensively oriented version of the theory of international politics (realism) and a defensively oriented version (neo-realism) of the same over-arching theoretical framework [62, p. 150]. Rational choice theory [63], or neo-institutionalism [64], which explains many of the phenomena discussed in neo-realism in terms of the institutional environment and institutional behavior can also be regarded as an offshoot of this group of theories. Neo-institutionalism has many elements in common with the complex systems approach to economics and economic history [64–66].

Leaving apart the differences between these two somewhat differing versions, both realist theories explain the United States' forceful military response to the September 11 terrorist attacks, as terrorism is countered by the use of force. Commenting on those attacks, Jack Snyder [67, p. 56] argues: "Despite changing configurations of power, realists remain steadfast in stressing that policy must be based on positions of real strength, not on either empty bravado or hopeful illusions about a world without conflict." In other words, this means that terrorist threats are clearly defined and should be always dealt with relevant resolute forces, while no other approaches should be taken into account.

Liberalism in international relations theory/security theory is derived from an assumption that international politics is not a "jungle." Liberals see world politics as a cultivatable "garden," which combines a state of war with the possibility of a "state of peace" [68, p. 19]. Reflecting the aims of the individual, liberal states view security not only in military terms, but also in terms of the protection and promotion of individual rights. In this approach, combating terrorism focuses far more on the application of legal instrumentalities than on the use of military force (see, for example [69]).

The third theoretical concept of security is based upon constructivist and post-modernist foundations of social science. In this case, security is understood in a broadened sense, going beyond political and military issues and is viewed as an intersubjective result of discourse. The concept of the broadened interpretation of security theory developed by the Copenhagen School takes its name from the Copenhagen Peace Research Institute, where new ideas of security were developed in the 1990s [70]. The constructivist approach not only concerns itself with a broadened interpretation of security but also seeks a deepened interpretation, which means that the individual essentially becomes a reference object. This approach gives rise to the concept of *human security* and embodies various aspects of life, e.g., food security, water security, and obviously, the threat of *terrorism to individuals. Curiously, this broader and deeper school of thought arising out of peace studies shares a great deal of common philosophy with the Petraeus Doctrine of counter-insurgency [71, 72].

The essence of the concepts contained in the Copenhagen school's theory of security can be summarized as follows. Security is not treated as a traditional "objective" concept referring primarily to military and political threats. Under the influence of constructivism, post-modernism, and post-structuralism, it is perceived to arise as a result of social discourse, an "act of speech"—performative utterance, and an outcome of "securitization." In this case, security is understood in a broadened sense, going beyond political and military issues, and is viewed as an intersubjective result of discourse. Security refers to the following sectors: military, economic, political, environmental, and societal. Following Buzan et al. [70] the concepts of existential threat and securitization are employed by this approach. Any public issue can be securitized, meaning the issue is presented as an existential threat, requiring emergency measures and justifying actions outside the normal limits of political procedure. Security is thus a self-referential practice, because it is in this practice that the issue becomes a security issue—not necessarily because a real existential threat exists, but because the issue is depicted as such a threat.

Discourse that takes the form of presenting something as an existential threat to a referent object does not by itself create securitization. It is solely a securitizing move and the issue is securitized only if and when the audience accepts it as such. Securitizations in different sectors frequently cannot be separated. Securitization studies aim to gain an increasingly precise understanding of who securitizes, on what issues (threats), for whom (referent objects), why, with what results, and under what conditions [70]. It is then evident that securitization may be influenced by various factors, including political power. Thus, security as a result of the securitization discourse about threats may be biased and prone to distortions deriving from the interests of the dominant securitizing actor.

There exists a specific link between complex systems studies and securitization, which concerns many issues beyond terrorism. Securitization is a self-reflexive idea since it includes the reflection of the observer/participant about the process of defining a threat to security. In terms of complex systems studies, it means that all securitized aspects of social life, including terrorism, can be analyzed as characteristics of modern complex society. In such case the idea of "soft" complexity emerging when conscious actors/observers face challenges of reflection, self-reflection and self-reference in studying terrorism—similar to other threats—is part of social reflection. In this sense, securitization connects to soft complexity through its epistemology.

Buzan [73], one of the founders of Copenhagen School, treats the approach to terrorism in the USA after September 11, 2001, frequently labeled as the "global war on terrorism" (GWoT), as an example of macro-securitization comparable with the Cold War. Macro-securitization in this case means that terrorism is treated as a global threat to all countries, not only to the developed Western world. The sense of macro-securitization is expressed in the statement: "The war on terror is like a new Cold War where everything is subordinated to a single purpose" 73, p. 1106.

Terrorism on a global scale is securitized according to the above pattern and is linked to a large number of other phenomena, which have previously been securitized. As examples, Buzan recalls major political declarations by NATO, the European Union, and the US Government, arguing about potential security threats resulting from the links between terrorism and organized crime, especially in drug trafficking, human trafficking including illegal labor, prostitution, and slavery as well as weapons sales, the proliferation of weapons of mass destruction (WMD), regional conflict, and state failure. In relation to the securitization of WMD, the macro-securitization includes a strong concern that not only "rogue states," but also terrorist organizations might acquire nuclear weapons or other WMD [73, p. 1105].

The application of the idea of macro-securitization to the "Global War on Terrorism" is not only an intellectual exercise, but is also a valuable instrument for the analysis of the consequences of terrorism for modern society. In comparison to other viewpoints, such as neo-realism, it allows for a deeper explanation of the full range of the dimensions of counter-terrorism policy and its associated activities. What is especially important in the constructivist approach to security is the consideration of the possibility of an unjustified securitization, which may be used as an instrument for the implementation of extraordinary actions with negative consequences to society. The self-reflexiveness of the constructivist approach might best be characterized by the classical question "*Quis custodiet ipsos custodes*?" (Who guards the guardians?) One can also see another commonality with soft complexity science at this point and that is the recursive feedback loop between act of securitization and its self-conscious evaluation.

The main conclusion of the constructivist view (i.e., that the macro-securitization approach to terrorism is comparable to the ideological foundations of the Cold War) may be summarized as follows: "It thus becomes clear that terrorism poses a double threat to liberal democratic societies: open direct assaults of the type that have become all too familiar, and insidious erosion as a consequence of the countermeasures taken. It is easy to see how this dilemma drives some towards seeking a solution in total victory that will eliminate both the terrorists and the contradiction. But if it is impossible to eliminate terrorists, as is probably the case, then this drive risks the kind of permanent mobilization that inevitably corrodes liberal practices and values" [73, p. 1117].

Broadly speaking, then, the intersubjective interpretation of security proposed by the Copenhagen School gains significant ground for post-modernist, critical approaches to security theory. The main assumption of the critical studies derived from the Frankfurt School is that post-modernism and constructivism are examinations and critiques of society and culture, constructed by applying knowledge from the social sciences and the humanities. Critical security studies, by exposing the intersubjective character of security, put in doubt classical ideas such as rationality, prediction, etc. Although already visible in initial works concerning broadened security theory [70], constructivism has exerted a strong impact on the development of recent security theory, especially in Europe.

Critical terrorism studies can be then viewed as a part of critical security studies. Some ideas about terrorism presented by the critical approach were perceived as controversial, e.g., the difficulties in defining terrorism, the role of state-sponsored terrorism and state terrorism, and cultural biases in terrorism studies. From the point of view of the application of complex systems studies to research on terrorism, the main contribution of the critical approach lies in extending reflection on terrorism to the problems of culture, cognition, and perception [74, 75]. These considerations are in a natural way connected with "soft" complexity [14].

As a synthesis of approaches to terrorism as a threat to the security of contemporary society and reflecting the complexity of terrorism, we should like to introduce new tools and bring about an updated paradigm which will allow the various targeted state and NSAs to deal with terrorism more effectively. The first thing to do is to recognize that there is a problem, and then recognize that it is a new problem and it calls compellingly for a new kind of solution. To that end we must be prepared to rename and re-man some of our traditional institutions. However, this kind of activity has to be driven by market pull (that is it must already have the support of the general populace, and not just from Congressmen and Congresswomen, Senators, and present and past high level officials). We suggest that this modern, multi-faceted organization should focus exclusively on terrorists and terrorist activity, in order that institutional memory is preserved intact, and to provide for an orderly succession of DCI's and DNI's.

With respect to the change in linguistic paradigm, we shall refer to the complex problems of terrorism in a complex global society as the "New Security Dilemma." That name, hopefully indicates to the reader some of the self-reflexivity which characterizes the new discipline. Similarly the new nomenclature should be a semiotic indicator that the problems and the levels of analysis reflect the shifting of the focus in the discipline from state security to security of other reference objects, including individuals, thus creating a security theory, with the important caveat that classical realism can still provide a new approach to terrorism. Following Cerny [76, p. 30]: "In this environment, civil wars, ethnic wars, cross-border wars, warlordism, terrorism, and the like must be addressed not as military questions but rather as social, economic, and political ones. What is needed is not so much a war on terror as a political, economic, and social war on the causes of terror—uneven development, inequality, injustice, and, perhaps most importantly, the incredible frustrations engendered by the revolution of rising expectations in a globalizing world."

## Systemic Interpretation of Terrorism in Modern Complex Society

Terrorism can be viewed as an example of collective behavior and the communication of a group of people operating in a broader social environment. Therefore terrorism may be viewed as a subsystem of variously defined social systems. It is also evident that terrorism is a complex phenomenon and the term complexity applied together with terrorism may have all meanings, beginning from a broadly interpreted metaphor to more rigorous understandings. When searching for the links between complexity studies and terrorism it is necessary to ask the following questions:

1. How can terrorism *treated as a collective behavior* be defined in systemic terms?
2. What are the specific features of modern society which make it more vulnerable to terrorism than the traditional societies of the past?
3. What are the new forms of terrorism, which are particularly able to affect contemporary complex society?
4. What might be the negative consequences of countermeasures addressing this type of terrorism?

Partial answers to these questions may be found in various other discussions about the "old" and the "new" terrorism and in other attempts at periodizing terrorism. Rapoport [77] presented first periodization, distinguishing four waves of terrorism (anarchist, nationalist, 1960s leftist, and the current religious wave). Each wave had a precipitating event, lasted about 40 years before receding, and, with some overlap, faded as another wave rose to take center stage.

This typology was supplemented by the fifth wave identified by Kaplan [78]. The fifth wave of terrorism is characterized by the existence of groups that began on an international wave but which for some reason (clearly these are organizations which have turned inward), cut ties to their international benefactors or ideological/religious bedfellows, and sought to realize a utopian vision of a radically perfected society on the *local* level. The goal of such groups is the creation of a new man and a new woman comprising an ethnicity or tribal society that is the reconstitution of a lost "Golden Age" model or an entirely new world *in a single generation*. There have been such movements emerging from the various waves of Rapoport's theory, and they share a sufficient *zeitgeist* to constitute a kind of wave of their own. From 11 features of the 5th wave, the ideology of creating new men and women, self-isolation from other terrorist organizations, and excessive violence against women seem most specific [78, p. 13].

The impact of terrorism on modern complex society can be better captured with the idea of "new" terrorism partly relating to the above periodizations. The divide between the "old" and the "new" terrorism, proposed initially in the 1990s, is broadly discussed and contested in theoretical arguments drawing largely upon history. The separation between both types of terrorism should be distinguishable by clear criteria of their division, rather like Hegelian *Thesis* (Greek θέσις) and Antithesis (Greek ιντίθέσις) and, obviously, this is not what has been observed. Several marks of distinction or discontinuity have been proposed by a number of scholars but nothing has yet really come out of this effort. According to Laqueur [79, p. 4], the new terrorism is characterized by a greater accessibility to WMD and unchanged human nature. This kind of terrorism can also be depicted using alternative features proposed by a number of other authors [49, pp. 80–87].

The compilation of the key features of the new terrorism, which is obviously not complete, includes the following features. First, the new terrorism is conducted with, or with the threat of, WMD—chemical, biological, radiological, or nuclear (CBRN). Second, the lethality and indiscriminate nature of the current attacks is rising. The violence used by terrorists in the past is seen as a rational with a discriminating selection of targets in order to send a message to their opponents as well as to their constituents. Third, in contrast to the old secular terrorism, the new terrorism is often religiously inspired. The prospect of seeing WMD employed by religious fanatics only adds an extra level of threat to modern terrorism. Fourth, the network structure of the organization defines much of its behavior [80–82], and is the most specific feature attributable to the new terrorism. In contrast to previous generations of terrorists and guerillas, the new organizations generally do not have a clear, hierarchal organizational structure with an ideological, strategic, and tactical leadership. Such loose network structures make the terrorist groups difficult to detect and infiltrate. Similarly, as in other areas of the economy and social life, such a "networkization" is a result of development of the Internet, which provides access to terrorist methods, weapon fabrication, and like-minded individuals with whom ideology and operational information can be exchanged in an anonymous setting, which has become a specific battlefield between all those who try to hide and those who track them.

According to the previously referenced study [49], related to the fourth feature is a lack of state sponsorship that the new terrorism seems to have. The new terrorists provide themselves with funds through a variety of criminal means, such as smuggling weapons or drug trading and trafficking. In addition, the new terrorists are for the most part amateurs who operate on a peculiar part-time basis [34, 35]. They are often living normal lives in the societies against which they are fighting.

This collection can be supplemented with other characteristics proposed by Banks et al. [83, pp. 17–18]. Incidentally, the set of characteristics for modern terrorism exhibits virtually no overlap, so it perfectly reflects the existing discrepancies which characterize the discussions of terrorism which we previously reviewed. Maintaining the continuity of enumeration, the fifth characteristic of terrorism is international, or even transnational, that is, which has been enabled by the development of the media.

Attacks against political, historic, and cultural symbols can be treated as the sixth attribute of the new terrorism. Finally, contemporary terrorism is increasing due to the development of various modern media, beginning from the classical and ending with the Internet [84]. As observed by Nacos [85, p. 10], the "act of terrorism is in reality an act of communication." On the one hand, commercial media terrorist attacks are the "breaking news" to increase their audience, but at the same time, terrorists know that the modern media constitute a kind of "proxy" weapon.

The above typologies constitute a point of departure for the search of "complexity-era" terrorism. However, they also avoid directly addressing the increasing vulnerability of contemporary society. Only by examining both aspects of the problem (i.e., by addressing changing patterns of terrorist ideology, instruments (weaponry), methods and communications and then examining their impact on specific new characteristics of modern society, including electrical power grids, transportation infrastructure, internet commerce, and medical treatment centers, all of which can be exploited as vulnerabilities by both the "old" and the "new" terrorism, can a more comprehensive picture of terrorism in the "complex society" be accurately elaborated. It must be remarked, however, that those new characteristics are not only vulnerabilities, but have a dual character and may constitute advantages in counter-terrorist activities, for example the development of Internet has completely changed the nature of open source intelligence.

The threat of terrorism in a modern complex society can be treated as one of the attributes of a "risk society" as depicted by Anthony Giddens [86] and Ulrich Beck [87–89]. A risk society develops or emerges as a consequence of changes in contemporary society on a global scale, which, for the most part, can be summarized with a single word—"modernization." One of the causes of the risk is the interdependency between major elements of the global system. The modern complex world provides better living conditions for the average citizen, who in two generations typically has products and services which were inconceivably costly or else simply "inconceivable" in his/her grandparents' day. However, this kind of rapid growth simultaneously creates risks and vulnerabilities resulting from complexity and the unreliability of the technology as well as emergent factors which may negatively impact basic social structures (manufactured risk). As Beck [89, p. 8] puts it: "A core contradiction in contemporary society is the fact that advanced modernity, with the aid of its scientific instruments and its mass mediated communication, is forced to accord the priority to the mega-threats it itself has generated, although it is clear that it lacks the necessary concepts to observe or impute, let alone 'manage,' them adequately—at any rate, not as long as the institutional status quo is absolutized and held constant in an ahistorical manner."

In addition to the above core paradox of the "risk society," one dominant feature of modernity reflexivity means that as a society examines itself, it in turn changes itself in the process [26, 28, 90, 91]. In Beck's [89, p. 11] interpretation, the risk is also a reflexive phenomenon. It directly leads to reference to social *autopoiesis* and makes "soft" complexity studies the fundamental method of studying the risk of social systems. The interpretation of risk shown below only strengthens the argument that ideas developed within complex systems research are the only instruments capable of helping to understand the deepest meaning of terrorism.

Focusing on the place of terrorism in the "risk society," Beck [89, p. 15] describes the paradoxes of terrorist activities. Contrary to economic and environmental crises, which can be understood as side effects of radicalized modernization, terrorist activities must be understood as intentional catastrophes. More precisely, they conform to the principle of the intentional triggering of unintentional side effects. The principle of deliberately exploiting the manifest vulnerability of modern civil society replaces the principle of chance and accident. What is striking is how the global anticipation of terrorist attacks is ultimately "manufactured" in an involuntary interaction with the power of the Western mass media, Western politics, and the Western military. To put it pointedly, the belief in "global terrorism" springs from an unintended self-endangerment of modern Western society.

Terrorism in modern risk society brings about another paradox often exposed in theory and policymaking discussions as an argument against too far-reaching counter-terrorism activities. Beck [92, p. 330] puts it as follows: "… in order to protect their populations from the danger of terrorism, states increasingly limit civil rights and liberties, with the result that in the end the open, free society may be abolished, but the terrorist threat is by no means averted. The dark irony here is that, while very general, risk-induced doubts on the part of average citizens regarding the alleged benevolence of the promises of governments to protect them leads to criticisms of the inefficiency of scholarly and state authorities, critics are blind to the possibilities of erecting (or expanding) the authoritarian state on this very inefficiency."

There may be different descriptions of the characteristics of contemporary society which make it simultaneously more vulnerable and more resilient to terrorism. An example of a typology of such characteristics is given in Table 2.1. Due to the

**Table 2.1** Systemic characteristics of complex contemporary society facilitating/hampering terrorist activities

| Attribute of modern society | Perceived as a vulnerability | Perceived as advantage in protecting against terrorism |
|---|---|---|
| Reliance on modern sophisticated technology—high speed trains, air transportation, energy supply, biotechnology | Modern technologies in all areas of life (non-IT-based control systems) are becoming more sophisticated and vulnerable, e.g. high speed trains, air transportation, energy supply, threats of use of biological weapons | Modern technologies of that category can be used for protection against terrorism, e.g. more advanced methods of tracking terrorists traffic |
| Reliance on IT technology in private and professional life, predominantly the systems of control (non-communication technology) | Systems of control dependent upon sophisticated and subsequently more complex devices, which can become more exposed to targeted attacks (cyber-terrorism) | Sophisticated systems of control can become more reliable due to multiple-level (loop) protection mechanisms |
| Development of monitoring technologies, including data gathering, storing, and tracking | Possibility of hacking and destroying the systems (cyber-terrorism) | Facilitation in identification of terrorist activities |
| Development of communication technology—traditional and the Internet | Possibility of clandestine communication and networking (cyber-terrorism) | Possibility to transfer information and track communication |
| Expansion of the media, including the Internet | Media as the means of communication is one of most important facets of terrorism | Publicizing acts of terrorism may contribute to negative perception of terrorist acts |
| Increased standards of living and changing patterns of behavior leading to more openness and trust in interpersonal relations | This feature of modern developed societies has been significantly damaged by terrorist attacks in the 1990s and especially after September 11, 2001. An example—changing patterns of social behavior in the USA | In an open society of people trusting each other, suspicious behavior can be more easily identified |
| Radical religious inspiration | Increased determination of terrorists and possibility of manipulation | Facilitates disclosure if publicly demonstrated |
| Increasing individualization and changes (weakening) of traditional social ties—family, friends, acquaintances | Facilitates activities of clandestine organizations as well as rise of the "lone wolf" terrorists—ideologically/religiously motivated and/or mentally imbalanced individuals | Weakening social coherence makes less plausible and more difficult terrorist activities of groups united by ideology and political goals |

*Source*: Author's personal research

specificity of cyber-terrorism's characteristics (which can only be associated with that form of terrorism) cyber-terrorism is herein treated very cursorily.

Reflexivity and self-reflexivity, the core ideas of "soft" complexity, bring about a consideration of the meta-role of reflections about terrorism. It is evident that the secrecy of terrorist organizations and the secrecy of counter-terrorist activities are their most fundamental attributes. Therefore, a question should be asked: what is the role of scientific publications in which terrorism is described and analyzed? What if such publications are aimed at improving the knowledge of terrorism, including the amelioration of the methods for identifying and combating terrorist? Last, but not least, if one of the aims of this chapter, and of the entire volume, is to deepen the understanding of terrorism with the assistance of complex systems concepts, in such cases, will those publications serve present and future generations of terrorists as a source of information about the actions of their enemies? If such publications are an element of deception, a question that arises is who is to be deceived—society as a whole, policymakers, or terrorists? It can be easily seen that the arguments above lead to multiple levels of hierarchical considerations about knowledge, and common knowledge (in the sense used in game theory, in the theory of communications, and in modern linguistic philosophy).[2]

This example is just a preliminarily look at how the clandestine character of terrorism and counter-terrorism action reminds one of the roles of reflexivity, self-reflexivity, and self-reference in public discourse on terrorism-related issues. Reflexivity and self-reflexivity have been always inherent attributes of social systems but at present, due to the development of science and technology allowing for higher self-awareness, they are regarded as representative features of modern complex society [26, 90].

## *The Role of Complex Systems in Understanding, Preventing, and Combating Terrorism*

Due to the intricate character of terrorism as a sociopolitical phenomenon, it has become natural that in addition to modern technology, the social sciences and psychology have been applied to studying and combating terrorism. As described by Weinberger in [95], in a *Wired* article with the very thought provoking title, "Can Social Scientists Win the War On Terrorism?," the possibilities that can be achieved in that area, thanks to computational social science (as complexity-based models are called in that article), would significantly enhance the capability of predicting who might join a terrorist or insurgent group.

It seems only natural that, as in other domains of social life, terrorism itself and the activities connected with it are attracting the attention of researchers in complex systems. Since terrorism is part of society seen as the "complexity of complexities," it is natural that approaches relating to the utterance of "complexity" have become common in the theoretical considerations of terrorism and in the context of counter-terrorist policymaking. Terrorism as a complex phenomenon can be also be seen as

[2] The concept of common knowledge was proposed by Lewis [93]. In an interaction of two agents A and B common knowledge exists when A knows that B knows and B knows that A knows that B knows … ad infinitum. It can be extended to n agents. A similar idea was used by Schelling [94] in the Department Store Problem.

a "wicked problem" characterized, among other things, by the absence of a single definition, by a multitude of different and sometimes contradictory approaches, dependent upon unique conditions, as well as having solutions which may be characterized by very grave consequences [96, pp. 4–6; 97].

By applying our earlier consideration of complex systems as a point of departure which avoids the superficial discourse in which terrorism, systems, and complexity are applied as bywords, if not buzzwords, it is possible to enumerate what are and what could be the applications of ideas drawn from modern complex systems studies to better understand and combat terrorism. The characterization of all of the features of terrorism which demand the application of systems thinking exceeds the scope of this study. Herein a kind of analogy with the application of systems thinking in management and international relations results (security studies) can be obtained by showing only the characteristics of terrorism that can be analyzed with the use of ideas drawn from complex systems studies as broadly defined.

As we mentioned earlier, in the application of complex systems mathematical models, analogies and metaphors in the social sciences, and subsequently to all of the various kind of studies of terrorism and in the study of all counter-terrorism activities, the following approaches can be identified: descriptive, explanatory, predictive, anticipatory, normative, prescriptive, retrospective, retrodictive (backcasting), control, and regulation.

In particular, the following aspects of terrorism are and should be a subject of interest to complex systems studies:

- The identification of the sociopolitical context of terrorism,
- The understanding of terrorism as a sociopolitical phenomenon,
- The identification of threats of terrorism,
- The explanation of the causes of terrorism,
- The possible responses to terrorism,
- The prevention of terrorism,
- The range of possible incident responses,
- Terrorism and social communication,
- The media and terrorism,
- The government and terrorism,
- The relationship between law enforcement and terrorism,
- Military responses to terrorism.

First and foremost, systems thinking, including complex systems studies, can be applied to gain a better understanding of terrorism as a sociopolitical phenomenon with all its psychological, historical, cultural, economic, and even philosophical factors. For these and other, similar relationships (e.g., the media and terrorism, politics and terrorism, law enforcement, etc.) the "soft" complexity studies seem to be more applicable, although mathematical models can be applied as well. For example, the study of terrorism by Grant [98] using the concepts of "soft" complexity developed by Luhmann [26, 27, 91] can be analyzed as a case of this type. Grant's study stems from assumptions about the self-referential qualities of the system and strives for a reduction of complexity in the system as it seeks to reduce dependence on uncertain communications.

In Grant's study, the complex communications between two complex social systems—terrorist organizations and counter-terrorist institutions—are scrutinized. Here two different types of systems, on the one hand, a fluid network which characterizes the terrorist organization, and on the other hand, the more hierarchically structured counter-terrorist institutions are communicating both directly and indirectly. In consequence, as the level of uncertainty is increased, so too does the level of complexity increase.

In a simplified way, this study can be used as an example which shows that the "soft" complexity approach to terrorism may not be directly useful for building models but rather, for making the language of the description and analysis of terrorism more sophisticated. It may help us to better understand terrorism, but it also gives us the grounds on which to build more advanced mathematical models ("hard" complexity) and allows for deepened heuristic interpretations of the results of mathematical modeling.

While general explanations are necessary for a better understanding of terrorism, and subsequently for the more efficient prevention of terrorism as well as for the more effective combating of terrorism, it is the mathematical models of CAS research, or "hard complexity," that are expected to be of practical use in counter-terrorism activities. In such cases, the models described in the remainder of this chapter are primarily designed for enhancing predictive capabilities in identifying, preventing, and combating terrorist activities.

## *Applications of Complex Systems Models to Counter-Terrorism Activities: Complexity and Prediction—The Key Issues in Dealing with Terrorism*

Our expectations of complex systems studies with respect to the theories of terrorism and related policymaking are similar to those in other areas of security policy. First, the elaboration of the theoretical foundations of terrorism should allow for a better understanding of terrorism as a sociopolitical phenomenon. Second, the applications of complexity science to the identification and prediction/anticipation of terrorist threats should assist in combating terrorism. Strategic surprise and prediction/anticipation are key issues in all security-oriented studies, but in the case of terrorism, they acquire a special weight since unexpected targets, methods, and time constitute the basic ingredients of terrorist attacks. As Sandler and Enders [99, p. 288] put it: "By making their acts appear random and dispersed, terrorists seek to cause anxiety to the widest possible audience." It may be concluded then, that the key elements of preventing and combating terrorism are vulnerability, surprise, and subsequently, prediction, prevention, and anticipation (pre-emption).

The question concerning the function of terrorist organizations and the occurrence of terrorist attacks is whether they are purely random or whether there are any regular patterns underlying them. The answer is relatively simple. Terrorism, like other social phenomena, demonstrates both random and non-random characteristics, and the only real techniques available for analyzing the latter are the models taken from complex systems studies—assuming that a sufficient amount of good quality data is available for building the models.

Similar to other social phenomena, especially in the normative sciences, expectations of an enhanced level of predictability in dealing with terrorism can be achieved, thanks to the application of complex systems models. However, there is then a new question which arises. If unpredictability, or a low reliability of prediction is the key feature of complex social phenomena, which ideas drawn from complexity studies can help the social sciences better understand the nature of prediction and of its limitations in such a context [100]? This question is of special significance for policy-oriented sciences dealing with social phenomena—economics, management, finance, and security studies—which aim not only at description and explanation, but also at providing guidance for action. In those areas, prediction is the one of main objectives of research.

The question of the limits of prediction has an additional significance in the application of complexity models in the social sciences in "generative social science" [101]. Undoubtedly, models drawn from "hard" complexity studies are helpful in prediction and policymaking but they are also prone to suffer from the general limits of the prediction of social phenomena. It thus becomes a matter of some debate as to the extent to which they can actually diminish the inherent structural impact of those limitations.

All social sciences are bounded by three limitations of prediction. The first stems from the very "objective" character of external reality, number of elements, their interactions, decomposition, fundamental limits of probability theory, and of mathematics in general, (computability, intractability), etc. The second results from limitations of human cognition—we frequently do not know what we do not know. The third derives from the political, cultural, and even the ethical constraints which shape social discourse on future events. In some cases, one may even consider "politically correct" [102] limitations on discourse as the worst-case scenario.

As a matter of fact, all of these barriers overlap and they are all composed of more elementary factors. The latter two barriers are the result of a mixture of psychological and sociocultural constraints. In any social and economic/financial prediction, a special challenge is also associated with "unthinkable" events—"the unknown unknowns." Without delving into an ontological discussion regarding the extent of possible futures and future states of prediction, the barriers to prediction for the behavior of social systems can be reduced to two interrelated groups—epistemological limits at the level of the individuals (observers and actors) and social limits resulting from interactions of those individuals.

## Epistemological Limits

- Limits resulting from constructivism and post-modernism (inter-subjectivity),
- Limits deriving from non-linearity as a distinctive facet of complex systems,

- The fundamental limits of mathematical models, which in some cases, e.g. nonlinearity and indeterminism, computational complexity, and computational (i.e., algorithmic) intractability, can be treated as an ontological limit. In essence, we are not only dealing with the limits of cognition, but we are now also limited by the very structure of nature,
- The subjectivity of our definitions of risk/threat/hazard, etc.,
- The limits to our abilities in identification (taxonomy and typology)
- The logical deficiencies of our methods (i.e., the fallacy of observing the consequent)
- Limitations with respect to our understanding of and ability to communicate about risk, (motivated bias)
- The inherent cognitive limits of the observer—limited physiological capability to identify and process variables (information) depicting the phenomenon (phenomena) under scrutiny; these limits are also the causes of "bounded rationality" [103], framing, and prospect theory [104], all cognitive distortions which affect our reasoning.

**Sociopolitical Limits**

- The sociopolitical consequences of complexity of social systems,
- Sociopolitical pressures (external pressure, conformism, political correctness),
- Sociocultural factors—cultural bias in prediction and anticipation,
- The inherent limits of subjectivity and inter-subjectivity exposed in postmodernist and constructivist approaches, e.g. definitions of meaning, deficiencies in transfer (negotiation) of meaning,
- Uneven access to information, including secrecy (such a situation corresponds with asymmetry of information in economics).

Since terrorism-oriented theory and policy are predominantly based upon mathematical modeling—"hard complexity"—it is worthwhile to remind the reader of some of the basic limitations on the applications of those models in predicting social phenomena.

The epistemological and social obstacles of prediction in social sciences have already been mentioned. The inherent limits of mathematical modeling go to the very foundations of probability theory and of mathematics, in general, described with reference to the works by Kolmogorov [105] and Chaitin [106]—descriptive complexity, Solomonoff–Kolmogorov complexity, Kolmogorov–Chaitin complexity, stochastic complexity, algorithmic entropy, or program-size complexity—and Wolfram [107] (Computational Irreducibility). Those barriers result from attempts to provide an explanation to the observation made by Kolmogorov: "…the epistemological value of probability theory is based on the fact that chance phenomena, considered collectively and on a grand scale, create a non-random regularity." Quoting and commenting on this statement, Mandelbrot and Hudson [108, p. 30] add that: "sometimes this regularity can be direct and awesome, at other times strange and wild."

Bearing in mind the above and other limitations of the prediction of social phenomena, it must be emphasized that the models based upon complexity studies and related to terrorism, which are presented in the remainder of this chapter are treated twofold. First, they allow for prediction, directly and indirectly, by showing observed behavioral regularities in the phenomena under study. Those regularities may not necessarily occur in reality, so prediction is a rather heuristic exercise. Second, awareness of the limitations of predictability of phenomena reflected in those models also, paradoxically, allow for better predictions—it is now known what will not happen.

The application of complex systems ideas to studying terrorism and counter-terrorism can be employed at several different levels, beginning with broad speculation and ending with specific technical considerations. In this enterprise there are three areas of prediction that need to be taken into account:

- Prediction in preventing terrorism, allowing for the elimination of its fundamental political, social, and economic causes;
- Prediction in combating terrorism (the prevention of terrorist attacks, and prediction and pre-emption in counter-terrorist operations);
- The identification of the vulnerabilities of modern complex socioeconomic-political-systems.

At the highest level of generality, complexity-related considerations do not provide any special insights into the possibilities of prediction in the study of terrorism. Rather, they aim to explain terrorism as a sociopolitical phenomenon as was presented in the earlier part of this chapter. At this level of analysis, the limitations of prediction do not play much of an important role. For example, take the case of Russell Ackoff, one of the founders of modern systems thinking. For Ackoff, the systems approach is a vital necessity for combating terrorism at all levels of the societal hierarchy, although he links the emergence of terrorism to the unequal distribution of wealth all over the world. If such injustices could be eliminated through the applications of a systems approach, then the causes of terrorist attacks would be, to a large extent, eliminated [109].

As we saw in the previous chapter, in considering generalized models of terrorism and complex systems, Fellman and Wright [110, p. 3] emphasize two approaches. In the first case, they argue that when trying to predict the trajectory of something, you have to understand all of the details and keep track of every little thing. This is like solving terrorism by the application of surveillance and physical security measures. It is fundamentally impossible to do it well, particularly when it involves a large number of independent actors, each of which is difficult to predict. The second approach is founded upon a more universal assumption. While interpreting terrorism as a complex phenomenon state in terms of formal properties of the system, terrorist behavior falls somewhere between the purely chaotic and the fully deterministic realms, which can be represented as a non-linear dynamical system, characterized by a low-order chaotic attractor. Therefore, the greatest room for improving the performance of organizations tasked with preventing or combating terrorism is at the mid-range level. Not at the level of state leadership, and not at the level of mapping and predicting the behavior of each individual terrorist, but rather at an intermediate or organizational level—"action at the mid-range."

## *Complex Systems and Terrorism: Typology of Models*

The following presentation of mathematical models is elaborated for the prediction of various aspects of terrorism. A detailed explanation of complex systems and their dynamics would require a book-sized study. However, we can review the kinds of models and concepts which would go into that typology:

- Forecasting based on statistical methods,
- Exploratory data mining,
- Game theory (simple models and evolutionary game theory)
- Strategic thinking,
- Application of CAS,
- Network models,
- Power-law models and distributional patterns in terrorist attacks,
- Models of collective phenomena directly and indirectly related to complex systems, e.g. statistical thermodynamics, etc.

The above typology requires some clarification. First, although game theory and statistical models are not directly assigned to the category "complex systems models," they are included in the survey since many of them constitute either the point of departure for more advanced modeling or are applied in complexity-based models. Game theory as a method of modeling interactive decisions is an adequate method of studying interdependent decisions in CAS models and/or in networks. Second, applications of CAS include networks, but not all network models can be treated as directly related to CAS. Therefore two of the typological categories partially overlap one another.

Table 2.2 includes representative models for the aforementioned categories. The typology can be treated as an introduction to the detailed presentations of the models and should help provide a better understanding of the meaning of complexity of terrorism and its theoretical and practical consequences.

## Conclusion

The above survey draws several conclusions arising from the application of complex systems models to theory and policy concerning terrorism. Deeper insights that stem from the application of complex systems concepts in studying terrorism result from a better understanding of the dynamics of social phenomena. This improved understanding is achieved when those concepts are applied as mathematical models, analogies, and metaphors. Two particular achievements resulting from the

**Table 2.2** Typology of complex systems models applied in terrorism studies

| Type of models | Selected example works |
|---|---|
| Forecasting based on statistical methods—identification of patterns in time series | Sandler and Enders [99] |
| Predictive data mining | Jonas and Harper [111] |
| "Big data" | Page [112] |
| Game theory (simple models and evolutionary game theory) | Arce and Sandler [113], Bueno de Mesquita [114], Enders and Sandler [115] |
| Strategic management thinking | Fellman and Strathern [116] |
| Terrorism as complex adaptive system | Elliott and Kiel [117], Ahmed et al. [118], Bousquet [119] |
| Terrorism, complex adaptive systems, multi-agent based models, networks | ISAAC (Irreducible Semi-Autonomous Adaptive Combat) [120]: Enhanced ISAAC Neural Simulation Tool (EINSTein): SOTCAC (Self-Organized Terrorist and Counter-terrorist Adaptive Co-evolutions)—in Ilachinski [121–123] |
| Networks models (social networks analysis [SNA]) and complex networks | Krebs [124, 125], Fellman and Wright [110, 126], Fellman and Strathern [116], Fellman et al. [81], Sageman [127], Bohannon [128] |
| Power-law and patterns of occurrence of terrorist attacks | Cederman [129], Clauset and Young [130], Johnson et al. [131, 132], Piepers [133], Clauset and Wiegel [134], Clauset et al. [135], Zhu et al. [136], Clauset and Gleditsch [137] |
| Models of collective phenomena—statistical physics (percolation, Ising model) | Galam [138–140], Zhu et al. [136], August et al. [141] |

*Source*: Author's personal research

application of complex systems research to the study of terrorism are especially noteworthy. First, complexity-related ideas, both "soft" and "hard," allow for a profound understanding of terrorism as a sociopolitical phenomenon. Secondly, thanks to complex systems models of advanced description and the use of tools like social network theory, agent-based modeling, general systems theory, nonlinear dynamical systems modeling, and NK-Boolean dynamic fitness landscapes; new causal links have been discovered, allowing for the exploration of otherwise concealed phenomena, which has directly improved the ability to describe and predict of terrorist activities ranging from bio-terrorism and maritime piracy to irregular warfare and ethnic violence.

Complexity models show that stable regularities in observed social, economic, and political phenomena do not really exist. So the usefulness of those models for prediction is twofold. Firstly, they help in capturing the subtleties of the phenomena under study—thus, in some way, enhancing predictive capabilities. Secondly, complexity models are helpful in disclosing new limitations of prediction. In consequence, this leads us to better predictions—we now know with far greater clarity what is, and what is not predictable. This change in the way in which we approach the problem of prediction can further assist us in the efficient use of resources by showing us what kinds of problems are amenable to the efforts we wish to make and which problems might be better left for the future. In this fashion we may better direct the utilization of manpower, capital, and information so as not to unnecessarily drain our capabilities in a futile attempt to solve impossible problems or in a time and resource consuming effort which is bound to yield only very modest results. All of these topics will be discussed in detail, along with their models and their empirical context in the chapters which follow.

A final aspect of the new insights which CAS can provide is the study of the emergent properties of social, economic, and political systems. On the one hand, complexity science allows for more reality-relevant mathematical models, which can illustrate the specific characteristics of "emergent properties," thus enhancing analysis and prediction and allowing us to "expect the unexpected" in a precise, scientifically defined fashion. Even where these models do not directly enhance predictive capabilities they may in many instances provide unexpected, counter-intuitive, heuristically valuable solutions. We shall see several examples of this type of behavior in the latter portion of the book where Dr. Bar-Yam has applied such models to regional and ethnic violence as well as to a novel and extremely powerful explanation of the "Arab Spring" and the current fluid political situation in the Middle East.

A less immediately evident direction for the development of the applications of complex systems ideas in theoretical and applied considerations for policymaking arises out of qualitative or "soft" complexity ideas. The value of these new inputs is especially visible when reflexivity and self-reflexivity must be taken into account in order to understand the behavior of the parts of the system involved in terrorism and thus can be very useful in formulating counter-terrorism activities. One conclusion which this line of research leads us to is that sophisticated mathematical models of terrorism should embody the roles of cognition, reflexivity, and self-reflexivity somewhere within their structure. Later in this book, we will see specific examples of this kind of functioning in the mathematical models developed by Fellman et al., Ezell and Parnell. It must be added that the incorporation of such reflexivity is not limited to the study of terrorism alone, but refers to all manner of studies of modern, complex society.

As a result of the foregoing discourse, it seems that for the future of research in terrorism, more attention should be given to the epistemological foundations and qualitative interpretations of complexity than has been done in the past. This attention will help in developing a broader comprehension of the context in which the applications of complexity-based mathematical models take place. Further, it may be added that a synthesis of "hard" complexity with "soft" complexity could become a new instrument of better understanding terrorism, prediction of terrorist activities and counter-terrorism.

The final argument for developing a qualitative–quantitative methodology for complexity-based modeling in terrorism studies stems from the increasing significance of the need to understand the consequences of the cognitive processes of the actors involved. Modern complex systems modeling, especially CAS Research, allows for more advanced investigations into the cognitive processes of agents—whether they be terrorists or policymakers. What remains unknown is how the patterns of collective behavior are influenced or determined by the cognitive processes of the agents, i.e., how individual learning determines collective learning and what kind of behavior may emerge from that learning. A mixed approach, based upon a quantitative–qualitative complexity methodology offers substantial promise for future developments in this area of inquiry.

## References


1. Bar-Yam, Y. (2005). *Making things work*. www.Amazon.com
2. Holland, J. D. (1995). *Hidden order. How adaptation builds complexity*. New York: Basic Books.
3. Axelrod, R., & Cohen, M. D. (1999). *Harnessing complexity. Organizational implications of a scientific frontier*. New York: The Free Press.
4. Prigogine, I., & Stengers, I. (1984). *Order out of chaos: Man's new dialogue with nature*. New York: Bantam.
5. Saperstein, A. M. (2002). Complexity, chaos, and national security policy: Metaphors or tools? In D. S. Alberts & T. J. Czerwinski (Eds.), *Complexity, global politics and national security* (pp. 44–61). Honolulu: University Press of the Pacific.
6. Waldrop, M. M. (1992). *Complexity: The emerging science at the edge of order and chaos*. New York: Simon & Schuster.
7. Weaver, W. (1948). Science and complexity. *American Scientist, 36*, 536–544.
8. Simon, H. A. (1962). The architecture of complexity. *Proceedings of the American Philosophical Society, 106*(6), 467–482.
9. Ashby, W. R. (1963). *An introduction to cybernetics*. New York: Wiley.
10. Lloyd, S. (2001). Measures of complexity: A nonexhaustive list. *IEEE Control Systems Magazine, 21*(4), 7–8.
11. Gleick, J. (1987). *Chaos: The making of a new science*. New York: Viking.
12. Gell-Mann, M. (1995). What is complexity? *Complexity, 1*(1), 16–19.
13. Kauffman, S. (1995). *At home in the universe: The search for laws of self-organization and complexity*. Oxford: Oxford University Press.
14. Bar-Yam, Y. (1997). *Dynamics of complex systems*. Reading: Addison-Wesley.
15. Biggiero, L. (2001). Sources of complexity. *Human Systems, Nonlinear Dynamics, Psychology and Life Sciences, 5*(1), 3–19.
16. Kwapień, J., & Drożdż, S. (2012). Physical approach to complex systems. *Physics Reports, 515*(3–4), 115–226.
17. Zovko, I., & Farmer, J. D. (2002). The power of patience: A behavioral regularity in limit order placement. *Quantitative Finance, 2*(5), 387–392. Taylor and Francis Journals.
18. Mesjasz, C. (2010). Complexity of social systems. *Acta Physica Polonica A, 117*(4), 706–715. http://przyrbwn.icm.edu.pl/APP/PDF/117/a117z468.pdf.
19. Peters, E. (1991/1996). *Chaos and order in the capital markets*. New York: Wiley.
20. Peters, E. (1994). *Fractal market analysis*. New York: Wiley.
21. Farmer, J. D. (1999). *Physicists attempt to scale the ivory towers of finance*. Computing in Science and Engineering (IEEE), November–December, 1999.
22. Smith, E., Farmer, J. D., Gilllemot, L., & Krishnamurthy, S. (2003). Statistical theory of the continuous double auction. *Quantitative Finance, 3*, 481–514.
23. Cambel, A. B. (1992). *Applied chaos theory, A paradigm for complexity*. Boston: Academic.
24. Richardson, K. A., & Cilliers, P. (2001). Special editors' introduction. What is complexity science? A view from different directions. *Emergence, 3*(1), 5–23.
25. Maturana, H., & Varela, F. (1980). *Autopoiesis and cognition: The realization of the living*. Dordrecht: D. Reidel Publishing Co.
26. Luhmann, N. (1990). *Essays on self-reference*. New York: Columbia University Press.

27. Luhmann, N. (1997). *Die Gesellschaft der Gesellschaft*. Frankfurt am Main: Suhrkamp.
28. Luhmann, N. (1993). *Risk: A sociological theory*. New York: Aldine de Gruyter.
29. Smith, P. (2007). *An introduction to Gödel's theorems (Cambridge introductions to philosophy)*. Cambridge: Cambridge University Press.
30. Von Foerster, H. (1982). *Observing systems*. Seaside: Intersystems Publications.
31. Lakoff, G., & Johnson, M. (1980). *Metaphors we live by*. Chicago: University of Chicago Pressy.
32. Wittgenstein, L. (2002). *Philosophical investigations [the German text with a revised English translation]* (3rd ed.). Oxford: Blackwell Publishers.
33. Sato, Y., Akiyama, E., & Farmer, J. D. (2002). Chaos in learning a simple two-person game. In *Proceedings of the national academy of science* (Vol. 99, No. 7).
34. Fellman, P. V. (2011). The complexity of terrorist networks. *International Journal of Networking and Virtual Organisations, 8*(1/2), 4–14.
35. Fellman, P. V. (2011). The Nash equilibrium revisited: Chaos and complexity hidden in simplicity. *Unifying Themes in Complex Systems V.* Berlin: Springer.
36. Hayek von, F. (1967). Studies in philosophy, politics, and economics. London: Routledge & Kegan Paul.
37. Wallerstein, I. (2000). *The essential Wallerstein*. New York: The New Press.
38. Sokal, A., & Bricmont, J. (1998). *Fashionable nonsense. Postmodern intellectuals' abuse of science*. New York: Picador.
39. Baez, J. (1992). *The crackpot index*. N-Category Café. http://math.ucr.edu/home/baez/crackpot.html
40. Fellman, P. V. (1999). Reviews: Life at the edge of chaos: Creating the quantum organization. In M. D. Youngblood (Ed.), *Emergence* (Vol. 1, Issue 2). Dallas, TX: Taylor and Francis.
41. Fellman, P. V., Post, J. V., & Carmichael, C. (2012). The fundamental importance of discourse in theoretical physics. *International Journal of Theoretical and Mathematical Physics, 2*(5), 116–121.
42. Dahl, R. A. (1991). *Democracy and its critics*. New Haven: Yale University Press.
43. Laqueur, W. (2003). *No end to war: Terrorism in the twenty-first century*. New York: Continuum International Publishing Group.
44. Laqueur, W. (2012). *A history of terrorism [7th printing, with a new introduction by the author]*. New Brunswick: Transaction Publishers (first 2001).
45. Schmid, A. P. (2004). Frameworks for conceptualizing terrorism. *Terrorism and Political Violence, 16*(2), 197–221.
46. Schmid, A. P. (2011). *The Routledge handbook of terrorism research*. New York: Routledge.
47. Schmid, A. P. (2012). The revised academic consensus definition of terrorism. *Perspectives on Terrorism, 6*(2), 158–159.
48. Schmid, A. P., & Jongman, A. J. (2005). *Political terrorism: A new guide to actors, authors, concepts, data bases, theories, and literature*. Revised, expanded and updated edition prepared under the auspices of the Center for International Affairs, Harvard University, Piscataway, NJ: Transaction Publishers.
49. Defining Terrorism. (2008). WP 3, Deliverable 4, October 1. *Transnational terrorism, security and the rule of law*. A project financed by the European Commission under the 6th Framework Programme. http://www.transnationalterrorism.eu/tekst/publications/WP3%20Del%204.pdf, 27 March 2012.
50. Gadek, R. (2010). *Terrorism: The best definition of terrorism in light of 9–11*. http://criminaljustice-onlineblog.com/12/terrorism-the-best-definition-of-terrorism-in-light-of-9-11/, 12 July 2012.
51. White, J. R. (2012). *Terrorism and homeland security* (7th ed.). Belmont: Wadsworth, Cengage Learning.
52. Schneider, F., Brück, T., & Meierrieks, D. (2010). *The economics of terrorism and counter-terrorism: A survey* (part I). Cesifo Working Paper no. 3011, Category 1: Public Finance. http://www.cesifo-group.de/portal/pls/portal/docs/1/1185552.PDF, 26 November 2012

53. Stepanova, E. (2008). *Terrorism in Asymmetrical Conflict. Ideological and Structural Aspects*. SIPRI Research Report no. 23, Oxford: Oxford University Press.
54. Denning, D. (2000). *Cyberterrorism*. Testimony before the Special Oversight Panel of Terrorism Committee on Armed Services, US House of Representatives, 23 May. http://www.cs.georgetown.edu/~denning/infosec/cyberterror.html
55. Bakker, E., & de Graaf, B. (2011). Preventing lone wolf terrorism: Some CT approaches addressed. *Perspectives on Terrorism, 5*(5–6), 43–50.
56. Baudrillard, J. (2001). *The spirit of terrorism* (Translated by Dr. R. Bloul, Le Monde, 2 November). http://www.egs.edu/faculty/jean-baudrillard/articles/the-spirit-of-terrorism/, 15 September 2012.
57. Baudrillard, J. (2002). *The spirit of terrorism and requiem for the Twin Towers*. London: Verso.
58. Schmid, A., & Jongman, A. (1988). *Political terrorism: A new guide to actors, authors, concepts, data bases, theories and literature*. Amsterdam: North-Holland.
59. Laqueur, W. (1977). *Terrorism*. London: Weidenfeld and Nicolson.
60. Keohane, R. (2005). *After hegemony: Cooperation and discord in the world political economy*. Princeton, NJ: Princeton University Press.
61. Keohane, R., & Nye, J. S., Jr. (2011). *Power and interdependence* (4th ed.). Boston: Longman.
62. Snyder, G. H. (2002). Mearsheimer's world—Offensive realism and the struggle for security: A review essay. *International Security, 27*(1), 149–173.
63. Libach, M. I. (2003). *Is rational choice theory all of social science?* Ann Arbor: University of Michigan Press.
64. Knott, J., & Miller, G. (1987). *Reforming bureaucracy: The politics of institutional choice*. Englewood Cliffs: Prentice Hall.
65. Arthur, W. B. (1994). *Increasing returns and path dependence in the economy* (2nd ed.). Ann Arbor: University of Michigan Press.
66. Aoki, M. (2001). *Toward a comparative institutional analysis*. Cambridge: MIT Press.
67. Snyder, J. (2004, November–December). One world, rival theories. *Foreign Policy, 145*, 52–62.
68. Doyle, M. W. (1997). *Ways of war and peace*. New York: W.W. Norton.
69. Oseth, J. (1985). *Regulating U.S. intelligence operations, A study in definition of the national interest*. Lexington: University Press of Kentucky.
70. Buzan, B., Wæver, O., & De Wilde, J. (1998). *Security. A new framework for analysis*. Boulder: Lynne Rienner Publishers.
71. Kaplan, F. (2013). *The insurgents: David Petraeus and the plot to change the American way of war*. New York: Simon and Schuster.
72. Petraeus, D., Amos, J. F., & McClure, J. C. (Eds.). (2009). *U.S. Army/U.S. Marine Corps counterinsurgency field manual*. Kissimmee: Signalman Publishing.
73. Buzan, B. (2006). Will the 'global war on terrorism' be the new cold war. *International Affairs, 8*(25), 1101–1118.
74. Gad, U. P. (2012). Preventing radicalisation through dialogue? Self-securitising narratives versus reflexive conflict dynamics. *Critical Studies on Terrorism, 5*(3), 391–407.
75. Jackson, R., Breen Smyth, M., & Gunning, J. (Eds.). (2009). *Critical terrorism studies. A new research agenda*. London: Routledge.
76. Cerny, P. G. (2005). Terrorism and the new security dilemma. *Naval War College Review, 58*(1), 11–33.
77. Rapoport, D. C. (2004). The four modern waves of terrorism. In A. K. Cronin & J. M. Ludes (Eds.), *Attacking terrorism: Elements of a grand strategy* (pp. 46–73). Washington: Georgetown Press.
78. Kaplan, J. (2008). Terrorism's fifth wave: A theory, a conundrum and a dilemma. *Perspectives on Terrorism, 2*(2), 12–24.
79. Laqueur, W. (1999). *The new terrorism: Fanaticism and the arms of mass destruction*. Oxford: Oxford University Press.

80. Carley, K., Lee, J.-S., & Krackhardt, D. (2001). Destabilizing networks. *Connections, 24*(3), 31–34. INSNA, 2001.
81. Fellman, P. V., Clemens, J. P., Wright, R., Post, J. V., & Dadmun, M. (2007). *Disrupting terrorist networks – A dynamic fitness landscape approach.* http://arxiv.org/ftp/arxiv/papers/0707/0707.4036.pdf, 28 November 2007.
82. Fellman, P. V., Frutos, D., Thanakijsombat, T., & Teekasap, P. (2011). The complex landscape of maritime piracy. *Unifying Themes in Complex Systems, Volume VIII,* New England Complex Systems Institute Series on Complexity, Cambridge, MA: NECSI Knowledge Press, http://necsi.edu/events/iccs2011/ICCS2011_Proceedings.pdf, 1397–1410.
83. Banks, W. C., de Nevers, R., & Wallerstein, M. B. (2008). *Combating terrorism. Strategies and approaches.* Washington: CQ Press.
84. Tuman, J. S. (2010). *Communicating terror. The rhetorical dimensions of terrorism.* Thousand Oaks: Sage.
85. Nacos, B. L. (2002). *Mass-mediated terrorism: The central role of the media in terrorism and counterterrorism.* Lanham: Rowan and Littlefield.
86. Giddens, A. (1999). Risk and responsibility. *Modern Law Review, 62*(1), 1–10.
87. Beck, U. (1992). *Risk society. Towards a new modernity.* London: Sage.
88. Beck, U. (2002). The terrorist threat: World risk society revisited. *Theory, Culture & Society, 19*(4), 39–55.
89. Beck, U. (2009). Critical theory of world risk society: A cosmopolitan vision. *Constellations, 16*(1), 3–22.
90. Giddens, A. (1991). *Modernity and self-identity. Self and society in the late modern age.* Cambridge: Polity.
91. Luhmann, N. (1995). *Social systems.* Palo Alto: Stanford University Press (Original work published in German in 1984).
92. Beck, U. (2006). Living in the world risk society. *Economy and Society, 35*(3), 329–345.
93. Lewis, D. K. (1969). *Convention: A philosophical study.* Oxford: Blackburn.
94. Schelling, T. C. (1960). *The strategy of conflict.* Cambridge: Harvard University Press.
95. Weinberger, S. (2007, August 8). Can social scientists win the war on terrorism? *Wired.* http://www.wired.com/dangerroom/2007/08/in-short-what-a/, 22 January 2012.
96. Celebi, E. (2006). *Systems approach to terrorism: Countering the terrorist training subsystem.* Monterey: Naval Postgraduate School, https://www.hsdl.org/?view&did=469542, 21 October 2012.
97. Horn, R. E., & Weber, R. P. (2007). *New tools for resolving wicked problems: Mess mapping and resolution mapping processes.* MacroVU(r), Inc. and Strategy Kinetics, LLC. http://www.strategykinetics.com//New_Tools_For_Resolving_Wicked_Problems.pdf, 22 April 2011.
98. Grant, C. B. (2004). Uncertain communications: Uncertain social systems. *Soziale Systeme, 10*(2), 217–232.
99. Sandler, T., & Enders, W. (2007). Applying analytical methods to study terrorism. *International Studies Perspectives, 8*(3), 287–302.
100. Taleb, N. N. (2007). *The black swan. The impact of the highly improbable.* New York: Random House.
101. Epstein, J. M. (2007). *Generative social science: Studies in agent-based computational modeling (Princeton studies in complexity).* Princeton: Princeton University Press.
102. Mesjasz, C. (2008). Security as attributes of social systems. In H. G. Brauch, J. Grin, C. Mesjasz, P. Dunay, N. C. Behera, B. Chourou, U. Oswald Spring, P. H. Liotta, & P. Kameri-Mbote (Eds.), *Globalisation and environmental challenges: Reconceptualising security in the 21st century* (pp. 45–62). Heidelberg: Springer.
103. Simon, H. A. (1997). *Models of bounded rationality: Empirically grounded reason* (Vol. 3). Cambridge: MIT Press.
104. Kahneman, D., & Tversky, A. (1979). Prospect theory: An analysis of decision under risk. *Econometrica, 47*(2), 263–292.

105. Li, M., & Vitányi, P. M. B. (2008). *An introduction to Kolmogorov complexity and its applications (Texts in Computer Science)*. New York: Springer Science + Business Media.
106. Chaitin, G. J. (2001). *Exploring randomness*. London: Springer.
107. Wolfram, S. (2002). *A new kind of science*. Champaign: Wolfram Media Inc.
108. Mandelbrot, B., & Hudson, R. L. (2004). *The (mis)behavior of markets. A fractal view of risk, ruin, and reward*. New York: Basic Books.
109. Knowledge Wharton (2002). *To combat terrorism systems approach is vital*. Knowledge@Wharton, http://knowledge.wharton.upenn.edu/article.cfm?articleid=595&CFID=151186460&CFTOKEN=39430185&jsessionid=a83060ea8c2b119b6aad35633bf412446391, 20 April 2011.
110. Fellman, P. V., & Wright, R. (2004). Modeling terrorist networks. *The Intelligence: Journal of U.S. Intelligence Studies, 14*(1), 59–66.
111. Jonas, J., & Harper, J. (2006). Effective counterterrorism and the limited role of predictive data mining, *Policy Analysis*, no. 584, Cato Institute, December 11. http://www.cato.org/sites/cato.org/files/pubs/pdf/pa584.pdf, 15 June 2010.
112. Page, D. (2012). *Big data: The next big deal in anti-terrorism*. Homeland1, November 1. http://www.homeland1.com/homeland-security-products/technology/articles/1362256-big-data-the-next-big-deal-in-anti-terrorism, 11 December 2012.
113. Arce, D. G., & Sandler, T. (2003). An evolutionary game approach to fundamentalism and conflict. *Journal of Institutional and Theoretical Economics (JITE), 159*(1), 132–154.
114. Bueno de Mesquita, E. (2008). The political economy of terrorism: A selective overview of recent work. *The Political Economist, 10*(1), 1–12.
115. Enders, W., & Sandler, T. (2012). *The political economy of terrorism* (2nd ed.). Cambridge: Cambridge University Press.
116. Fellman, P. V., & Strathern, M. (2007). *The symmetries and redundancies of terror: Patterns in the dark*. http://arxiv.org/abs/0707.4028, 28 November 2007.
117. Elliott, E., & Kiel, L. D. (2004). A complex systems approach for developing public policy toward terrorism: An agent-based approach. *Chaos, Solitons and Fractals, 20*(1), 63–68.
118. Ahmed, E., Elgazzar, A. S., & Hegazi, A. S. (2008). *On complex adaptive systems and terrorism*. http://arxiv.org/PS_cache/nlin/pdf/0501/0501032v1.pdf, 7 August 2011.
119. Bousquet, A. (2012). Complexity theory and the war on terror: Understanding the self-organising dynamics of leaderless jihad. *Journal of International Relations and Development, 15*(3), 345–369.
120. ISAAC (Irreducible Semi-Autonomous Adaptive Combat). (1997). http://www.cna.org/research/tags/isaac, 7 September 2010.
121. Ilachinski, A. (2004). *Artificial war: Multiagent-based simulation of combat*. Singapore: World Scientific Publishing.
122. Ilachinski, A. (2005). *Self-organized terrorist-counterterrorist adaptive coevolutions, part I: A conceptual design*. Aleksandria: CNA Corporation.
123. Ilachinski, A. (2007). Complex adaptive systems, multiagent-based models, and some heuristics regarding their applicability to URW. In R. R. Luman (Ed.), *Proceedings on combating the unrestricted warfare threat: Integrating strategy, analysis, and technology* (pp. 199–215). Unrestricted Warfare Symposium 2007, The Johns Hopkins University Applied Physics Laboratory. http://www.jhuapl.edu/urw_symposium/previous/2007/pages/proceedings/2007/papers/Ilachinski.pdf, 7 September 2010.
124. Krebs, V. E. (2002). Uncloaking terrorist networks. *First Monday*, *7*(4). http://firstmonday.org/htbin/cgiwrap/bin/ojs/index.php/fm/issue/view/144, 10 September 2007.
125. Krebs, V. E. (2007). Connecting the dots—Tracking two identified terrorists, orgnet.com. http://www.orgnet.com/prevent.html, 10 September 2007.
126. Fellman, P. V., & Wright, R. (2003). *Modeling terrorist networks – Complex systems at the mid-range*. Paper prepared for the Joint Complexity Conference, London School of Economics, September 16–18. http://www.psych.lse.ac.uk/complexity/Conference/FellmanWright.pdf. Accessed 15 June 2007.

127. Sageman, M. (2008). *Leaderless Jihad terror networks in the twenty-first century*. Philadelphia: University of Pennsylvania Press.
128. Bohannon, J. (2009). Counterterrorism's new tool: Metanetwork analysis. *Science, 325*, 409–411.
129. Cederman, L.-E. (2003). Modeling the size of wars: From billiard balls to sandpiles. *American Political Science Review, 97*(1), 135–150.
130. Clauset, A., & Young, M. (2005). *Scale invariance in global terrorism*. http://arxiv.org/abs/physics/0502014/, 10 October 2007.
131. Johnson, N. F., Spagat, M., Restrepo, J. A., Bohórquez, J. C., Suárez, N., Restrepo, E. M., & Zarama, R. (2005). *From old wars to new wars and global terrorism*. arXiv:physics/0506213v1, 19 November 2011.
132. Johnson, N. F., Spagat, M., Restrepo, J. A., Becerra, O., Bohórquez, J. C., Suárez, N., Restrepo, E. M., & Zarama, R. (2006). *Universal patterns underlying ongoing wars and terrorism*. arXiv:physics/0605035v1, 19 November 2011.
133. Piepers, I. (2006). Casualty dynamics in wars and terrorism and the scale-free organization of social systems. http://arxiv.org/ftp/nlin/papers/0606/0606054.pdf, 19 November 2011.
134. Clauset, A., & Wiegel, F. W. (2010). A generalized aggregation–disintegration model for the frequency of severe terrorist attacks. *Journal of Conflict Resolution, 54*(1), 179–197.
135. Clauset, A., Young, M., & Gleditsch, K. S. (2007). On the frequency of severe terrorist events. *Journal of Conflict Resolution, 51*(1), 58–88.
136. Zhu, J. -F., Han, X. -P., & Wang, B. -H. (2009). *Scaling property and opinion model for inter-event time of terrorism attack*. Department of Modern Physics, University of the Science and Technology of China, Hefei. arXiv:0910.3985v2, 15 August 2010.
137. Clauset, A., & Gleditsch, K. S. (2011). *The developmental dynamics of terrorist organizations*. http://arxiv.org/PS_cache/arxiv/pdf/0906/0906.3287v2.pdf, 15 August 2011.
138. Galam, S. (2002). The September 11 attack: A percolation of individual passive support *European Physical Journal B, 26*, Rapid Note, 269–272.
139. Galam, S. (2003). Global physics: From percolation to terrorism, guerilla warfare and clandestine activities. *Physica A, 330*, 139–149. http://arxiv.org/abs/cond-mat/0404265. Accessed 18 July 2011.
140. Galam, S. (2012). *Sociophysics. A physicist's modeling of psycho-political phenomena*. New York: Springer Science + Business Media.
141. August, F., Blanchard, P., Delitzscher, S., Gerald, H. G., & Krueger, T. (2010). *Passive supporters of terrorism and phase transitions*. arXiv:1010.1953v2, 4 December 2012.

# Chapter 3
# The Psychology of Terrorism

Elena Mastors

## Introduction

Psychologists and psychiatrists have long been interested in the study of terrorism. This is not surprising since terrorism fundamentally involves the study of behavior. As Victoroff [1] explains, "it would seem appropriate for the scholarly disciplines of psychology and psychiatry to bring their intellectual resources to bear on the political problems of terrorism, a problem that—stripped to the basics—is one of atypical human behavior" (p. 4). Psychologists and psychiatrists are not the only ones that apply psychological concepts to this complex area of study. Other discipline experts are using psychology to inform their understanding. While psychology cannot answer all questions regarding terrorism, it does provide significant insights that aid in its understanding. Ultimately, it is a blending of psychology with other disciplines that will aid us in providing the best path of analysis.

What is evident from a review of the literature discussed below is there are many disparate parts of terrorism, and an equal number of disparate concepts, theories, and models, levels of analysis and depictions of related processes. Some of these are interdisciplinary, others are not. This complicates providing a clear picture with policy and operational prescriptions. Even so, there are too many facets of terrorism to assume there can be one or even several overarching theory or model. Furthermore, at times, the assertions being made about terrorism are based on "theoretical presumptions couched as facts" [1, p. 4]. Finally, there is an antagonistic approach of researchers to the study of terrorism, stemming from the close relationship between terrorism studies and counterterrorism agencies [2]. Adding to this problem is the stigma associated with the perception of researchers going "native." To better understand terrorism, overcoming stigmas and speaking to them can indeed enhance our understanding.

E. Mastors (✉)
American Public University System, Charles Town, WV 25414, USA
e-mail: emmastors@gmail.com

 
P.V. Fellman et al. (eds.), *Conflict and Complexity*, Understanding Complex Systems,
DOI 10.1007/978-1-4939-1705-1_3

As expressed, the psychology of terrorism literature is vast. The purpose of this chapter is to provide some insight into the various concepts, theories, and processes. This is by no means an exhaustive discussion of all the work being done. For convenience, the literature is divided into those approaches covering terrorist psychopathologies, the counterpoint work on the "normality of terrorists, examples of work that examines motivations, characteristics and other attributes of terrorists, the notion of terrorist "profiling," and contributing group and organizational factors. The chapter concludes with a discussion of the reality of conducting research on terrorism, the necessity of interviewing terrorists to inform our understanding, testing theory in real world settings, and pursuing interdisciplinary approaches to fully explain this complex area of study.

## Terrorist Psychopathology

The earlier study of terrorism focused on the individual terrorist, in particular terrorists with abnormal personalities. Prior thinking was that those who turned to terrorism could be distinguished psychologically from the general population and these differences were pathological. As a result, many studies focused on these purported pathologies. For example, Cooper [3] and Pearce [4] argued that terrorists were sociopaths. Others focused on narcissism as a key attribute [5–7]. Post [8] contended that psychological mechanisms of externalization and splitting are found "with extremely high frequency in the population of terrorists, particularly among the leadership" (p. 129). In another study, Billig [9] provided a detailed account of the participation in radical activities of lawyer Horst Mahler, who eventually became the leader of the Red Army Faction in Germany. Citing Mahler's dissatisfaction with fighting the system through legal battles, the terrorist leader began to participate in street demonstrations in the mid-1960s. Billig maintained that Mahler, subject to an authoritarian father who was also a member of the Nazi party, was a product of these circumstances. According to Billig,

> Mahler, who was born nearly 4 years prior to World War II, had an ambivalent relationship with his Nazi father. He identified with his father's masculine image when he disapproved of his comrades who "cried with they were hurt." He labeled it weakness when they went on hunger strikes in their desire to be given the privileged status of political prisoner. He objected to being treated differently from other prisoners and refused to be exchanged for a prominent German political figure who has been kidnapped by his comrades.
>
> Like other terrorists, he objected to the traumatizing assertiveness of his father's authority and displaced his conflicts on the workings of 'total society.' He compared society to a private enterprise in which the boss demanded that 'everybody must obey him without opening his mouth,' just as a child must obey his father. His conflictual feelings toward his father are suggested further by his expressed respect for his parent as a 'good man' and by his identification with the father's authoritarian role (p. 45).

In other studies focused on pathologies, Feuer [10] and Kent and Nichols [11] drew similar conclusions; that is, participation in terrorist activities resulted from feelings of hostility toward parents stemming from childhood abuse or rebellion [12].

Thus, "the terrorist's hostile focus is so great during childhood and adolescence that it continues into adulthood and becomes very narrow and extreme, ostensibly explaining the terrorist's absolutist mindset and dedication" [12, p. 16].

In his work, Meloy [13] examined the characteristics of what he termed the "violent true believer." According to him, a violent true believer has the following 12 attributes:

1. He has *tactical understanding* of suicide as a weapon of terrorism.
2. He may have a strong *envious impulse* to damage or destroy the goodness that he sees in his enemy since he cannot possess it himself.
3. As a result of personal suffering or indoctrination, he comes to feel that his is *helplessly dependent* on the object of his envy.
4. He attains a *sense of omnipotence* in the moments before his suicidal–homicidal act which overcomes other intolerable feelings and is the last sense of self while alive.
5. He likely has had a *history of despair and depression* which leaves him vulnerable to recruitment.
6. He develops a *sense of entitlement*—the specific belief that he has a right to kill others, even civilians, to further his cause—which compensates for his sense of hopelessness and helplessness, at least in this life.
7. He entertains *grandiose fantasies* suffused with expectations of pleasure, power, and knowledge in an afterlife. These fantasies are buttressed by his anticipated joining of a select group of martyrs that have preceded him.
8. If psychopathy is severe in the violent true believer, he is not likely to sacrifice himself, but will prepare others to do so.
9. He has a capacity for complete *emotional detachment* from himself and others, particularly in the hours prior to the terrorist act.
10. His violence is *predatory*: planned, purposeful, and emotionless.
11. He may be *clinically paranoid* in the sense that an irrational fear of imminent attach is psychologically present, or he has developed over the course of time a paranoid pseudo-community (e.g., infidels, Americans, unbelievers) upon which he can project blame for all the ills and misfortunes of his life.
12. And finally, he has *a sense of foreshortened future* as the date of his attack approaches.

Meloy argued that these characteristics provide a theoretical portrait of a terrorist. From here, personality can be indirectly assessed by gathering physical evidence, archival data, and observable evidence.

## The Normality of Terrorists

Challenging this perspective were studies demonstrating that terrorists were not unique in comparison to the general population. Often cited is the work on the German Baader–Meinhof Gang, where detailed records kept of the members indicated as such [14, 15]. Taylor and Qualye [16] argued that terrorists were not pathological in their study on terrorists in Northern Ireland. Clark [17] conducted interviews with Basque separatists (ETA) and found them to be "psychologically healthy." The same holds true for studies of the Italian Red Army and Palestinians [18].

Why is there not a preponderance of individuals exhibiting psychopathological personalities? For one, it isn't beneficial for a secretive group to recruit psychologically unfit members, in particular those with serious personality disorders, as they can severely hamper the functioning of the group. Further, psychopaths do not have the qualities sought after by terrorist groups. As Horgan [15] noted, "the presence of pathological egocentricity commonly found in psychopathic individuals seems to conflict with some of the required characteristics sought after by terrorist leaders and their members—high motivation, discipline and an ability to remain reliable and task-focused in the face of stress, possible capture and imprisonment" (p. 51). Furthermore, psychopaths are not necessarily prone to sacrifice on behalf of a group. Thus, as McCauley [14] asserts, "so far as I am aware, no one has ever suggested that a psychopath's moral blindness can take the form of self-sacrifice" (p. 15).

However, this is not to suggest that members of terrorist groups or those motivated to join them are all psychologically fit. After all, they willingly joined a violent group. The important point is that it is generally accepted these members are not the norm and do not stand out as unique against the general population. Still, despite evidence to the contrary, the perpetuation of this view of the terrorist group member as psychologically disturbed continues in the policy and academic communities. As Horgan [15], aptly noted, "bizarre as it may seem to the experienced psychologist, the profile of 'the terrorist' as a crazed fanatic, hell-bent on mindless destruction for its own sake still exists, with varying degrees of subtlety, within the literature" (p. 24). Experienced psychologists are amongst those who express this view, and in fact, while some have backed from their original claims, they do so by characterizing terrorists as not having major, but minor psychopathologies [15]. They are, however, psychopathologies nonetheless.

## Individual Motivations and Characteristics

While the perspective that terrorists suffer from pathologies has been heavily criticized, the studies themselves prompted the increased interest in examining alternative psychological views of why individuals turn to terrorism. Here there are several diverse areas of study.

Once such approach is examining the role of frustration, in particular what is termed the frustration-aggression hypothesis. This theory, associated with Davies [19], maintains that violence, in this case, revolutions, is a response to frustration. A criticism of this approach is that many would-be terrorist group members find themselves being frustrated, yet they do not turn to terrorism. Others purport that economic misery leads to violence [1]. Subsequent studies do not bear out a clear link between economic standing and terrorism (see, for example, Krueger and

Malecova [20] study of Palestinians, and Taylor and Qualye [16] assertions about Irish Republican Army members). But as Victoroff [1] noted, "their (Kreuger and Malekova's) analysis is based on socio-economic background, not on socioeconomic prospects" (p. 20). Victoroff also maintained that additional considerations relative to this perspective need to be investigated:

> Furthermore, nationalist–separatist and many religious fundamentalist terrorist tend to enjoy the support of their communities. In such cases, terrorism may be a prosocial activity *ostensibly* undertaken on behalf of all classes. If the *entire* in-group (that of the political actor) faces economic disparities relative to an out-group (that of a privileged target), participation in political violence would not be expected to be an economic class phenomenon but a group-of identity phenomenon. Further research will be necessary to determine the relationship between class of origin, economic expectations, individual factors, and terrorism (p. 20).

Moghaddam [22] points to the notion of perceived deprivation in explaining terrorism. For example, a person in Bombay can be extremely poor, but not feel unjustly treated and a person living comfortably in Riyadh could feel very unjustly treated. In other words, perception matters and this is a psychological process that socio-economic data may not be able to capture [23].

King and Taylor [23] described the differences between relative deprivation at the personal and group level. The concept of relative deprivation was originally brought about by studying the attitudes of US military personnel during WWII, specifically looking at discontentment. As they explained, noting the difference between personal and group deprivation,

> At the time of the survey, many more promotions were awarded in the air force than within military police. Yet on the survey, more air force personnel complained about the lack of promotion as compared to military police. The researchers attributed the greater discontent to the salience of promotions within the air force; for those who did not get promoted, the many promotions were a constant reminder of their lack of advancement. They key to understanding the troops' morale then, was not the objective quality of their circumstances, but rather their circumstances relative to their chosen target of social comparison.
>
> Personal deprivation was then applied to a group context, and thereafter discontent with personal circumstances—personal relative deprivation—was distinguished from discontent arising from comparing the circumstances of one's group—or group relative deprivation. This distinction was a major theoretical development, and has led to important nuances. For instance, personal relative deprivation has been linked to more inward emotions, such as self-esteem, delinquency, and depression, whereas group-based relative deprivation has been found to be a stronger predictor of collective action and prejudice toward other groups (p. 609).

Moral disengagement, as articulated by Bandura [24], is another perspective on terrorism. Bandura argued that individuals do have high moral standards and these are a product of socialization and prevents inhumanity toward others. However, individuals can disengage from this morality resulting in inhumane acts. Morality has been studied by others. For example, in their discussion of morality, Skitka [25] argued that moral conviction is a "strong and absolute belief that something is right and wrong, moral or immoral" (p. 36). Moral convictions have strong motivational components and "do not require reason or evidence" (p. 36). In turn, moral mandates are attitudes or stands derived from moral convictions. This has implications for justifying extreme behavior. In their work, Ginges et al. [26] contended that moral reasoning is a factor in decisions to engage in or support political violence.

In their study of motivation, Cottee and Hayward [27] put forth that terrorists are motivated by three things: the desire for excitement; the desire for ultimate meaning; and the desire for glory. As they noted, "what makes it attractive is its allure as a *life-mode* or *way of being*" (p. 965). Further, "terrorism, for those who practice and embrace it, can be profoundly thrilling, empowering and spiritually intoxicating, and that this particular aspect of it may inform, along with other key motivations, no doubt, the decision to engage in it" (p. 965).

Expanding on motivation theory, Stankov et al. [28] examined the psychological components of the extremist mind set, described as "a pattern of beliefs, feelings, thoughts, and motivation that tends to be mobilized under facilitating conditions—that may lead to violent behavior" (p. 70). Accordingly, the extremist mindset has three scales of measurement; Proviolence, Vile World, and Divine Power. Proviolence is the willingness to use violence to attain personal or social goals. Vile World involves the belief that a vile world is responsible for the suffering of ones group, and Divine Power refers to the fact that violence is sanctioned by a higher power. They further added that individuals who score high on Vile World and Divine Power will not certainly engage in violence unless they are also Proviolent. Those who are Proviolent can also be manipulated to become higher in Vile World and Divine Power. Together, high scores on all measures may lead to better understanding of terrorist behavior.

Psychologists also examine characteristics of leaders at-a-distance, in particular their need for power, locus of control, ethnocentrism, need for affiliation, cognitive complexity, distrust of others, self-confidence, and task-interpersonal emphasis [28–30]. In this methodology, content analysis is used to examine communications from the leaders by focusing on spontaneous content, to explain their behavior. This approach is very useful to the study of terrorist leaders and as such, over the years, has been used to explain the behavior of many leaders, including those heading armed groups [31].

Integrative complexity is another psychological area of investigation that can be useful to the study of terrorism. Integrative complexity "focuses on both *differentiation* (which are evaluatively distinct dimensions of a problem taken into account by decision makers) and *integration* (which are the connections made by decision makers among differentiated characteristics)" [32]. Integrative complexity has been applied to the study of terrorists, but not extensively. For example, Suedfeld [33] examined the integrative complexity of communications of Western, al-Qaeda and Taliban leaders in the case of Afghanistan.

Operational code involves the study of leaders by examining their philosophical beliefs (what the nature of the political universe is) and instrumental beliefs (what are believed to be the best strategies and tactics for achieving goals) [32]. This approach has its roots in the psychoanalytic work of Leites [34] and was later refined devoid of psychoanalytic elements [32]. The operational code has been used to study terrorists in Amal and Hezbollah [35] and was later applied to the operational code of groups including al-Qaeda and Hamas [36]. As in the case of integrative complexity, it has not been widely used.

The interplay between cognition and emotions is an area of study in psychology that has great potential in providing insight into a host of topics [32]. For instance, McCauley [14] asserted that emotions such as anger can be a strong motivating factor and is an area of study with great potential in explaining terrorism. Through his humiliation–revenge theory, Jurgensmeyer [37] argued that anger feeds violence. Moghaddam [22] suggested that revenge may be a motivating factor in suicide attacks in places such as Lebanon, Israel, Turkey, and Sri Lanka. Yet, in an examination of 1993–1998 Palestinian attacks, personal grudges were not found to be "a necessary factor and apparently not even a major one in initiating the wish to embark on a suicide mission, although in all probability it was a contributing factor in some of the cases" (p. 107).

Suicide bombers are frequently studied as a subset of terrorists. As Merari [38] explains, "the self-immolation element makes this form of terrorism substantially different in both its psychological foundations and potential consequences from other terrorist attacks that involve high risk for the perpetrators" (p. 101). Generally accepted is that the first suicide bombing occurred in 1981 in an attack on the Iraqi embassy in Beirut, Lebanon, "although as a methodical terrorist tactic, they were first used in Lebanon in 1983 by radical Islamic groups that later formed Hizballah" [38, p. 102]. Conclusions by those studying the personalities of suicide bombers echo the sentiments of the conclusions drawn regarding other terrorists; that is, for the most part, they do not suffer from pathologies and that there is not a single profile of them [38–42]. Yet, while demographic characteristics of this subset are available, personality traits and motivations are more difficult to come by. While the common conclusions are that they are "normal," Merari [38] cautioned that those are making these conclusions are not psychologists. Therefore, further study involving clinical interviews and personality tests is warranted.

In his own study that spanned data from 1993 to 1998 of Palestinian militants, Merari did not find evidence of psychopathology nor a common personality type. They were not suicidal, as traditionally understood. Group level factors such as indoctrination, group commitment, and a personal pledge were important in keeping the recruit focused on the suicide mission. Moghaddam [22] examined the case of Palestinian suicide bombers during the Second Intifada and offered a framework of analysis that included both individual (motivational) and organizational (organizational goals and methods of training and indoctrination) factors. As he explained of motives:

> Individual motives, which may include the desire to reap expected benefits in the afterlife, the urge to seek revenge for the death or injury of a close friend or family member, or the real or perceived humiliation brought about by Israeli occupation. The second set, which defines these goals and motives that leads organizations to plan suicide attacks, includes political aims of and tactical considerations for the use of suicide bombings. These two sets of motives converge at the recruitment state, when organizations identify and mobilize individuals who have professed a willingness to die (p. 68).

The institutional phase comes into play where the organization takes control over the volunteer. Here indoctrination and training takes place and commitment is solidified. Finally, Fields et al. [40] found varied motivations of Palestinian suicide bombers and some evidence of depression.

## Terrorist "Profiling" Be Gone

Many contend that because of the diversity of terrorists, there is no "terrorist profile" [15] and that profiling terrorists is futile. The latter has important policy and operational implications. For example, relying on profiling to detect would-be terrorists is not a prudent approach given that demographic profiles are varied and there are not particular constellations of personality attributes that are evident. However, this should not completely discount the "profiling" of terrorists or their groups, of which psychological approaches can provide some insight. Knowing your enemy leads to better operational and policy decisions. For example, Mastors and Norwitz [43] argued that examining the background, biographical elements, characteristics, motivation, and other factors of terrorist leaders can help devise strategies to influence or disrupt their activities.

## Group, Organizational, and Intergroup Level Factors

Looking at group and organizational factors is fundamentally important to the study of terrorism. For example, group level factors may be a factor in an individual's radicalization. Examining group dynamics can also lead to a better understanding of their functioning, and understanding them can provide insight into why members leave [15]. And, studying indoctrination strategies can help inform us of how individuals are recruited and remain committed.

Social identity theory, a social-psychological approach to studying intergroup behavior explains there are ingroups (those we belong to) and outgroups (those that others belong to) [44]. The theory is based on the principle of intergroup comparison. Individuals join groups to bolster their self-esteem and, by doing so, acquire an identity. When members compare their ingroup to relevant outgroups, they will evaluate their group more positively. Individuals can and do react to threatened or negative social identity. They can leave the group or a join a higher status group (social mobility). Alternatively, they may engage in social creativity whereby they compare their group to the outgroup on a different dimension, reevaluating for a more favorable outcome, or comparing their group to another lower status group, a strategy known as social competition. Competition between groups that result from an unfavorable comparison can explain why terrorist groups emerge and fight against their perceived enemies [32] and even why individuals choose to join terrorist groups.

Further, social identity theory explains rivalry between groups within organizations or networks. For example, Mastors [45] used social identity to explain rivalry between various groups in the al-Qaeda and the core of al-Qaeda run by Usama bin Laden and Ayman al-Zawahiri.

Important psychological concepts that provide keen insight into motivation and recruitment strategies and maintaining conformity [32] include dehumanization, where the enemy is portrayed as subhuman; scapegoating, where the enemy is portrayed as the root of all the problems of the ingroup and stereotyping. For example, Israel and the Jews are consistently dehumanized by Hezbollah, Hamas, and al-Qaeda and depicted as pigs. The United States is portrayed by al-Qaeda to be the source of Muslim ills worldwide. Additionally, Horgan [15] argued that language choice is a powerful tool and that the use of military language is also significant for it dehumanizes the enemy, shifting the enemy from one of a person to a "target."

Conformity in groups can also be understood by examining the work of Stanley Milgram (see for example [46]). Milgram conducted a serious of experiments in the early 1960s on obedience to authority. He was interested in why individuals in Nazi Germany could engage in such horrifying ways. The subjects in the study were informed they would test the ability of people to learn—a group labeled "the learners." The subjects were seated at a shock generator with 30 switches and voltage ranging from 15 to 450 V. The learners were asked to memorize words, and when they made a mistake, the subjects were told by a person in a lab coat (the authority figure) to administer shocks. Even as the shocks grew in intensity, 2/3rds of the subjects continued to administer them, some demonstrating sighs of severe stress in doing so. The subjects were not pathological, and their individual responsibility for engaging in harmful behaviors toward others became secondary to obeying the authority figure in the lab coat, who they viewed as ultimately responsible. This case study can be instructive in our understanding of how terrorists function in groups [15].

In explaining events that occurred in Abu Ghraib prison, Zimbardo [47] maintained that when a person joins a group, deindividuation occurs; that is, there is a loss or suppression of personal identity and personal responsibility is shifted from the individual to the group. Individuals then commit acts they wouldn't necessarily enact as individuals [32]. Thus, deindividuation can increase conformity in groups. Horgan [15] illustrates this with an example of the Provisional IRA and the group's dictates articulated in their *Green Book*. According to Horgan,

> the qualities of deindividuation are evident at various points within the terrorism process and...can play a role in strengthening the readiness of a PIRA Volunteer to follow orders and engage in an operation, but they can also come to bear on reassuring the Volunteer at any post-event phase and serve to minimize responsibility, doubt and lack of confidence in the moral justification of their what has just happened (p. 133).

Not all groups are bastions of conformity, where all members agree on strategy, tactics, and goals and occasionally, conflict erupts within groups. This can be a result of several factors including personal conflict, mistrust, and substantive procedural disagreements [45, 48]. Even the strain caused by deviant members that do not conform can threaten the cohesion of a group [45], and measures may be taken by members of the group to reassert conformity. For example, members of the loyalist Ulster Defense Association in Northern Ireland, governed by a leadership council, would actively contain non-conforming leaders. In certain instances, these leaders were expelled and even killed [49].

Members leave due to issues that erupt in groups and their departure can be voluntary or involuntary. In terrorist groups, however, leaving is often a difficult task and terrorists may be ostracized, live in fear of death, and even (or eventually) killed, as Taylor and Qualye [16] illustrate in the book on the IRA. The fear of being ostracized or killed often compels group members to stay. Others are permitted to leave after a monetary "buy out" (personal correspondence, March 2008). But Horgan [15] contended there are psychological implications of disengagement and highlights Brockner and Rubin's [50] concept of psychological traps. Psychological traps are

> situations where an individual, having decided upon some course of action that he or she expects will return a reward (in the broadest possible sense), for example, joining a terrorist group or remaining in such a group, finds that the actual process of goal attainment requires a continuing and repeated investment in some form over some degree of time. This repeated investment, in a psychological sense, will probably be required of that individual to sustain his or her involvement, but still the eventual goal may continue to be a very distant realization…Somewhere in this process is an inevitable stage when people find themselves in a decisional no-man's land, the realization that he or she has made quite a substantial investment but still not yet achieved his or her expected goal (p. 146).

This is a juncture point because the individual has made considerable investment, but to leave means abandoning his/her goal. "The ensuing entrapment encompasses the 'spiraling of commitment,' so frequently seen in members of terrorist groups" (p. 146).

Bjorgo and Horgan [51] list five factors of psychological, as opposed to physical, disengagement. Psychological disengagement is defined as disillusionment, which may be voluntary or involuntary.

1. Disillusionment arising from incongruence between the initial ideals and fantasies that shaped a person's initial involvement and their subsequent experiences with the reality of what is entailed by involvement—in other words, the mismatch between the fantasy and the reality
2. Disillusionment arising from disagreement over tactical issues
3. Disillusionment arising from strategic, political, or ideological differences
4. Becoming burned out
5. Changing personal priorities

The authors draw upon examples from several terrorist groups to illustrate disillusionment. On the other hand, physical disengagement can encompass the following:

1. Voluntary exit from the movement
2. Involuntary exit from the movement
3. Involuntary movement into another role
4. Voluntary movement into another role
5. Involuntary exit from the movement altogether
6. Experiences stemming from psychological disengagement that acts as a catalyst for physical disengagement across factors 1–4 above.

Social identity also portends that individuals can leave a group in search of a higher status group as a result of a negative or threatened identity. Additionally, some terrorists do move on to form other groups, emerging as a rival group. For example, the Armed Islamic Group (GIA) in Algeria split and a new group, the Salafist Group for Call and Combat (GSPC) emerged in 1992, which later merged with al-Qaeda to form al-Qaeda in the Maghreb.

## Radicalization and Recruitment Processes

Others seek to explain the processes involved in radicalization and many concepts discussed so far in this chapter are highlighted in this approach. Through a series of psychological processes, Moghaddam [22] explains radicalization as a narrowing stairway. "As people climb the stairway, they see fewer and fewer choices, until the only possible outcome is the destruction of others, or oneself, or both" (p. 70). On the ground floor, is the perception of fairness; when there is a perception of being unjustly treated, individuals look for ways to address their grievances. On the second floor, perceived possibility of personal mobility and perceptions of procedural justice are two psychological factors that affect behavior. On the second floor, leaders also highlight threats from "dissimilar outsiders." On the third floor, recruits are encouraged to become disengaged from morality. The fourth floor is where a recruit enters the secretive world of the terrorist group, and on the fifth and final floor, violence is committed against others.

Mastors and Deffenbaugh [52] see recruitment of individuals into the various groups in the al-Qaeda network as a convergence of motivation and attractiveness to the group message. They argue that individuals are motivated for a variety of personal, economic, social, and political reasons, although these are not mutually exclusive, and more than one motivation can apply. This is the push to join, but is not compelling enough on its own. When a motivated recruit meets a resonating message (the pull), a recruit is poised to join a terrorist group. In his book, Wiktorowiz's looked at recruitment to al-Muhajiroun through the lens of four factors; cognitive opening, religious seeking, frame alignment, and socialization. While there are similarities to the two approaches, Wiktorowicz is a process-orientated model, while Mastors and Deffebaugh are not depicting a process of recruitment but isolating possible variables. In another discussion of radicalization, Sageman [53] depicted a process whereby individuals learn of events or experience them personally, they become outraged, seek outlets for their outrage, and then this becomes of interest to a terrorist group.

In their comparison of Sageman, Moghaddam, Wiktorowicz [54], Borum's [55] "pathways to terrorism" (stages of "it's not right," "it's not fair," "it's your fault," and negative stereotyping) and Sliber and Bhatt's [56] four stage model (stages of pre-radicalization, self-identification, indoctrination, and jihadization), King and Taylor [22] argued that there are several commonalities amongst the approaches.

> First and foremost, the models converge on the assumption that radicalization is a transformation based on social-psychological processes. All five models describe emotions, cognitions, and social influences that, when operating in right order and combination can lead

> someone to endorse and engage in terrorism. To be precise, Wiktorowicz's model stops short of predicting actual terrorism, but nevertheless portrays a radicalization process equivalent in other models (p. 609).

With similarities in the models also come differences. The authors noted that each has a different number of stages of radicalization, and offer either factors leading to radicalization, or as the process as linear with stages. Further, the extent to which religion is applicable differs from approach to approach.

## Conclusion

Clearly, as demonstrated by the literature, a one-size-fits-all approach to studying terrorism does not exist, nor should it because of the many facets at play. Even with basic component parts, terrorism is a complex phenomenon with many higher level elements and processes. There is disagreement over the proper way to study terrorist groups and their members. Clearly, the psychological literature yields much about individual and group processes, but more research must be conducted on the ground, versus in laboratory settings. There also should be less theorizing about terrorism, and more application of strategies and possible solutions. Finally, psychology, while instructive, does not have all the answers. Studying terrorism is an interdisciplinary effort and frameworks, theories, and models must reflect this approach to be most effective. Perhaps a way forward is to map the many component parts relevant to the study of terrorism, isolating variable for further study. While certain concepts and theories may inform this mapping, they must be thoroughly tested in real world settings to truly demonstrate their applicability to terrorism. An alternative is to begin to develop more generalities and theories based on real world cases and field study.

Another area of concern in the study of terrorism is the data from which conclusions are drawn. Researchers agree that interviewing terrorists can be extremely beneficial in understanding their behavior and advancing related studies. This personal interaction comes with challenges. As Brannan et al. [2] argued in their hermeneutic of crisis management,

> the lack of interaction with actual 'terrorists' is evidenced by the literature, and *not* talking to terrorists seems to have become established as a source of scholarly credibility. Based on the notion that the perpetrators of 'terrorist' activities gain legitimacy as well as a propaganda outlet if researchers engage them in dialogue and allow them to make their own cases, terrorism studies seems to have virtually placed a premium on *avoiding* first-hand contact with the subjects of their research (p. 7).

Others have expressed views that researchers fear that by engaging terrorists, they could be branded as terrorist sympathizers [15]. But the simple fact is that information gleaned directly through conversation with terrorists about their behavior, group and organizational processes, and other avenues of research is essential and there exists a small amount of this type of open source information compared to the wider body of literature on the topic. However, this is not to suggest that all academics, by virtue of being academics, are skilled at eliciting information and have the necessary skills, sort through fact and fiction, while maintaining objectivity (either pro or against the cause or the individual). Just as intelligence agencies provide training on interviewing techniques and offer briefings on threat environments, perhaps young researchers could benefit from the same education.

Access to terrorists is another consideration but according to Horgan [15] access is not always as complicated as it seems. For instance, he demonstrates access has not been an issue for researchers seeking interaction with groups in Northern Ireland. Naturally, other areas of the world may present more challenging circumstances, and dangerous situations such as kidnapping and the beheading of journalists should inform these research field activities. However, interviewing incarcerated terrorists is certainly another avenue of gaining insight.

Interviews are not the only avenue of data collection, and we can glean much by examining what terrorists and terrorist groups write and say. For example, examining statements, communiques, manuals, and books can provide significant insight into strategy, tactics, goals, and approaches to the enemy, strategies of recruitment, indoctrination, conformity, and so forth. Secondary sources of information have long been relied upon for research. If used, however, it is important to understand the agenda of those providing such information and the possibility of misinformation [2].

Overall, the psychology of terrorism literature is robust and there are many individuals engaged in scholarly research and analysis. The prospects are good for expanding our knowledge, and there is certainly much room for the sharing of ideas and collaboration.

## References

1. Victoroff, J. (2005). The mind of a terrorist. *Journal of Conflict Resolution, 49*(1), 3–42.
2. Brannan, D. W., Esler, P. F., & Strindberg, N. T. A. (2001). Talking to terrorists: Towards an independent analytical framework for the study of violent substate activism. *Studies in Conflict and Terrorism, 24*(1), 3–24.
3. Cooper, H. H. A. (1978). What is a terrorist: A psychological perspective. *Legal Medical Quarterly, 2*(4), 16–32.
4. Pearce, K. (1977). Police negotiations: A new role for the community psychiatrist. *Canadian Psychiatric Association Journal, 22*(4), 171–174.
5. Crayton, J. W. (1983). Terrorism and the psychology of the self. In L. Z. Freedman & Y. Alexander (Eds.), *Perspectives on terrorism* (pp. 33–41). Wilmington: Scholarly Resources.
6. Lasch, C. (1979). *The culture of narcissism*. New York: Norton.
7. Pearlstein, R. M. (1991). *The mind of a political terrorist*. Wilmington: Scholarly Press.
8. Post, J. (2004). Terrorist psycho-logic: Terrorist behavior as a product of psychological forces. In W. Reich (Ed.), *Origins of terrorism: Psychologies, ideologies, theologies, states of mind* (pp. 25–40). Washington: Woodrow Wilson Press.
9. Billig, O. (1985). The lawyer terrorist and his comrades. *Political Psychology, 6*(1), 29–46.
10. Feuer, L. (1969). *The conflict of generations: The character of and significance of student movements*. New York: Basic Books.
11. Kent, L., & Nichols, W. (1977). The psychodynamics of terrorism. *Mental Health and Society, 4*(1–2), 1–8.
12. Ruby, C. L. (2002). Are terrorists mentally deranged? *Analyses of Social Issues and Public Policy, 2*(1), 15–26.

13. Meloy, J. R. (2004). Indirect personality assessment of the violent true believer. *Journal of Personality Assessment, 82*(2), 138–146.
14. McCauley, C. (2007). Psychological issues in understanding terrorism and the response to terrorism. In B. Bonger, L. M. Brown, L. E. Beutler, J. Breckenridege, & P. G. Zimbardo (Eds.), *Psychology of terrorism* (pp. 13–31). Oxford: Oxford University Press.
15. Horgan, J. (2005). *The psychology of terrorism*. London: Routledge.
16. Taylor, M., & Qualye, E. (1994). *Terrorist lives*. London: Brasseys.
17. Clark, R. (1983). Patterns in the lives of ETA members. *Terrorism: An International Journal, 6*(3), 423–454.
18. Kruglanski, A. W., & Fishman, S. (2006). The psychology of terrorism: "Syndrome" versus "tool" perspectives. *Terrorism and Political Violence, 18*(2), 193–215.
19. Davies, J. C. (1973). Aggression, violence, revolution and war. In J. N. Knutsen (Ed.), *Handbook of political psychology* (pp. 234–260). San Francisco: Jossey Bass.
20. Krueger, A. B., & Malecova, J. (2002). Education, Poverty, Political Violence, and Terrorism: Is There a Connection? Working Paper No. w9074, http://papers.nber.org/papers/w9074. Accessed 25 June 2007.
21. Moghaddam, F. M. (2007). The staircase to terrorism. In B. Bonger, L. M. Brown, L. E. Beutler, J. Breckenridege, & P. G. Zimbardo (Eds.), *Psychology of terrorism* (pp. 69–80). Oxford: Oxford University Press.
22. King, M., & Taylor, D. M. (2011). The radicalization of homegrown jihadists: A review of the theoretical models and social psychological evidence. *Terrorism and Political Violence, 23*(4), 602–622.
23. Bandura, A. (2004). The role of selective moral disengagement in terrorism and counterterrorism. In F. Moghaddam & A. Marsella (Eds.), *Understanding terrorism*. Washington: American Psychological Association.
24. Skitka, L. (2002). The dark side of moral conviction. *Analyses of Social Issues and Public Policy, 2*(1), 35–41.
25. Ginges, J., Atran, S., Sachdeva, S., & Medin, D. (2011). Psychology out of the laboratory. *American Psychologist, 66*(6), 507–519.
26. Cottee, S., & Hayward, K. (2011). Terrorist (e)motives: The existential attractions of terrorism. *Studies in Conflict and Terrorism, 34*(12), 963–968.
27. Stankov, L., Knezevic, G., & Saucier, G. (2010). Militant extremist mindset: Proviolence, vile world, and divine power. *Psychological Assessment, 22*(1), 70–86.
28. Hermann, M. G. (1980). Explaining foreign policy behavior using the personal characteristics of political leaders. *International Studies Quarterly, 24*(1), 7–46.
29. Winter, D. G., Hermann, M. G., Walker, S. G., & Weintraub, W. (1991). The personalities of Bush and Gorbachev measured at a distance: Procedures, portraits, and policy. *Political Psychology, 12*(2), 215–243.
30. Hermann, M.G. (1987). Assessing the foreign policy role orientations of sub-Saharan African leaders. In S. Walker (Ed.), *Role theory and foreign policy analysis* (pp. 70–95). Durham: Duke University Press.
31. Mastors, E. (2000). Gerry Adams and the Northern Ireland peace process. *Political Psychology, 21*(4), 839–845.
32. Cottam, M. L., Mastors, E., Preston, T., & Dietz-Uhler, B. (2010). *Introduction to political psychology*. New York: Psychology Press.
33. Suedfeld, P. (2003). Integrative complexity of western and terrorist leaders in the war against the Afghan terrorist regime. *Psicología Política, 27*, 79–91.
34. Leites, N. (1951). *The operational code of the Politburo*. New York: McGraw Hill.
35. Schbley, A. (1990). Religious terrorists: What they aren't going to tell us. *Conflict and Terrorism, 13*(3), 237–241.
36. Picucci, P. M. (2008). Terrorism's operational code: An examination of the belief systems of Al-Qaeda and Hamas, Unpublished dissertation.
37. Jurgensmeyer, M. (2000). *Terror in the mind of god*. Berkeley: University of California Press.

38. Merari, A. (2007). Psychological aspects of suicide terrorism. In B. Bonger, L. M. Brown, L. E. Beutler, J. Breckenridege, & P. G. Zimbardo (Eds.), *Psychology of terrorism* (pp. 101–115). Oxford: Oxford University Press.
39. Atran, S. (2003). Genesis of suicide terrorism. *Science, 299*(5612), 1534–1539.
40. Fields, R., Elbedour, S., & Abu Hein, F. (2002). The Palestinian suicide bomber. In C. Stout (Ed.), *The psychology of terrorism: Clinical aspects and responses* (pp. 193–223). Westport: Praeger.
41. Pape, R. A. (2003). The strategic logic of suicide terrorism. *American Political Science Review, 97*(3), 343–361.
42. Post, J., Ali, F., Henderson, S. W., Shanfield, S., Victoroff, J., & Weine, S. (2009). The psychology of suicide terrorism. *Psychiatry, 72*(1), 13–31.
43. Mastors, E., & Norwitz, J. (2009). Disrupting and influencing the leaders of armed groups. In J. Norwitz (Ed.), *Pirates, terrorists and warlords* (pp. 323–340). New York: Skyhorse Publishing.
44. Tajfel, H. (1970). Experiments in intergroup discrimination. *Scientific American, 223*, 96–102.
45. Mastors, E. (2014). *Breaking al-Qaeda: Psychological and operational techniques*. Boca Raton, FL: CRC Press.
46. Milgram, S. (1963). Behavioral study of obedience. *Journal of Abnormal and Social Psychology, 67*(4), 371–378.
47. Zimbardo, P. (2007). *The lucifer effect*. New York: Random House.
48. Borum, R. (2004). *Psychology of terrorism*. Tampa: University of South Florida.
49. Mastors, E. (2008). Can the Ulster Defence Association transition into mainstream politics? *Journal of Policing, Intelligence and Counter Terrorism, 3*(1), 7–30.
50. Brockner, J., & Rubin, J. Z. (1985). *Entrapment in escalating conflicts*. New York: Springer.
51. Horgan, J. (2009). Individual disengagement: A psychological analysis. In T. Borgo & J. Horgan (Eds.), *Leaving terrorism behind: Individual and collective disengagement* (pp. 17–29). London: Routledge.
52. Mastors, E., & Deffenbaugh, A. (2007). *The lesser jihad: Recruits and the Al-Qaeda network*. Lanham: Rowman & Littlefield.
53. Sageman, M. (2008). *Leaderless jihad*. Philadelphia: University of Pennsylvania.
54. Wiktorowicz, Q. (2005). *Radical Islam rising*. Lanham: Rowman & Littlefield.
55. Borum, R. (2003). Understanding the terrorist mindset. *FBI Law Enforcement Bulletin, 72*(7), 7–10.
56. Silber, M. D., & Bhatt, A. (2007). *Radicalization in the west: The homegrown threat*. New York: NYPD Intelligence Division.

# Chapter 4
# A Framework for Agent-Based Social Simulations of Social Identity Dynamics

**M. Afzal Upal**

## Introduction

Over the last few decades social scientists have accumulated considerable evidence in support of the hypothesis that people like to think well of themselves. For instance, most business people believe themselves to be more ethical than average [1], 90 % of business managers rate their performance as superior to their average peers [2], and most surgeons believe their patients' mortality rates to be lower than average [3]. Furthermore, the desire to self-enhance appears to be cross cultural [4]. Psychologists believe that this desire to think that we are better than our peers in desirable attributes is driven by people's deep seated need to feel good about themselves. Psychologists argue that the need for a positive self-esteem is a fundamental human need similar to people's need for food and shelter. Social psychologists refer to our knowledge of who we are as *identity* and argue that people's knowledge of which groups they belong to (called their *social* identity) is a key component of people's identities around the world [5]. In fact, according to many social psychologists, people's perceptions of their groups are even more important than people's knowledge of their individual characteristics because group perceptions tell us about beliefs and behaviors of an ideal individual in that group. Thus social identity provides a mould for individual identity [6].

Self-esteem is the evaluative component of identity. People feel better if they believe that their individual characteristics are valued but also if the groups that they belong to are also valued. Many social psychologists argue that the need for a positive self-esteem is what causes people to strive for a positive *social* identity and can cause some people to engage in activities such as outgroup derogation and ingroup favoritism [7, 8]. Social psychologists have also found that perceptions of threats to

M.A. Upal (✉)
Defence R & D Canada, 1133 Sheppard Ave W, Toronto, ON Canada, M3K2C9
e-mail: Afzal.Upal@forces.gc.ca

 
P.V. Fellman et al. (eds.), *Conflict and Complexity*, Understanding Complex Systems,
DOI 10.1007/978-1-4939-1705-1_4

people's social identity are more likely to motivate people to engage in violence against outgroup members than threats to personal identity or threats to non-identity related beliefs [9]. As such, this research suggests that social identity beliefs are key to understanding intergroup conflict in general and insurgent behavior in particular [9]. Thus a more predictive model of how people's identity beliefs are formed and modified over time would be extremely valuable to military commanders interested in understanding insurgent beliefs and behavior and designing influence operations to affect them.

## Predictive Sociological Models

While scientific papers written in human languages (such as English) have been the mode of representation overwhelmingly preferred by social scientists to represent their models of people's thoughts and behavior, some social scientists have criticized these models for being too imprecise and in some cases so much so that they are not even falsifiable. This has prompted some researchers to strive for closed-form mathematical formulations to represent societal behaviors. Thus crowd movements have been represented using complex differential equations originally developed to represent fluid dynamics [10]. The problem preventing a wider use of mathematical models is that before closed-form equations can be developed, we need to identify all the relevant variables and fully understand how they interact with each other. This is the knowledge that is typically not available for most social science phenomena of interest. Bottom-up computational modeling approaches such as agent-based social simulation have been suggested as an alternative as they are more precise than verbal models but do not require a complete understanding of the social phenomena.

Most agent-based (as well as other computational) models of human decision making are based on rational choice theory [11] which suggests that when faced with a number of possible actions, people select the action that seems to them to offer them maximum utility in the given situation. Rational choice models are particularly suited to bottom-up agent-based social simulation approaches because they provide an individual a rationale for each decision.

While rational choice models of human behavior have had some success in explaining human behavior in a wide range of domains, they have also been criticized for their inability to account for people's seemingly irrational behavior in various situations. For instance, the 2005 bestselling book *What's the Matter with Kansas: How Conservatives Won the Heart of America* asks why do poor and middle class Americans "vote against their economic interests" [12]. Social identity theorists argue that these Americans vote for Republican candidates because they believe that they will do more to enhance the positive image of the groups they associate with, which in turn will improve their self-esteem. By acknowledging people's deep seated desire to feel good about themselves, social identity theory (SIT) allows us to see that such seemingly irrational decisions are rational after all.

## Self-enhancement, Self-verification, and Social Learning

It is rational for an agent motivated to view itself better than others to engage in motivated cognitive processes [13] that allow it to continue to maintain a positive view of itself and to enhance it further. Social psychologists have identified a number of strategies that people use to maintain an overly positive view of themselves. These strategies include:

- *Selectively focusing* on those features and events of our personality that make us look good,
- *Selectively remembering* events that show us in positive light and forgetting negative events,
- Public *self-affirmation* of positive aspects of our personality at the expense of negative or neutral aspects [14]. This includes prompting others around us for public affirmation by boasting about ourselves in front of others, and
- *Selective comparisons* with people we have a priori reasons to believe are worse than us.

If self-enhancement were the only motive driving people, we would expect people to be extreme narcissists, believing themselves to be the best at everything good. Since most people appear to think of themselves only slightly better than they really are, there must be other competing motives that balance out the drive to self-enhance. Psychologists have indeed identified the motivation for *self-assessment* as the desire to seek out new information about oneself to find out what sort of person one really is [15]. Self-verification motivates us to assess what others around us really think of us and incorporate that information into our self-concept. According to Bandura's social learning theory, our beliefs about ourselves as well as about other issues in life are affected by what others around us (in particular those who are like us) think about those issues [16]. Since we cannot maintain a positive view of ourselves without any regard for the views of those we like, the self-enhancement motive also results in self-presentation and impression-management strategies [17] designed to convince others that we are good people. These include the strategies of:

- *Self-promotion*—an attempt to persuade others that we are competent,
- *Ingratiation*—an attempt to get others to like us,
- *Intimidation*—an attempt to get others to think that we are dangerous,
- *Exemplification*—an attempt to get others to consider us a morally respectable, and
- *Supplication*—an attempt to get others to take pity on us as helpless and needy.

## Social Identity Management

Since, according to SIT, people (especially those who integrate their group's identity into their own identity) derive part of their self-esteem from their membership in groups, they also engage in various strategies to maintain and enhance their group's image in their own mind as well as in the eyes of others. By undertaking actions designed to improve their group's lot, group members hope to improve their own worth so that they can feel even better about themselves. The kind of identity management strategies that people are willing to engage in and the extent to which various strategies are effective in improving people's identity perceptions depends on a number of factors. The most important among these factors is the strength of one's *group identification*.

## Group Identification

*Group identification* has been variously defined as "the readiness to see oneself as a member of a particular social group" and "an awareness of similarity, ingroup identity and shared fate with others who belong to the same category" [18]. Not all group members feel equally strongly attached to a group. Some people integrate group's identity with their own and develop a stronger affiliation for the group than their fellow members. We define *group identity strength* as the degree to which a group member feels attached to a group. Studies suggest that those who are more strongly attached to a group are more strongly affected by a perceived change in the status of a group than those who are weakly attached to the group. Strongly attached group members are also more likely than weakly attached group members to engage in social actions designed to improve their group's status. Whether one perceives one's group to have a higher or a lower status than comparison groups is also a critical factor in a group member's selection of which identity management strategy to use in a given situation.

## Group Status Differences, Relative Deprivation, and Sociostructural Variables

In his wartime study of American soldier, Stouffer et al. [19] were the first to appeal to people's perceptions of *relative deprivation*. Relative deprivation refers to the sense of having less resources than what one is entitled to or how much one's competitors have [19]. Runciman [20] made a distinction between *egoistic relative deprivation* as an individual's sense of deprivation relative to other individuals while *fraternalistic relative deprivation* derives from comparisons of the members of our group with members of a comparison group [20]. Several studies have indicated that fraternalistic deprivation is a significantly more effective in prompting people to act than egoistic deprivation [21, 22]. People are also more concerned about well-being of their group in the future as compared to its past. SIT as well as social categorization theory specifies intergroup comparisons as key drivers of social identity management strategies. The dimensions of this comparison include both material resources such as wealth and skin color as well as symbolic resources such as honor and piety. Not all individuals receive an equal share of the group's resources. Those with higher status in a group can expect to receive a higher portion of the resources while those with a lower group status often get a smaller portion. According to SIT, a group member's perceptions of the value of a group are also affected by sociostructural variables of group *permeability*, *legitimacy*, and *stability*.

- *Permeability* of a group is the sense of how easy it is for outsiders to enter the group. A high permeability indicates a low cost of entry into the group while a low permeability indicates a high cost of entry to a group. If a group member perceives that it is difficult for outsiders to join the group and thus gain a share in group's resources then her expectations of future rewards would be higher than those of a group member who perceives that it's easy for anyone to join the group thereby diluting the share of group rewards for existing group members.
- *Legitimacy* of a status structure is defined as the people's sense of how legitimate the status hierarchy is. A legitimate status hierarchy indicates that relative positions of different groups are just and deserving while an illegitimate status structure indicates that relative positions of various groups are unfair and unjust. If a group member perceives her group's higher status relative to a comparison group to be legitimate then she can expect larger future rewards than a group member who perceives the higher status to be illegitimate.
- *Stability* of a status structure is defined as the people's sense of how likely the status hierarchy is to last into the future. A highly stable status structure is expected to last well into the future while an unstable status hierarchy is likely to go through changes. If a group member perceives her group's higher status relative to a comparison group to be highly stable then she can expect larger future rewards than a group member who perceives the higher status to be unstable and in danger of the status difference being eliminated or even being reversed.

Stability and legitimacy have a similar impact on individual group member's identity management strategies and are often considered together.

## Social Identity Management Strategies

SIT describes that people consider both collective strategies that can improve the individual's self-esteem and contribute to the improvement of their group's status as well as individual strategies that only improve the individual's own self-esteem without affecting the group's status.

## Individual Strategies

When continued social comparison yields undesirable results (i.e., when favoritism of one's ingroup in relation to other outgroups is not achieved), lower status group members are motivated to seek change. For instance, exploring connections with other groups—which offer more possibility for positive self-evaluation—may be pursued. Accordingly, *social mobility* refers to the extent to which the pursuit of alternate social identities is feasible. It assumes that group boundaries are *permeable* and that individuals can change social groups in order to bolster social status. If social mobility is not possible for lower status group members, then they may consider *individuation* [23] to separate themselves from the devalued group by assigning more weight to the individual self-esteem as compared with social self-esteem. Higher status group members are also known to occasionally engage in individuation when they believe their group's higher status to be illegitimate and unfairly obtained and maintained.

## Collective Strategies and Group Leaders

Members who integrate a group's identity into their own and as a result strongly identify with a group have more to lose from a potential decrease in their group's status and more to gain from a potential increase in their group's status. This higher stake gives them an increased motivation to seek an improvement in group's status rather than engaging in individual strategies. Those who fuse their own identity and the group identity, and are therefore seen as being more prototypical members of the group by others, are also likely to have a higher influence on fellow group members. Thus when they engage in collective strategies such as collective action against an outgroup or redefinition of social identity so that it is seen in more positive light, they are more likely to succeed in their efforts. Furthermore, those who are seen by group members to engage in efforts to improve the status of a group are rewarded by group members resulting in an even higher status for these individuals, at least within the group. This feedback loop can create vast discrepancies in status among group members with a few group leaders having vastly higher status than the large number of lay group members.

When lower status group members believe that the group boundaries are impermeable and intergroup relations are not secure (i.e., they are perceived to be *unstable* and *illegitimate*), then lower status groups are likely to engage in intergroup competition with the high-status group [24]. That is, they are likely to engage in collective action and conflict that is intended to change the social structure in their favor, e.g., by stealing outgroup resources [25, 26]. Therefore, lower status groups are very likely to engage in conflict with higher status groups when access to the higher status group is not possible through mobility and when the higher status group's position seems illegitimate.

If lower status group members believe group boundaries to be impermeable and they believe intergroup relations to be secure (i.e., stable and legitimate), then lower status group members are expected to engage in *social creativity* in order to redefine elements of the comparative context. Thus rather than changing groups (mobility) or attempting to alter the status of the ingroup (collective action/competition), group members will instead use different strategies to assert group identity. For instance, members of lower status groups may express ingroup preference in various ways: by introducing new dimensions of comparison; by emphasizing the importance of ingroup traits; and by turning to new comparison groups (e.g., Ellemers et al. [27]; or review the strategies of social creativity in Cheung et al. [28]]. They might do this by embracing new beliefs, such as "we may not be as powerful as them, but we're friendlier" (this thought might come to mind when Canadians contemplate their relation to their powerful neighbors to the south, the United States!). In organizational research, for instance, these self-enhancement dynamics help illustrate how lower status professional groups (e.g., garbage collectors, exotic dancers) respond to their social dilemma. That is, they tend to cope by defining their occupation along positive dimensions, such as having flexible hours, more work-life balance, the ability to meet new people, and the chance to work outside. Social creativity allows people to retain identification with their group, while redefining what that identity means [29].

Because a major focus of SIT has been to explain why lower status groups engage in collective action and conflict against higher status groups, the theory makes relatively few predictions regarding the identity management strategies of higher status or dominant groups. Nonetheless, SIT asserts that such groups are motivated to maintain and extend their socially dominant position. To do so, they engage in *expressive biases*—which often involve boasting and justifying dominance. Moreover, much like non-threatened lower status groups, *equal-status groups* are motivated to carve out positive distinctiveness by using creative biases. Again, these entail creating a different social reality in which the ingroup is perceived more favorably.

Below is Tajfel and Turner's [8] original model of the effects of sociostructural beliefs (i.e., perceived stability, legitimacy, and permeability) on social identification and social identity management.

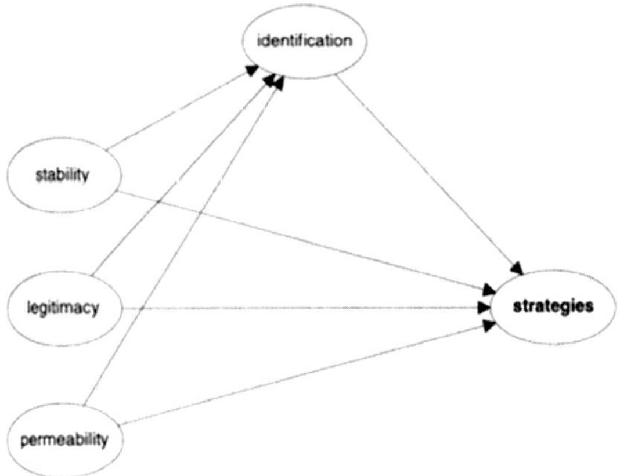

## Social Identity Management in Afghanistan

Mahmood, Rahmatullah, and Khalid are three fictional Afghans who were born a few months apart in 1980 in the Shah Wali Kot (SWK) District of Kandahar Province in Afghanistan. We will illustrate how differences in their intergroup perceptions have led them to adopt radically different social identity management strategies. We will primarily focus on dynamics of affiliation for the social groups of "pro-Karzai government Afghans" and the Taliban.

## Background

Similar to most people, the three Afghans have multiple identities including their gender, regional, tribal, and professional identities. While all three are Pashtun males that hail from the SWK region, their tribal affiliations differ. Mahmood is a Muhammedzai Pashtun. The Mohammedzai clan belongs to the Barakzai subtribe, one of four subtribes that make up the Zerak Durrani tribe and have the highest social status in the SWK region. Rahmatullah is a Noorzai who, and although they also consider themselves to be Durrani Pashtuns, they are suspected by Zerak Durrani of not being legitimate Durranis. Khalid identifies with the Hotak Gilzai tribe who has the lowest social status of the three social groups under discussion (Fig. 4.1).

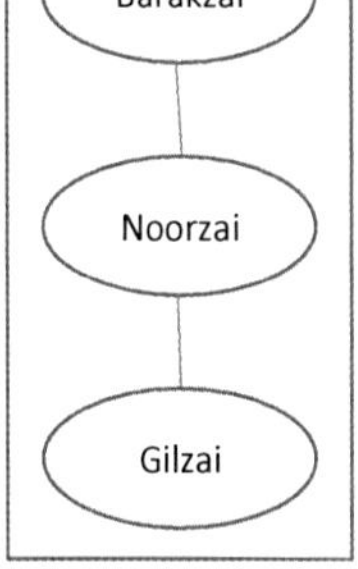


**Fig. 4.1** Kandahar Pashtun tribal hierarchy. Barakzais occupy the highest social status while Gilzais have the lowest status

## Individual Perceptions

### *Mahmood's Life: From Taliban to Pro-Karzai Government*

Like other typical Muhammedzai families, Mahmood's family owns considerable agricultural land which produces enough income to allow them to afford luxuries such as education and the ability to travel inside and outside Afghanistan. Muhammedzai's family was able to flee Afghanistan and escape much of the destructive aspects of the Soviet occupation in the 1980s but he still remembers seeing many of the village's mud-brick homes and family orchards obliterated by napalm. Two of Mahmood's cousins were pulled alive from bomb rubble; an aunt was not so lucky, another of the estimated 1.5 million people killed during the 10-year Soviet occupation. Mahmood went to a private English-medium school in Quetta, Pakistan where he learned Western notions of liberalism and leftwing ideas percolating among Quetta's elite. Mahmood also learned from his parents the glorious history of founding of modern Afghanistan by Ahmad Shah Durrani as an empire ruled by Zerak Durranis of Kandahar. He learned that because the kings epitomized the Afghan mixture of brain and brawn, they were loved by all people of Afghanistan. He learned about the modernization efforts of the last Rabbani King, Zahir Shah, who in 1963 had introduced free elections, universal suffrage, and women's rights to Afghanistan, only to be overthrown by his cousin Doud Khan.

Although Mahmood blamed the communist Khalq party in Afghanistan for inviting Soviets and for forcing his family to leave his birthplace, he had some sympathy for their modernization program for Afghanistan. He wished for a re-establishment of the Zerak Durrani monarchy through a return of the exiled King Zahir Shah to the Afghan thrown. He was disappointed when the communist government fell and the infighting among the Mujahideen factions began to wreak havoc in Afghanistan's cities. As the fighting dragged on for years and the factions fought with each other, he became hopeless for peace ever returning to Afghanistan and his family being able to return home. Although, like most well-educated Afghans in Quetta, he was suspicious of Taliban as a ragtag group of madrassa students supported by Pakistan's intelligence services, he grew to view them more positively as they finally brought peace to most of Southern Afghanistan. Listening to the largely positive accounts of Taliban's conquests in Pakistani media and criticism of Taliban by his liberal friends, Mahmood felt disconnected from his friends whose proclamations of wanting peace in Afghanistan he now found to be increasingly hypocritical.

In summer of 1996, as Mahmood turned 16, 2 weeks after the Taliban took Kandahar, Mahmood asked his father for permission to join the Taliban. Using his contacts, Mahmood's father got Mahmood a job in the Taliban foreign office in Kandahar. There was hardly any paperwork and no membership card. What signaled his new status was an expensive black turban bought for him by his father. In August 1996, he took a paying job at UNICEF's Kandahar office paying him about \$150 a month. The work at the foreign office increased after Kabul fell to the Taliban in late September, and eventually Mahmood got a full-time job translating.

The Taliban might still be in power today but for the hospitality they extended to Osama bin Laden. Bin Laden returned to Afghanistan in May 1996 as the guest of Jalalabad warlords who were not members of the Taliban but quickly ingratiated himself with Mullah Omar. "I saw bin Laden after he was brought to Kandahar in 1997," Mahmood recalled. Mahmood heard bin Laden speak at a house in Kandahar in 1998, not long after Qaeda agents financed by bin Laden blew up two US embassies, in Kenya and Tanzania, and President Clinton retaliated by launching 75 cruise missiles at what were thought to be four terrorist training camps near the eastern Afghanistan town of Khost. "He spoke against the US presence in the Holy Land, and amazingly, he spoke against Saddam Hussein that day," Mahmood recalled.

Over the next few years, Mahmood's job as a translator frequently took him on trips abroad to UAE, Saudi Arabia, and even the United States where he had the chance to observe how Taliban's image had deteriorated over time. He recognized the need to change that image but found his views strongly opposed by the majority of his fellow Taliban. "I nearly got into a fight with the chief justice of the Supreme Court, Mullah Saqib, who had verified the edict to demolish the Bamiyan Buddhas," he recalls. "I said, 'Why can't we have women's education?' And he said, 'We'll have it later.' I said: 'There isn't any time. Why are we waiting?' He said to me, 'I think you were really indoctrinated by America.' That really ticked me off. I wanted something good for Afghanistan. I was saying what I was saying because it was for the good of Afghanistan, not because I was being paid by the CIA."

The ensuing months of Mullah Omar's reluctance to expel bin Laden did not dispel Rahmatullah's disillusionment. He seriously considered quitting the Taliban and taking up a friend's offer to work for CNN. In the wake of 911 events, when it became clear that a US attack was imminent, he moved his family back to Quetta and restarted his life in exile. The events that followed proved that his decision had been wise. The Taliban government was easily toppled the following months by their arch-enemy the Northern Alliance mostly made up of ethnic Tajiks, Uzbeks, and the Hazara with minimal US special operations support. Even though a few Taliban leaders had also been killed, most had escaped to Pakistan leaving behind hundreds of their foot-soldiers to be massacred in Mazar-i-Sharif while others were imprisoned and tortured in Baghram and Guantanamo Bay. While Mahmood felt some sympathy for Taliban and an alarm at subordination of Pashtuns by Tajik–Uzbek–Hazara coalition, he was weary of publicly speaking in favor of Taliban and risk exposing his past involvement with them in the post 911 atmosphere where mere suspicion of being Taliban was enough to land one in Jail even in Pakistan.

Mahmood was ecstatic when the Munich conference in December 2011 selected Hamid Karzai, a Popalzai and a fellow Durrani exile from Quetta, as Afghanistan's President. Mahmood joined the Taliban in 1996 because he believed them to be more legitimate and stable and therefore deserving of higher social status than his current group of liberal Afghan exiles in Quetta. However, over time his perceptions changed and by 2002, he had come to view them as less legitimate. Through mutual friends he offered his services and was offered a position in the Afghan foreign office in Kabul. He moved to Kabul in 2002 and has remained there since serving in various senior positions in the Karzai government.

## Khalid's Story: From a Destitute Orphan to a Taliban Commander

Khalid doesn't remember much about his father other than the fact that like his mother, he was a Hotak Gilzai Pashtun who died in a Soviet bomb that exploded over their village when Khalid was just 3 years old. His mother escaped with him to a refugee camp near Peshawar, Pakistan. The Jalozai refugee camp, a densely packed slum of tents and mud-brick shacks about 20 miles southwest of Peshawar, housed a hundred thousand refugees at its peak. The living conditions at the camp were extremely poor. The place was so crowded the families lived check by jowl with each other. The make shift tents were unable to stop the sunlight and persistent dust. In the winter, the tent offered little shelter from frigid temperatures and in the summer the heat became unbearable leading to several deaths due to dehydration alone. A few shared pit latrines provided the only sanitation, and when it rained the sewage flooded into the tents. Khalid's mother (and increasingly Khalid, as he grew older) had to fight with other refugees to get some flour, cooking oil, and potatoes delivered by the UNHCR.

Unable to afford even the funds needed to buy uniform and nominal school fee for the public school in Hayatabad, Khalid's mother sent him to the madrassa run by Ittehad-I Islami Party at the Jalozai camp. Khalid excelled at the Quran and Hadith memorization tasks at the madrassa and he looked forward to going to school. He learned about how Muslims had conquered the world when they followed the true Islam and how they had lost power and prestige to the West since they stopped doing so. He learned that the efforts of Khalq communists to ape Western notions of modernity were doomed to fail because Muslims can only succeed if they return the true path of Islam. Khalid also enjoyed listening to the stories of old men at the camp who were full of nostalgia for their mountain homes with fresh air, wilderness, and the sense uniqueness and independence that they had lost in this miserable crowd of the refugee camp. "It'd be better to be dead than living here in a foreign jail," they often proclaimed. He also learned from them about the long history of enmity between Gilzais and Rabbanis; how Rabbanis had sided with the seventieth century Iranian King Nadir Shah who in turn rewarded them by giving them lands that had rightfully belonged to the Gilzais. He also learned about how Afghanistan had been the graveyard of empires from Alexander the Greek to the wicked British and how the Godless Soviets were also destined to be forced out by the brave Islamic Mujahideen.

Khalid vividly remembered the jubilation felt by the whole camp following the Soviet withdrawal from Afghanistan in 1989 with everyone dreaming of returning in peace to their homes. Over time, however, the enthusiasm turned into bitterness as the Mujahideen leaders wanting all power for themselves fought each other. "Power corrupts the best of them," said the elders. Khalid believed that all the Mujahideen leaders been corrupted by lust for power and money and he prayed to God to free Afghans from their grip. The rise of the Taliban seemed to Khalid to be an answer to his prayers. Like Khalid, Taliban were fellow madrassa students and teachers hailing from the refugee camps in Pakistan who wanted to free Afghanistan from the corrupt warlords. The Taliban also looked attractive to him with their distinctive black turbans, their discipline, and their access to modern weapons. Khalid's chance to join the Taliban came when a Taliban commander visited Jalozai camp in the spring of 1996 to seek help for the battle of Kabul. By the time, Khalid arrived in Kabul, the battle was over as Ahmad Shah Massoud's Northern Alliance withdrew northward, he was, however, involved in the battle of Mazar-i-Sharif in 1998. Following the death of the commander of his group, he was appointed as the commander of his group of 20 men.

In 2001, following the Northern Alliance re-taking of Kabul, he fled to Southern Waziristan from where he has been involved in cross border raids on US troops in Jalalabad. He feels little affiliation for the Karzai government whom he regards as a puppet government installed and supported by the US. He believes that just like the Greeks, the British, and Soviets, Americans will soon be forced by Afghans to leave their homeland and a truly Islamic government of Afghans will rule Afghanistan according to the Sharia law.

## Rahmatullah: A Pashtun Farmer on the Fence

Rahmatullah comes from a long line of farmers. Their love of land is what prevented them from leaving their fig, pomegranate, and apricot orchards when most of their neighbors fled following the Soviet invasion in 1979. Rahmatullah's grandfather, who passed away when he was just 2 years old, lived his whole life without visiting any major city and without having ridden a bicycle or having set foot in an automobile! Rahmatullah's father was the village Malik ad respect by the other Noorzai who were the dominant group in their village. Rahmatullah grew up with a strong sense of being a brave Pashtun who had chosen to stay close to his land and defend it rather than running away like many others when it became too difficult. He wishes that all foreigners (including Taliban whom he accuses of being Pakistanis) would leave Afghanistan and stop dividing Afghans by ethnicities and tribes and let them live in peace. That said, he has some sympathy for both the Taliban aims of rooting out corruption by imposing Sharia law in Afghanistan and for the pro-Karzai government's attempts to rebuild the country's infrastructure (such as repairing the nearby Dahla Dam) to help farmers like him. He just wishes that the two sides could sit down and negotiate a power-sharing agreement rather than have Afghans kill each other.

## The Dynamics of Sociostructural Perceptions and Shifting Affiliations

Although we have described them as individual perceptions, the thoughts and beliefs of the three Afghans represent stereotypical beliefs of the three dominant tribal groups in the SWK region of Kandahar Province of Afghanistan. They help us understand that most SWK Zerak Durrani Pashtuns (especially the Muhammedzai and the Popalzai) are largely supportive of the Karzai government because they believe that Zerak Durranis have a higher social status than other tribal groups and that their tribe is the legitimate ruler of Afghanistan and that this will likely be the case in the future. According to SIT, Barakzai and Popalzai Taliban are likely to adopt the social identity management strategy of mobility. This is because as members of a lower status group who believe that the higher status of their comparison group (pro-Karzai Government) is legitimate and stable, they are likely to shift their affiliations to the comparison group if it's open to them. Gilzai Pashtuns, with their historical grievances against the Durranis, however, feel that the higher social status enjoyed by Durranis is illegitimate and that while Durranis currently have more resources because of the US help, this situation is not likely to last into the future. According to SIT, when members of a lower status group perceive that their lower status is illegitimate and that the status hierarchy is unstable, they are likely to engage in collective action to improve the social standing of their group. This can involve violent conflict with higher status group members. Noorzai Pashtuns with their status in the middle of the pack occupy the middle ground and their support for either group may tilt the balance in its favor. On the one hand, Noorzais believe that they are Zerak Durranis and that Durranis are rightful rulers of Afghanistan. On the other hand, they know that their legitimacy as Durranis is questioned by the higher status Muhammedzais and Popalzais and they have some sympathy for those who are seeking a change in the status quo.

Since social identity management actions are designed to enhance the social self-esteem of the individuals, it's not surprising that these actions also affect those individual's sociostructural perceptions on which the social self-esteem depends. This sets up a feedback loop since the choice of social identity management actions is a function of the individual's sociostructural perceptions. For instance, the violent insurgent actions taken by the SWK Hotak Gilzai may enhance perceptions of instability of Hotak Gilzai as well as other Pashtuns and thus may make them more likely to believe that the higher social status of pro-Karzai government may be unstable and illegitimate causing even more people to support Taliban thus turning this into a self-fulfilling prophecy. Similarly, successful defections by Barakzai Taliban resulting in their joining the Karzai government increase the perceptions of permeability of the pro-Karzai government for former Taliban members, thus making Karzai government more attractive to any disaffected Barakzai Taliban.

To summarize, people are motivated to act to improve their group's status to fulfill their self-enhancement needs. What action people take depends on their beliefs about the relative status of their group compared with other groups, their sociostructural beliefs regarding the stability and legitimacy of the relative intergroup status, and their perceptions of the resources and permeability levels of the two groups. The social identity management actions in turn affect people's sociostructural beliefs and their intergroup perceptions. This feedback loop between sociostructural beliefs and agent actions makes it difficult to analyze the dynamics of social identity using closed-form equations. Agent-based social simulation is an excellent tool for understanding such dynamic systems. Next section outlines the framework for the agent-based social simulation (ABSS) system we are designing to better understand social identity dynamics.

## Agent-Based Simulation Framework for Social Identity Dynamics

Agents in our framework are rational, i.e., they evaluate the cost and benefits of various actions in a given situation and select the action that promises to maximize the *(benefit—cost)* differential. Agents have twin objectives of self-enhancement and self-assessment. They seek to enhance their self-esteem while also engaging in social learning to ensure that their beliefs are still anchored in with those of their peers. An agent has a personal identity as well as a social identity. The personal identity defines an agent's unique distinguishing abilities and resources while the social identity describes the agent's beliefs about the groups in its information environment. An agent's s*elf-esteem* also has two components: personal and social. The personal component of self-esteem depends on one's evaluation of one's personal resources and attributes as compared to resources and attributes of other individuals, whereas the social component depends on an agent's evaluation of future rewards it expects to receive as a result of its membership in that group. The social rewards an agent can expect to receive as a result of its affiliation with a group is a function of the strength of the agent's affiliation for the group, agent's status within the group, and the agent's sociostructural beliefs about the portion of resources likely to come to the individual during the individual's lifetime. The portion of the resources the agent gets now and into the future depends on:

- Per capita group resources, i.e., absolute level of resources the whole group possesses divided by the number of agents in the group.
- Permeability of the group. If the group has a low cost of entry for members of other groups, then members of groups with lower resources are likely to enter into the group and lower the per capita group resources [30, 31].
- The stability and legitimacy of the group indicate how likely the group status differences are to persist into the future and therefore determine the future rewards that the individual members are likely to get by virtue of their membership in the said group.

For instance, in a world with two groups, an agent may have affiliation $A_{g1}$, stability $S_{g1}$, and legitimacy $L_{g1}$ with Group G1 while an affiliation $A_{g2}$, stability $S_{g2}$, and legitimacy $L_{g2}$ with Group G2. If Group G1 has $N_{g1}$ members and $R_{g1}$ resources while Group G2 has $N_{g2}$ members and $R_{g2}$ resources then the agent's social esteem would be

$$= A_{g1} \times \left(S_{g1} + L_{g1}\right) \times R_{g1} / N_{g1} + A_{g2} \times \left(S_{g2} + L_{g2}\right) \times R_{g2} / N_{g2}$$

Group affiliation values range from −1 to +1. A negative value indicates an adversarial relation with the group while a value of 0 indicates that one is indifferent to the group. A positive value indicates a positive affiliation with the group.

Depending on whether they perceive group One's status to be higher or lower than the group Two and depending on their perceptions of the values of sociostructural variables, the individuals engage in various identity management strategies (for a summary see: [32]).

## Identity Management: Individual Strategies

a. *Mobility:* Mobility refers to joining new social groups in order to improve one's status, and relies on intergroup permeability. When individual agents perceive that the expected benefit of weakening their affiliation with a lower status group and strengthening their affiliation for a higher status group outweighs the expected costs, they will change groups. An individual's estimate of the cost of leaving the ingroup and joining the outgroup will be inversely proportional to the perceived permeability of the outgroup. The estimate of the benefits is proportional to the status difference between the two groups. That is, individuals "disassociate from the group and pursue individual goals designed to improve their personal lot rather than that of their ingroup" [33].
b. *Individuation:* Individuation refers to an individual's attempt at preserving self-esteem by psychologically separating oneself from the devalued group, and instead focusing on personal identity rather than on group membership. When individuals perceive that (a) the expected benefits of moving closer to another group are less than its expected costs and (b) the expected benefits of collective strategies are less than the expected costs, they will attempt to decrease the weight of the social component of self-esteem and increase the weight of the individual component of their self-esteem. This involves a shift from social to personal self-categorization [23], such that individuals no longer define themselves as group members but as unique individuals who are not affected by group evaluations (Ng 1989). This individual strategy allows people to escape from their own negative social identity while the evaluation of other group members remains unchanged.

## Identity Management: Collective Strategies

An individual should engage in collective strategies (such as *intergroup conflict* or *social identity entrepreneurship*) if the benefits of staying in the group outweigh the benefits of leaving the group.

c. *Intergroup Conflict/Collective Action*: An individual should seek social change or engage in conflict with outgroup members (and encouraging other ingroup members to do so) if the expected benefits of doing so outweigh the costs. The benefits to the individual would be both individual benefits such as war booty and gain in one's reputation (and resulting homage from fellow group members) as an upstanding defender of the group interests as well as social identity benefits that would be expected to follow because of one's membership in a group with additional resources and the resulting higher status [34].

d. *Social Identity Entrepreneurship*: When individuals perceive that the expected benefits of advocating a change in group identity are greater than the expected costs of advocating a change, they will engage in social identity entrepreneurship (e.g., [35]). Benefits include both individual benefits such as gains in one's reputation (and resulting homage from fellow group members) as an upstanding defender of the group interests as well as social identity benefits that would be expected to follow because of one's membership in a group with a higher status (to the extent that such efforts are successful). The costs of advocating a change to group identity include being perceived as a trouble maker for not following intergroup norms. An individual should engage in redefining their perceptions of how to compute group status and then propagating these perceptions to other ingroup and outgroup members in an attempt to reshape other's perceptions. The following list describes various strategies of social identity entrepreneurship in more detail:

   i. *Re-evaluation of Comparison Dimension* refers to tendencies of members of low status groups to reverse the evaluation of the poles of the relevant, i.e., the status defining comparison dimension, without actually changing the positions of the two groups on this dimension.
   ii. *New Comparison Dimension*: attempt to improve ingroup evaluation by rejecting comparisons on dimensions implying negative outcomes for the ingroup and preferring comparisons on which the ingroup holds a higher status position than the relevant outgroup [36].
   iii. *Superordinate Re-categorization* [37]: ingroup and outgroup merge into a new common ingroup. Both groups define themselves in terms of a common, higher level, ingroup and strive for positive social identity by means of social comparisons with other higher level outgroups.
   iv. *Subordinate Re-categorization*. In this case, self-evaluation is improved by splitting the former ingroup into two or more subgroups with the new, lower level, ingroup being perceived to be of higher status relative to the new outgroup(s).
   v. *New Comparison Group*. According to this strategy, members of low status groups select a new outgroup which holds a lower status position relative to their ingroup on relevant comparison dimensions. By downward comparisons [38, 39] the evaluation of the own group can be improved collectively.
   vi. *Temporal Comparison*. Individuals refrain from comparing with other social groups and rather prefer comparisons with their own group at other points in time. For instance, if the ingroup's position appears to have improved over time, subjects achieve a positive comparison outcome by comparing their ingroup's current position with its past position on relevant comparison dimensions.

vii. *Comparison with Standard* [40]: Individuals refrain from comparing with other groups and compare their ingroup with standards instead—standards which are socially shared goals or norms. When there is a satisfying approximation to these goals, subjects are able to evaluate their ingroup positively regardless of their disadvantaged position relative to the outgroup.
viii. *Boosting*. Boosting (aka basking; [41]; see also Wann 1994) involves the re-evaluation of a negative comparison dimension caused by ingroup defeat. This is done by accentuating the future probability of success of the group (e.g., "even if we lose today, we are still the best and we will win next time").

e. *Boasting*: Boasting is the legitimation of group dominance. It is elicited by feelings of pride when group status is perceived as legitimate and mainly constitutes the demonstration of ingroup bias and preference. Boasting appears to serve two main functions:

   i. The expression of group identity and value, and;
   ii. The legitimation of ingroup superiority and outgroup derogation.

In the context of the current model, boasting should ensue when individuals perceive that the expected benefits of raising the profile of a status-affirming event (e.g., a sports win over their opponent group's team) outweigh its expected costs. Because the intergroup context is not static, group members must continuously monitor events in their environment to update their perceptions of group resources. When events occurs in which group members have the opportunity to showcase their strengths and talents, they are expected to do so. In other words, when events occur which exemplify ingroup dominance, high-status group members raise the profile of those events (in the eyes of others and for themselves) and engage in boasting. Individual benefits might include gains to one's reputation (and resulting homage from fellow group members) as an upstanding defender of group interests. Social identity benefits would be expected to follow as a result of the group's higher status (to the extent that boasting efforts are successful). Other benefits include reduction in the feeling that high status might be taken away (i.e., boasting has a reassuring, threat-reducing function). Costs would include being perceived as overly egoistic and aggressive. Boasting is also likely to increase intergroup tension and animosity [42, 43] and could potentially trigger collective action by the victimized lower status group members.

f. *Justification*: Group justification (e.g., denial of responsibility) is a response to the potential *threat* of a loss of status in the future [44]. An unstable and illegitimate context is especially threatening for high-status group members because it lowers self-esteem, suggesting that the high status is undeserved [45]. When the status of a dominant group becomes dubious or uncertain, high-status group members will attempt to reassert their status. Group justification in response to threat can re-establish self-esteem and group identification, but it also increases discrimination against outgroups (e.g., [45]).

When individuals perceive the expected benefits of advocating a change to legitimacy and stability perceptions to outweigh the costs of seeking a change to these perceptions, they will do so. The benefits to the individual would be both individual benefits such as gain in one's reputation (and resulting homage from fellow group members) as an upstanding defender of the group interests as well as social identity benefits that would be expected to follow because of one's membership in a group with a higher status (to the extent that such efforts are successful). The costs of advocating a change to group identity include being perceived as a trouble maker for not following group norms. An individual should engage in redefining their perceptions of how to compute group status and then propagating these perceptions to other ingroup and outgroup members in an attempt to reshape other's perceptions.

g. *Raising Group Boundaries*: Raising group boundaries is an attempt by group members to decrease perceptions (of in and outgroup members) of the permeability of their group. When individuals perceive the expected benefits of advocating a decrease in their group's permeability to be greater than the cost of advocating a decrease in group permeability, then they will do so. The benefits to the individual would be both individual benefits such as a gain in one's reputation from those group members who adhere to hierarchy-enhancing myths (and resulting homage from such group members) as an upstanding defender of the group interests as well as social identity benefits that would be expected to follow because of one's membership in a group with additional per capita resources and the resulting higher status. The costs of higher boundaries are also social as well as individual. Individual costs would be loss of reputation in the eyes of those fellow members who adhere to hierarchy-attenuating myths. Social costs would include loss to the group of the potential individual resources that the potential new entrants to the group would have brought to the group.
h. *Lowering Group Boundaries:* Lowering group boundaries is an attempt by group members to increase perceptions (of in and outgroup members) of the permeability of their group. When individuals perceive the expected benefits of advocating an increase in their group's permeability to be greater than the expected cost of advocating an increase in group permeability, then they will do so.

## Agent-Based Social Simulation as a Tool for Social Science Theory Development

ABSS being bottom-up models are designed by using existing social science theories to specify agent's key internal reasoning processes and key modes of interaction with other agents. Since it is neither desirable nor possible to specify all cognitive processes and agent-interaction rules, all ABSS systems rely on a number of parameters. Simulations are then run by assigning values to these parameters within a reasonable range (the reasonable range identified by subject matter experts). If social patterns similar to the ones that occur in real world human societies emerge from a certain range of parameters then the simulation can help us understand and explain the real world social patterns by making explicit the micro–macro causal links that explain that certain social patterns are caused by certain ways of thinking and/or certain rules of inter-agent interactions. These hypothesized links can help us revise socio-cognitive models and provide direction for further quantitative and qualitative experiments with human participants. The improved social theories and models can then be used to design the next generation of ABSS models. Thus a close interaction between computer scientists designing the ABSS models, theoretical social scientists working on social science theory, and experimental social scientists working on testing theories and collecting data from real world human participants is necessary for the development of predictive models that can be useful to decision makers.

## Conclusion

The Western militaries need to better understand social identity dynamics of target populations (especially when operating in culturally different environments) to better assess the impact of various kinetic and non-kinetic actions before excuting them. Agent-based social simulation can be a valuable tool in this endeavor both for socio-cognitive theory development and for what-if analysis of various courses of action. This chapter develops an agent-based framework based on synthesis of Social Identity Theory [21], rational choice theory [11], and social identity entrepreneurship model of leadership[1] that can serve as a starting point in this much needed effort.

**Acknowledgment** The author would like to thank Dr. Lisa Legault for her valuable feedback on an earlier draft of this article.

## References

1. Brenner, S. N., & Molander, E. A. (1977). Is the ethics of business changing. *Harvard Business Review Catalog, 19*, 48–54.
2. French, J. R. P. (1968). The conceptualization and the measurement of mental health in terms of self-identity theory. In S. B. Sells (Ed.), *The definition and measurement of mental health*. Washington: Department of Health, Education, and Welfare.
3. Gawande, A. (2002). *Complications: A surgeon's guide to an imperfect science*. New York: Macmillan.
4. Kitayama, S., Markus, H. R., et al. (1997). Individual and collective processes in the construction of the self: Self-enhancement in the United States and self-criticism in Japan. *Journal of Personality and Social Psychology, 72*, 1245–1267.

[1] While the three characters are fictional, we based them on various news reports of the Afghanistan war. In particular, the description of the Mahmood character draws heavily on the 2006 New York. Times Magazine article by Brown [46].

5. Tajfel, H. (1972). Social categorization. English manuscript of 'La catégorisation sociale'. In S. Moscovici (Ed.), *Introduction à la Psychologie Sociale*. Paris: Larousse.
6. Taylor, D. (2002). *The quest for identity: From minority groups to generation xers*. New York: Praeger.
7. Tajfel, H., & Turner, T. J. (1979). An integrative theory of intergroup conflict. In W. Austin & S. Worchel (Eds.), *The social psychology of intergroup relations*. San Francisco: Brooks-Cole.
8. Tajfel, H., & Turner, T. J. (1986). The social identity theory of intergroup behavior. In S. Worchel & L. W. Austin (Eds.), *Psychology of intergroup relations*. Chicago: Nelson Hall.
9. Sidanius, J., & Pratto, F. (2001). *Social dominance theory: An intergroup theory of social hierarchy and oppression*. Cambridge: Cambridge University Press.
10. Kachroo, P., Al-Nasur, S. J., Wadoo, S. A., & Shende, A. (2008). *Pedestrian dynamics*. Berlin: Springer.
11. Gilboa, I. (2010). *Rational choice*. Cambridge: The MIT Press.
12. Frank, T. (2005). *What's the matter with Kansas: How conservatives won the heart of America*. New York: Holt Paperbacks.
13. Kunda, Z., Miller, D., et al. (1990). Combining social concepts: The role of causal reasoning. *Cognitive Science, 14*, 551–577.
14. Steele, C. M. (1988). The psychology of self-affirmation: Sustaining the integrity of the self. In L. Berkowitz (Ed.), *Advances in experimental social psychology*. New York: Academic.
15. Sedikis, C. (1993). Assessment, enhancements, and verification determinants of the self-evaluation process. *Journal of Personality and Social Psychology, 65*, 317–338.
16. Bandura, A. (1976). *Social learning theory*. New York: Prentice Hall.
17. Leary, M. R. (1996). *Self presentation: Impression management and interpersonal behavior*. Boulder: Westview Press.
18. Brewer, M. B., & Gardner, W. (1996). Who is this we? Levels of collective identity and self representations. *Journal of Personality and Social Psychology, 71(1)*, 83–93.
19. Stouffer, S. A., Suchman, E. A., et al. (1949). *The American soldier, I: Adjustment during army life*. Princeton: Princeton University Press.
20. Runciman, W. G. (1966). *Relative deprivation and social justice*. London: Routledge and Kegan Paul.
21. Abrams, D., & Hogg, M. A. (1990). *Social identity theory: Constructive and critical advances*. London: Harvester Wheatsheaf.
22. Kelly, C., & Breinlinger, S. (1996). *The social psychology of collective action*. London: Taylor and Francis.
23. Turner, J. C., Hogg, M. A., Oakes, P. J., Reicher, S. D., & Wetherell, M. S. (1987). *Rediscovering the social group: A self-categorization theory*. Oxford, UK: Blackwell.
24. Doosje, B., Spears, R., & Ellemers, N. (2002). The dynamic and determining role of ingroup identification: Responses to anticipated and actual changes in the intergroup status hierarchy. *British Journal of Social Psychology, 41*, 57–76.
25. Wright, S. C. (1997). Ambiguity, social influence and collective action: Generating collective protest in response to tokenism. *Personality and Social Psychology Bulletin, 23*, 1277–1290.
26. Wright, S. C., & Taylor, D. M. (1998). Responding to tokenism: Individual action in the face of collective injustice. *European Journal of Social Psychology, 28*, 647–667.
27. Ellemers, N., Spears, R., & Doosje, B. (1997). Sticking together or falling apart: Ingroup identification as a psychological determinant of group commitment versus individual mobility. *Journal of Personality and Social Psychology, 72*, 617–626.
28. Cheung, I., Dewit, Y., Filardo, E., Thomson, M. H., & Adams, B. (2012). *A review of social science literature on social identity dynamics and scientific fundamentalism*. Defence Research & Development Canada Toronto Report CR 2012–077.
29. Elsbach, K. D., & Kramer, R. M. (1996). Members' responses to organizational identity threats: Encountering and countering the Business Week rankings. *Administrative Science Quarterly, 41*, 442–476.

30. Harth, N. S., Kessler, T., & Leach, C. W. (2008). Advantaged Group's emotional reactions to intergroup inequality: The dynamics of pride, guilt, and sympathy. *Personality and Social Psychology Bulletin, 34*, 115–129.
31. Levin, S., Federico, C., Sidanius, J., & Rabinowitz, J. (2002). Social dominance orientation and intergroup bias: The legitimation of favoritism for high-status groups. *Personality and Social Psychology Bulletin, 28*, 144–157.
32. Mummendey, A., Kessler, T., Klink, A., & Mielke, R. (1999). Strategies to cope with negative social identity: Predictions by social identity theory and relative deprivation theory. *Journal of Personality and Social Psychology, 76(2)*, 229–245.
33. Haslam, S. A., & Reicher, S. (2007). Identity entrepreneurship and the consequences of identity failure: The dynamics of leadership in the BBC prison study. *Social Psychology Quarterly, 70(2)*, 125–147.
34. Sturmer, S., & Simon, B. (2004). The role of collective identification in social movement participation: A panel study in the context of the German gay movement. *Personality and Social Psychology Bulletin, 30*, 263–277.
35. Bernach-Assollant, I., Laurin, R., Bouchet, P., Bodet, G., & Lacassagne, M. (2010). Refining the relationship between ingroup identification and identity management strategies in the sport context: The moderating role of gender and the mediating role of negative mood. *Group Processes and Intergroup Relations, 13(2)*, 639–652.
36. Tajfel, H. (1978). *Social categorization, social identity and social comparison*. New York: Psychology Press. Reprinted in T. Postmes, & N. R. Branscombe (Eds.). (2010). Rediscovering social identity: Key readings in social psychology (pp. 119–128).
37. Gaertner, S. L., Dovidio, J. F., Anastasio, P. A., Bachman, B. A., & Rust, M. C. (1993). The common ingroup identity model: Recategorization and the reduction of intergroup bias. In W. Stroebe & M. Hewstone (Eds.), European review of social psychology (Vol. 4, pp. 1–26).
38. Wills, T. A. (1981). Downward comparison principles in social psychology. *Psychological Bulletin, 90*, 245–271.
39. Wills, T. A. (1991). Similarity and self-esteem in downward social comparison. In J. M. Suls & T. A. Wills (Eds.), *Social comparison*. Hillsdlae, NJ: Erlbaum.
40. Masters, J., & Keil, L. (1987). Generic comparison in human judgement and behavior. In J. Masters & W. Smith (Eds.), *Social comparison, social justice, and relative deprivation* (pp. 11–54). Hillsdale, NJ: Erlbaum.
41. Finch, J. F., & Cialdini, R. B. (1989). Another indirect tactic of (self-) image management: Boosting. *Personality and Social Psychology Bulletin, 15*, 222–232.
42. Levin, S., Sidanius, J., Rabinowitz, J. L., & Federico, C. (1998). Ethnic Identity, legitimizing ideologies, and social status: A matter of ideological asymmetry. *Political Psychology, 19(2)*, 373–404.
43. Turner, J. C., & Brown, R. J. (1978). Social status, cognitive alternatives and intergroup relations. In H. Tajfel (Ed.), *Differentiation between social groups*. London: Academic Press.
44. Doosje, B., Spears, R., & Ellemers, N. (2002). Social identity as both cause and effect: The development of group identification in response to anticipated and actual changes in the inter group status hierarchy. *British Journal of Social Psychology, 41*, 57–76.
45. Branscombe, N. R., & Wann, D. L. (1994). Collective self-esteem consequences of outgroup derogation when a valued social identity is on trial. *European Journal of Social Psychology, 24(6)*, 641–657.
46. Brown, C. (2006). *The freshman*. New York Times Magazine. http://www.nytimes.com/2006/02/26/magazine/26taliban.html?

# Chapter 5
# DIME/PMESII Models

**Dean S. Hartley III**

The Baader–Meinhof Gang in Germany (late 1960s) used terror tactics. The group styled itself the Rote Armee Fraktion (RAF) and in its 30 years of activity (until 1998) was held responsible for 34 deaths and many injuries [1]. The counter-strategy for this group was legal in nature: they broke laws and should be held accountable. The RAF members may have cared about their ideology; however, few others did. The group ultimately died out, whether because of this approach or despite it. In today's (2013) terms, this was a small group—small enough that Wikipedia lists the names of all of the known members (about 60).

Today's terrorists will probably also die out, ultimately. In the meantime, they have killed considerably more than 34 people and have shown the ability to recruit large numbers of new members, despite very high losses of their own. Purely legal strategies have proved inadequate and purely military strategies have proved inadequate. Current strategies include combinations of legal, economic, military, informational—in short, whatever someone can think of that might work. However, these strategies have impacts on those who are not current members of the terrorist organizations, some good and some bad. Counter-terrorism strategies must then account for collateral effects (a standard term from military campaign planning of the Cold War). Further, the recruiting efforts—and their successes—have led to the understanding that the situation of the general populace surrounding the terrorist groups should be understood and perhaps addressed directly, beyond the focused efforts against the terrorists.

Three broad approaches to the terrorist problem are available: do nothing and wait for the threat to die out; react to each event as it seems most appropriate at the time; create and test strategies, using and adapting the strategies as the threat changes. The first two options appear to be "strawman" options, presented to make

D.S. HartleyIII (✉)
Hartley Consulting, Oak Ridge, TN, USA
e-mail: DSHartley3@comcast.net

 
P.V. Fellman et al. (eds.), *Conflict and Complexity*, Understanding Complex Systems,
DOI 10.1007/978-1-4939-1705-1_5

the third option look to be the obvious choice. However, historically, the first option has been immensely popular for other threats: consider Hitler and Bosnia. The second option is actually a good choice for new threats: the intensity, scope, and nature of the threat are unclear when it is new. Our current terrorist threat is no longer new and the 12 years since the attack of 9/11/2001 and the fall of the towers have provided us with a good appreciation of the intensity, scope, and nature of the threat. We have adopted the third choice.

Creating and testing strategies does not require the use of a model. Each strategy could be tested in the real world against the real threat. However, it is cheaper in terms of lives lost to build models and test the strategies there—first. The ones that seem promising can be tested in the real world. There are also other uses for models in this domain, such as predicting events and understanding the nature of the domain. Whatever their purposes be, useful models in this domain require more than military or legal effects. They require a full spectrum of social effects.

## Social Modeling Concepts

Modern military operations cannot be described in purely kinetic effect terms (damage and kills). These operations include Peace Operations (PO) of several kinds, Humanitarian Assistance (HA) and Disaster Relief (DR), Counter Insurgency (COIN), Counter Terrorism (CT), and Operations Other Than War (OOTW), which include the preceding and other types of operations. Later operations have been called Stability and Support Operations (SASO) and Stability, Support, Transition, and Reconstruction operations (SSTR or SSTRO). Some of these operations are considered to be part of Irregular Warfare (IW).

The popular perception of security, the level of support for the indigenous government, the economic stability of the country, and other non-tangible variables are important indicators of success. Some tangible non-kinetic effect variables, such as the state of infrastructure reconstruction and the turnout at elections, are important. The technical approaches that have been used to model these operations have generated a new acronym: DIME/PMESII. The acronym PMESII refers to the Political, Military, Economic, Social, Information, and Infrastructure variables that describe the status of a situation (state vector). There have been arguments that other categories should be included in the taxonomy; however, for our purposes, we will use PMESII to refer to all state vector variables, regardless of taxonomy details. The acronym DIME refers to the levers of power that a (nation) state has to influence the PMESII state, Diplomatic, Information, Military, and Economic. As with PMESII, we will use DIME to refer to all such interventions, regardless of taxonomy details. Collectively, these concepts will be referred to as DIME/PMESII or simply as PMESII for brevity.

The performance of operations that required more than kinetic effects (a polite term for destructive effects through physical means) drove the development of DIME/PMESII models. Similarly, the development of DIME/PMESII models is

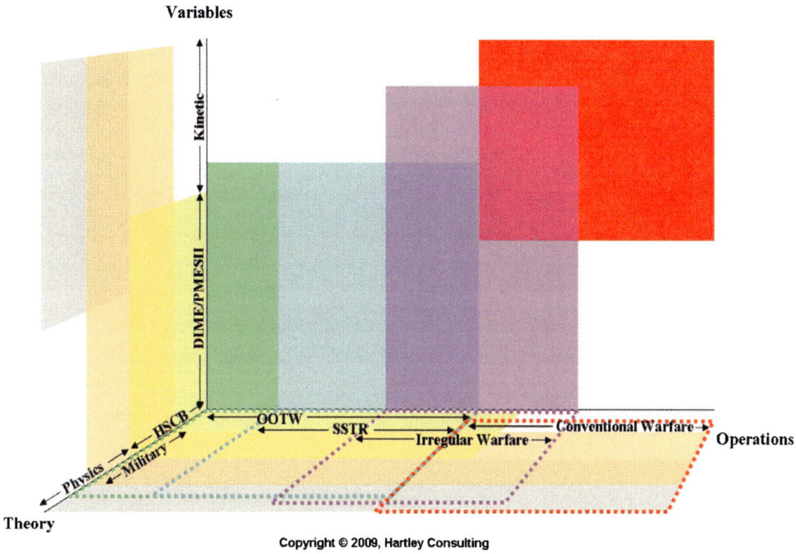


**Fig. 5.1** Theory vs variables vs operations

driving a need to understand and apply social science theories. Therefore we have a new acronym and term, HSCB Modeling, which stands for Human Social Culture Behavior modeling.

When using the acronym, HSCB, the focus is on the theoretical basis of a model; whereas, DIME/PMESII (or PMESII for short) focuses on the technical details needed to implement a model. When the focus is on the operations being modeled, models may be cited as OOTW, SASO, etc., models. These definitions are not synonyms, as shown in Fig. 5.1; however, it has become clear that most of the operations listed above will require DIME/PMESII modeling techniques, supported by a firm HSCB basis.

PMESII models are like other models; they have inputs, logic, and outputs, all mediated by variables. However, the historical reluctance of military modelers to use "soft factors" in their models suggests that there is some difference.

One difference can be described as dimensionality. Military computer combat models of the late 1960s and early 1970s typically had one-dimensional outputs, movement of the Forward Edge of the Battle Area (FEBA), and some only had two-dimensional inputs: total force values for each side. Consider Fig. 5.2 in which the PMESII categories have been subdivided. The entire complement of those early military models would be contained in the Military/Conflict subcategory.

The implicit assumption of Fig. 5.2 is that each of these 27 subcategories is independent of all the others. However, some of these subcategories are probably correlated. For example, good health care might not be found where education levels are too low. Thus, there might actually be fewer than 27 independent dimensions.

- Political
  - Government
  - Politics
  - Rule of Law
  - Security
- Military
  - Conflict
  - Government
  - Other
- Economic
  - Agriculture
  - Crime
  - Energy
  - Finance
  - Government
  - Jobs
  - Other
- Social
  - Basic Needs
  - Education
  - Health
  - Movement
  - Safety
  - Other
- Information
  - General
  - Information Operations
- Infrastructure
  - Business
  - Energy
  - Government
  - Transportation
  - Water

**Fig. 5.2** PMESII dimensions

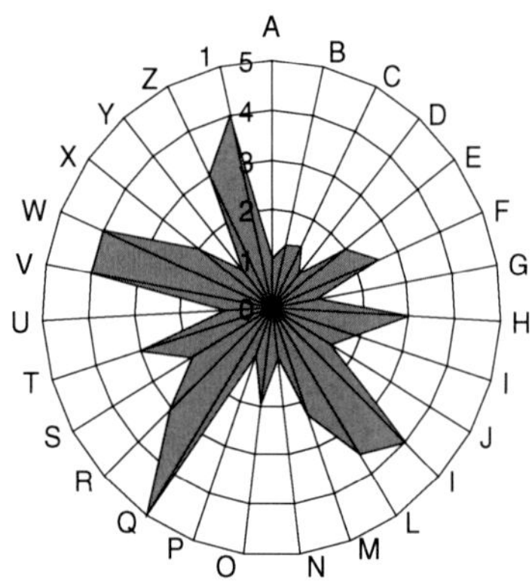


**Fig. 5.3** 27-dimensional radar chart

On the other hand, some of the subcategories are probably composites, increasing the actual number of dimensions. The true dimensionality can only be approximated; however, it is clearly larger than the dimensionality of those early combat models.

The dimensionality issue causes problems with understanding the outputs of a PMESII model, even assuming that it is a perfect model. Figure 5.3 displays a 27-dimensional PMESII snapshot of the state of some geopolitical area. A simulation model would generate a series of such snapshots over its simulated time domain.

A second difference between kinetic models and PMESII models lies in the nature of our understanding of their underlying realities. Kinetic models rely on physics, which at the gross level is well-understood. The non-kinetic parts of PMESII models rely on economics, sociology, psychology, and other very poorly understood sciences. Thus not only should the size of the implicit error intervals for each variable be increased, but also some of the assumed interaction relationships are probably wrong.

A third important factor is actually common to both types of models, although only recently recognized. During the early times of military computer modeling, all relationships were assumed to be essentially deterministic, although some required stochastic modeling due to inherent measurement errors. For example, artillery round trajectories were theoretically computable; however, the precise atmospheric conditions, barrel heating conditions, etc., would not be known in advance. Thus, the location of impact (and hence damage) was represented either by a distribution of impact points or an expected value-based computation. However, it was assumed that results were basically "nice" functions of inputs. For example, if 15 artillery pieces created X amount of damage and 25 artillery pieces created Y amount of damage, then 20 artillery pieces would create a damage amount between X and Y. (Actually, more complex relations were allowed; however, they had to be "sensible.")

Unfortunately, it was discovered that some simple computer codes actually had chaotic regions in which the results were extremely sensitive to very small input variations (very "un-nice" functions). At first it was thought that this was only a problem with the models. Later, it was decided that the real world might have chaotic regions; however, deciding whether there was a valid connection between the chaotic regions of a computer model and reality poses a problem. There is every reason to believe that PMESII models and the reality they model are also subject to these complexity and chaos problems.

## PMESII Model Characteristics

The PMESII domain is complex and there have been numerous attempts to characterize it.

## Mathematical Characterization of PMESII Systems

John Allen [2] of the Defense Advanced Research Projects Agency (DARPA) characterized PMESII system models as:

- *Non-linear*—the response to multiple stimuli is not the same as the composite of responses to individual stimuli;
- *Non-reversible*—don't always return to the origin (or don't return along the same path—hysteresis);
- *Non-deterministic*—future states are uncertain; the process is stochastic;
- *Non-stationary*—statistical properties vary with time;
- *Non-ergodic*—unexpected behavior may evolve or emerge because the statistical descriptions of individuals or sub-groups over time are not the same as the

statistical descriptions of the whole group at a given time or the whole group over time; and

- *Non-invertible*—the inputs and/or initial conditions cannot be inferred from knowledge of the output.

More properly, it should be said that this is a characterization of PMESII systems and that good PMESII system models should also embody these characteristics.

Allen continues that model variable inputs and outputs are values of, or changes in:

- *Probability of Event*: $P\{E_i\}$ at time $t_j$ (e.g., Terrorist attack);
- *Index*: value of $I_k$ at time $t_h$ (e.g., Popularity);
- *Physical quantities*: value of $P_1$ at time $t_f$ (e.g., Watt-hours or $/day);
- *Relationships*: strength $S_m$ at time $t_n$ (e.g., "strong");
- *Concept*: value of $C_p$ at time $t_s$ (e.g., Political Capital); and
- *Location* (not position): $L_d$ at time $t_v$ (e.g., Sadr City);

and that the variables are heterogeneous or even disjoint in terms of units and time references.

Allen's view of the PMESII modeling process is as follows:

- Behavior paradigms are created from theories in sociology, anthropology, political science, military science, economics, and psychology. Modeling paradigms translate these behaviors using logic/algorithms, equations, inference networks, system dynamics, and/or agent/object simulation. The best of these is selected for implementation.
- Knowledge engineering is used to create the databases, model parameters, and initial conditions for a particular scenario.
- Finally, the actions and effects of real-world users must be translated into definitions that correspond to the model. At that point, the actions can be input as independent variable values/inputs/stimuli; the model can be run; and the dependent variables/outputs/responses can be observed as effects.

## Early Descriptions of PMESII Model Characteristics

The first comprehensive review (in 1996) of the requirements for PMESII modeling is given in *Operations Other Than War: Research Requirements for Analysis Tools* [3]. This document reports on the results of several workshops that were attended by experts in Operations Other Than War (OOTW) and analyzes these results. The analysis includes a taxonomy of OOTWs, with categories of operations, attributes of the operations, and tasks requiring analytical support. The tasks are tied to the Universal Joint Task List (UJTL). The analysis also includes frequency distributions by category and responsible Combatant Command of historical OOTWs. The document includes a synthesis, consisting of a list of requirements for analytical tools and definitions of the requirements, with a roadmap.

The second major document (in 1999, reporting on a 1997 workshop) is the report of the MORS Workshop on OOTW Analysis and Modeling Techniques [4]. This report describes the plenary talks and the working group briefings and gives overall recommendations. The working groups were designed to span the types of tools needed for OOTW and were named as follows:

- Mission Definition & MOE/MOP/ROE Determination for Analysis
- Force Planning Tools
- Logistic and Mobility Planning Tools
- Effectiveness Measurement and Course of Action Analysis
- C4ISR Tools and Methods
- Operational Cost Estimation
- Impact Analysis
- OOTW Data Bases and Data Availability
- Executive Planning Review

## PMESII Modeling Constructs

Several approaches may be useful in developing PMESII models. Among these approaches are segmentation by agent or object types, segmentation by PMESII category, use of influence networks, and segmentation by activity types. Each of these approaches is discussed below.

## Segmentation by Agent/Object

In 2001, the Defense Modeling and Simulation Office (DMSO) began to consider the needs for PMESII models. DMSO commissioned a study on building an OOTW Toolbox [5]. The OOTW Toolbox report includes, among other things, a concept Hartley created in the early 1990s that categorizes the actors/objects of interest and describes the PMESII variables associated with each. Figure 5.4 illustrates the concept: a few significant persons and groups operate as independent actors, with geographically distributed demographic categories of people operating as opinion repositories. Events ($E_1$, $E_2$, ...) impact each of the actors/objects (in appropriate ways), with impacts propagating to external countries, including active interveners, and active NGOs. The active interveners and the NGOs post events to the event list, representing their actions. Actors with negative intent are not explicitly pictured; however, in the sense of the diagram their representation would be simple.

Figure 5.5 provides an object-oriented picture of the classes and variables (including some representative enumerations of values) for one of these agent-objects (the distributed demographic object).

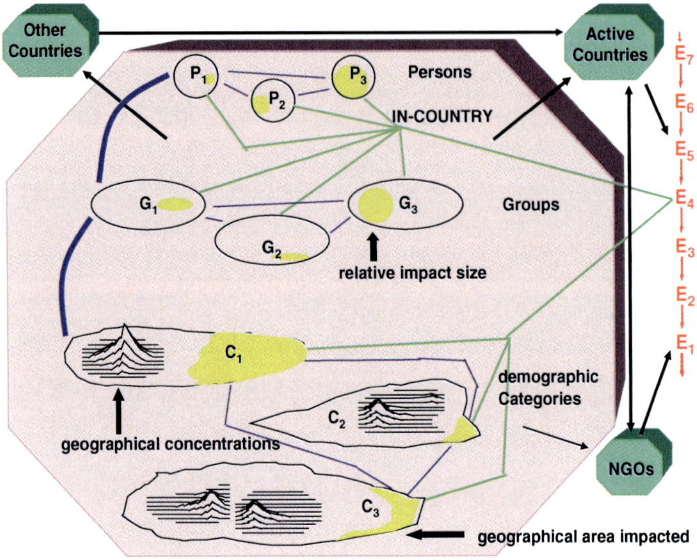


**Fig. 5.4** PMESII agent-objects

Chaturvedi's description of the very large-scale SEAS model {a model that can be used to test strategies} in a 2003 report [6] has more sophisticated diagrams and more detailed descriptions; however, the concept is virtually identical to that of Figs. 5.4 and 5.5. Because SEAS is an implemented model, rather than a conceptual description, it contains extensions that were not explored in Hartley's report to DMSO.

## Segmentation by PMESII Category

Headquarters, Department of the Army [7] describes the PMESII system in terms of links and nodes within and among the six PMESII categories, as shown in Fig. 5.6. Each category is considered to be a system and must be analyzed as such. Further, these systems interact with each other.

The nodes in the HQ DA picture are specific physical, functional, or behavioral entities within each system. The nodes can be facilities, forces, people, information, or other types of system components. The links are the connections between the nodes, which can be physical, functional, or behavioral in nature. In addition, the links have a strength-of-connection attribute. Node and link analyses will determine potential "decisive points," which help in identifying the centers of gravity.

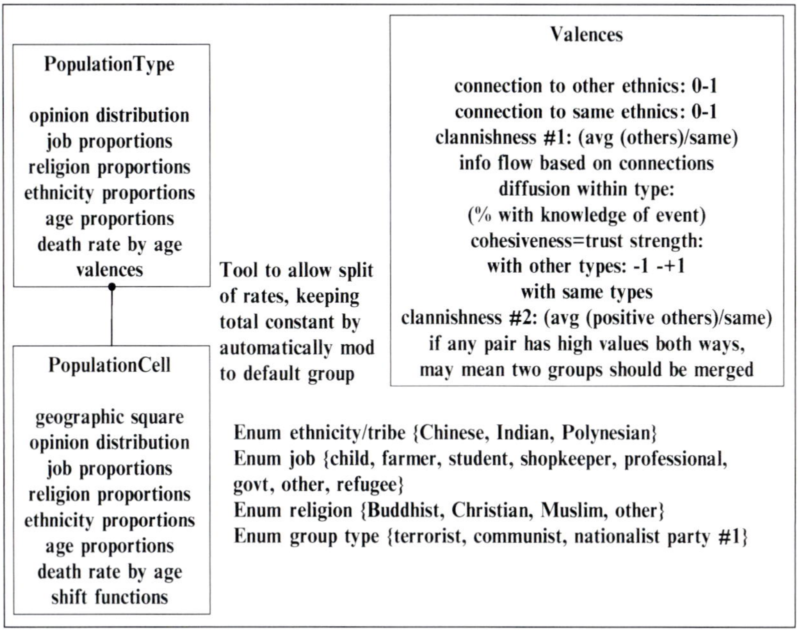


**Fig. 5.5** Potential PMESII distributed (population) object variables

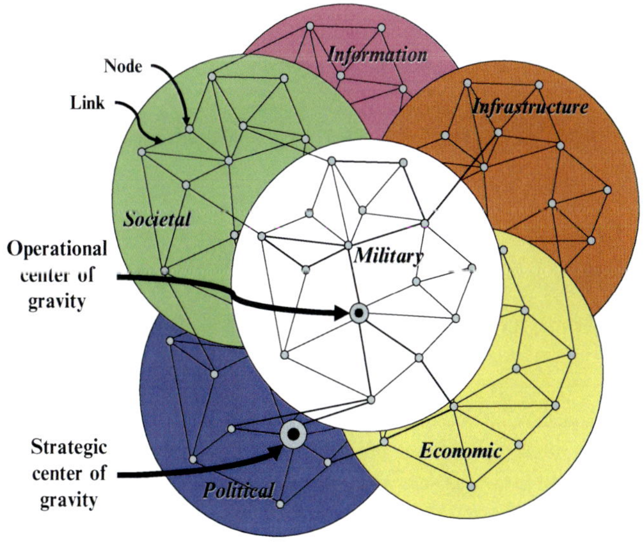


**Fig. 5.6** HQ DA view of PMESII system

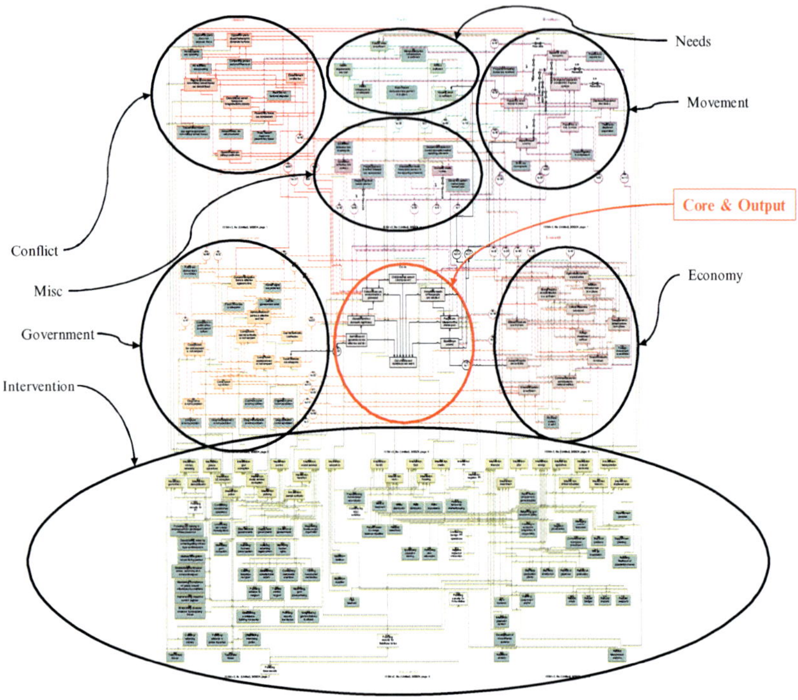


**Fig. 5.7** Logic diagram for the ISSM

## Influence and Causal Networks

The Analysts' Guide for the Interim Semi-static Stability Model (ISSM) {a model that supports understanding the evolution of a situation} [8] describes a node and link system that is divided into sectors with connections among the nodes in each sector and between the sectors (Fig. 5.7). There is visual similarity to the HQ DA concept; however, the nodes and links have different meanings. In the ISSM, the nodes are DIME and PMESII variables and the links are inferential links, some timeless and some time delayed. The Analysts' Guide contains detailed discussions of the nature and potential sources for the input variables.

There are several levels of variables. While this is a relatively simple model, it still has a large number of variables and requires some decomposition to support human understanding. The input and intermediate variables are divided into six sectors: conflict, economy, government, miscellaneous, movement, and needs. These variables feed into the variables of the final output, which is divided into the core sector and output variable. The relationships among these PMESII variables are shown in the upper two-thirds of the inference-diagram in Fig. 5.7. The lower third of Fig. 5.7 consists of the DIME variables, that is, the variables that represent the interventions in the situation being modeled. The upper portion is largely derived from *Doing Windows* [9]; however, that work did not include interventions and the lower portion has been added by the Hartley.

The input variables are indicated by light blue node interiors. The inputs in the top portion are called "observational" variables, because their values are derived from observation of the situation. Their relationships (weighted links) with the intermediate variables to which they connect are inferential and timeless. The canonical form of this relationship is, "I observe X, therefore I infer that Y is true." On the other hand, the intervention inputs have explicit gaps between the time of observation and the time of impact on the variables in the upper half. For example, a bridge building intervention has an immediate impact on the employment situation, but a delayed impact on the transportation situation.

These nodes and links contain the structurally invariant part of the model. Not shown is the user-created Preprocessing and Postprocessing logic that allows the user to customize the model to a particular scenario. The Preprocessor permits the user to create logic that converts data that are available into the inputs required by the invariant part of the model and the Postprocessor permits the user to create new measures of effectiveness, based on the standard variables. This custom logic allows the user to tailor the ISSM to the particular scenario of interest.

Wagenhals and Levis, in their description of the Pythia model, concentrate on the inferential connections among PMESII variables, both timeless and time delayed [10], with evident similarities to the inferential connections of the ISSM. Pythia is a computational support structure for implementing DIME/PMESII models. The following steps describe the process illustrated in Fig. 5.8:

- The Effects Based Operations modeling approach starts with the definition of desirable Effects on the Adversary (Red) and *the desired end state of Blue*;
- Then the user works backwards (from right to left) to the Centers of Gravity (COGs) of Red that influence the desired Effects—the arrows show the "Cause to Effect" (or in some cases "influence to effect") relationships (left to right);
- Then the user identifies the Operational Functions of Red that affect the COGs, which in turn influence the Effect(s);
- The user continues "unfolding" backwards in a "Cause or influence to Effect" chain till arriving at Actionable Events that can be carried out by Blue;
- Then, the user includes other external events, not controlled by Blue, that influence the achievement of the desired Effects on Red;
- There are also "Cause or influence to Effect" relationships that affect the *strength* of the influences (e.g., selected Information Operations); and
- Finally, time delays are assigned to the links and nodes, converting the influence net to a timed influence net.

Wagenhals and Levis do not explicitly divide the nodes into sectors; however, their layered approach of causal or inferential links, variable strengths to the links, and timing information are similar to the node and link approach of the ISSM.

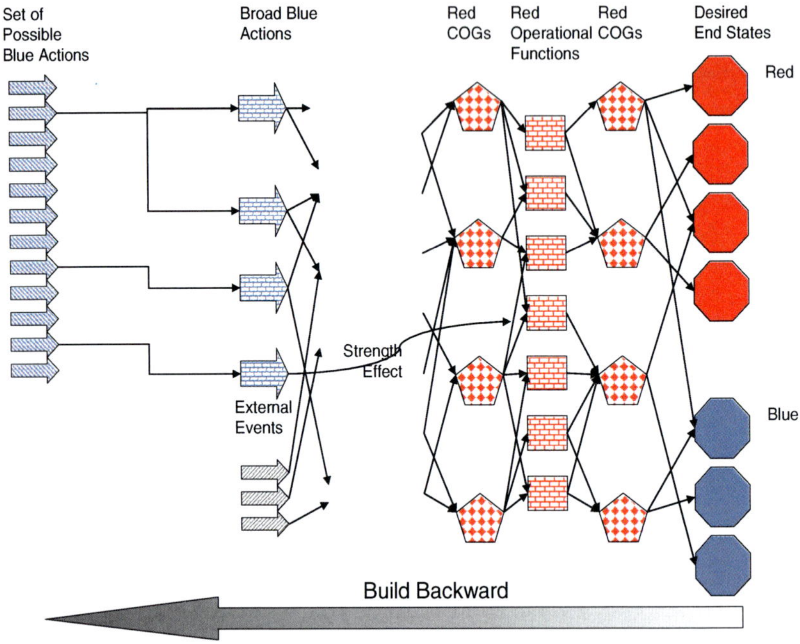


**Fig. 5.8** Pythia structure

## Segmentation into "Battles"

Hartley, Holdsworth, and Farrell [11] describe a segmentation of the PMESII system into activity types. The HQ DA segmentation is by the PMESII components and the ISSM segmentation is by national sectors. The HHF segmentation is labeled "Battles" and is derived from an unknown source within the Department of Defense.

Figure 5.9 shows these "Battles" in an irregular conflict, with their connections. (This is the part from the unknown source.) The diagram may be read by imagining a group of people who want to cause disruption or take power in the region. First they must create an organization, which involves an information "Battle" and a cycle of appearance and persuasion in which they are successful. Then they want to co-opt other organizations to their way of thought and to pursuing their aims or aims that support them. This also involves an information campaign. At some point their success is great enough to start looking for more control, the "Battle for the Streets," and to gain recruits, internally and externally. Both of these involve information campaigns. As the organization gains success, resistance sets in and there is a "Battle" for control of access to the country at the borders. The links described so far all involve converting popular and individual opinion to their side. The other links represent the physical flows of people (recruits) and power (usually destructive, but also services provided). Naturally, each of these "Battles" is (at least) two-sided, with resistance from the government and from other groups with their own ideas of

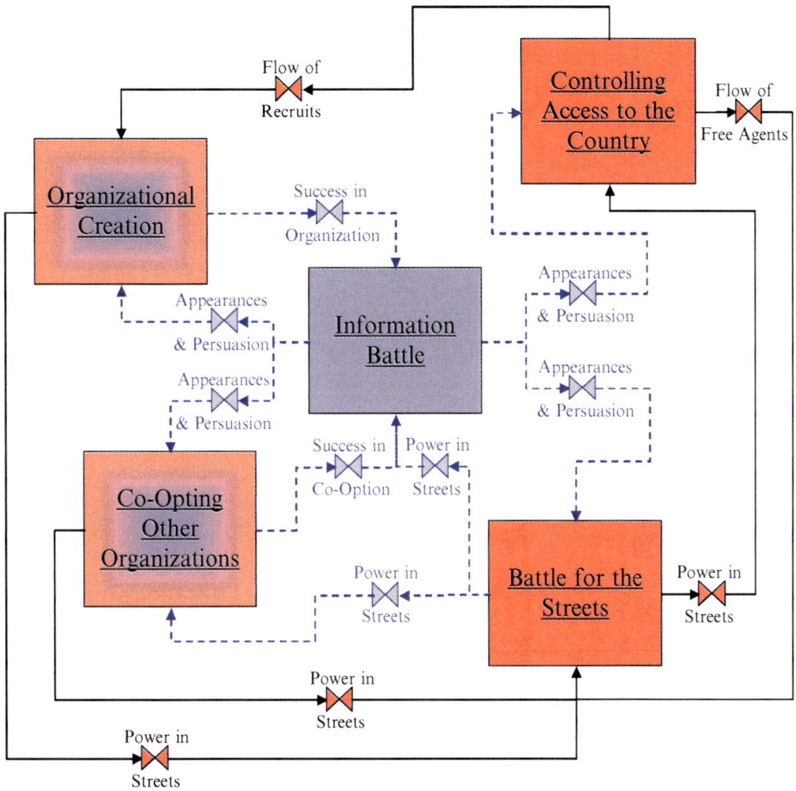


**Fig. 5.9** Irregular "Battles"

what should be done. Clearly, no kinetic (attrition and combat damage) model is adequate to the task of measuring all the outputs of these "Battles." A DIME/PMESII approach is required.

As complex as the picture in Fig. 5.9 is, it is inadequate. Something critical is missing and that is the effect this activity has on the daily life of the rest of the population—and the effect this daily life has on the "Battles" so far defined. Figure 5.10 adds the "Daily Life Battle." This "Battle" includes the economy (broadly defined), governmental activities, the basic needs of the populace, and the level of crime. It is the evaluation of "normalcy" that is the critical measure of success.

## Social Modeling Ontologies

An ontology is a description of the concepts and relationships that exist in a domain [12]. The PMESII taxonomy shown in Fig. 5.2 is an example of a simple ontology. Each of the subcategories of the top-level PMESII categories has an "is-a"

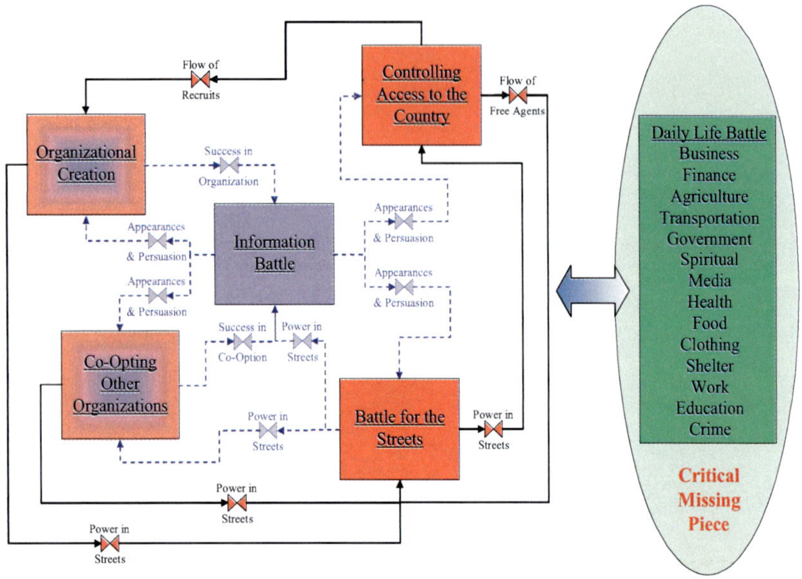


**Fig. 5.10** Irregular "Battles" plus Daily Life Battle

relationship with its category. That is, each instance in the "Conflict" subclass under the "Military" class *is an* instance of the "Military" class. In general, ontologies can be more complex than taxonomies. For example, multiple parents are allowed for each subclass and other relationships can be implemented, such as the "part-of" (or the reverse direction "includes") relationship that displays the components of a class.

Figure 5.11 shows the domain of interest: actors perform actions that affect the operational environment (OE) (including all actors, actions, and the rest of the environment), which is described by state variables, which in turn comprise the perceivable state of the world for the actors. The agent-objects discussed above are included in the "actors" component of the OE. The influences, causes, battles, DIME actions, etc., discussed above, are included in the "actions" component. The PMESII taxonomy, augmented with Kinetic and Environment categories and expanded down a level (with multiple parents allowed), comprises the state variable component.

Figure 5.12 provides a compressed view of the total ontology. The four individual class symbols that are connected by arrows in the figure represent literally hundreds of classes that have relationships with the four larger structures and with each other.

This ontology has been applied to model improvement programs and to validation of models [13].

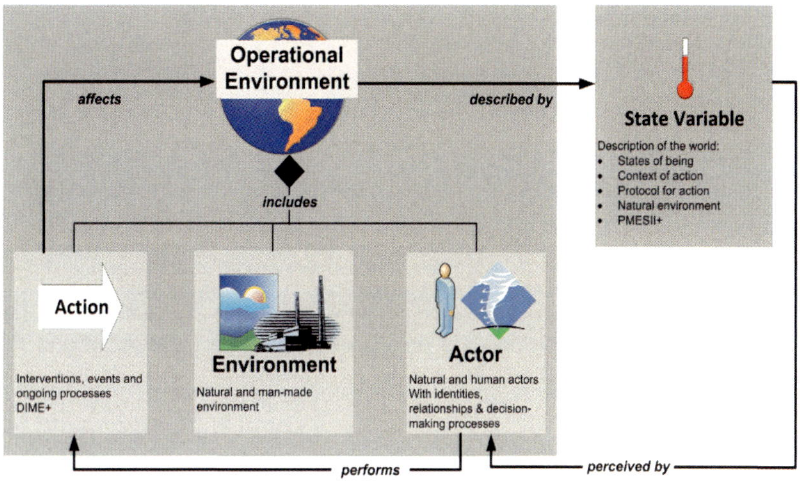


Fig. 5.11 The social domain

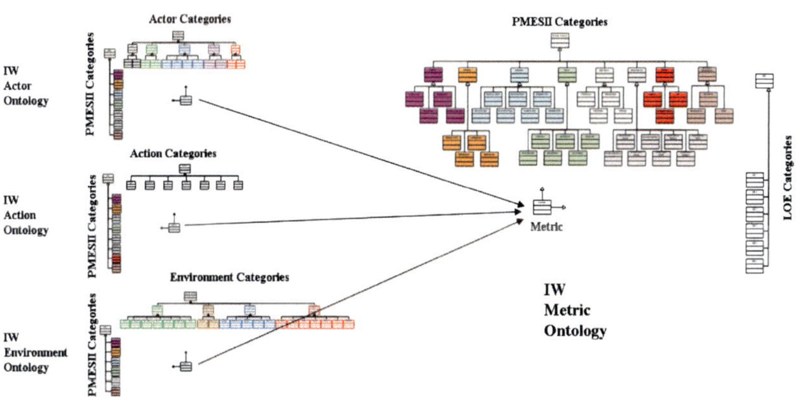


Fig. 5.12 The social ontology

## Technical Modeling Approaches

In addition to the modeling constructs discussed above, there are approach options that are relevant for models in general and PMESII models in particular. Several of these are discussed briefly.

## Proxy Variables vs. Causal Variables

Proxy variables are variables with good correlations to causal variables. Proxy variables may be used because the causal variables are unknown or because the causal variables are unobservable in practice. Proxy variables support large granularity models but cannot address the details that a coherent, validated theory with causal variables naturally addresses.

A defect of proxy variable models is that they can be "gamed." Because the proxy variable is not directly connected to the desired causal variable, it can be changed without affecting the causal variable (or with only minor effects). For example, suppose one were to use the price of cauliflower as a proxy variable for the cost of living. (One might do this if the price of cauliflower were readily available and not generally subject to price swings due to temporary fads.) Someone discovering this fact could cause estimates of the cost of living to change dramatically by influencing the price of cauliflower, a much easier proposition than actually changing the cost of living.

## Static Models, Time, and Feedback Loops

Time flows in the real world and actions cause reactions. These reactions are feedback loops, sometimes acting to damp the original actions (negative feedback) and sometimes adding to the effects of the original actions (positive feedback). Static models are representations of situations at a given instant and thus not capable of representing feedback loops. Modeling situations over time is directly addressed by simulation (e.g., discrete event simulation, time-stepped simulation, continuous simulation, or system dynamics). Simulation models can include feedback loops; however, the existence of feedback loops is dependent on what was programmed to be a part of the model, not the fact that it is a simulation.

## Closed Form vs. Human in The Loop

Closed form models are built to run without human intervention once the "start" button is pushed. Closed form models are easier to use and avoid the uncontrollable variability of human decisions inside the model run. Human-in-the-loop models allow for simpler formal models by substituting available humans' mental models for portions of the total model. Human-in-the-loop models require increased complexity in their use to mitigate human variability and simply because of the increased staffing demand.

## Stochastic vs. Deterministic

Deterministic models produce a single result (or set of results) for a given set of inputs. Stochastic models produce a single result for a given set of inputs—for a given run. A different run will (generally) produce a different result. Stochastic models are generally run many times with the same inputs to obtain a distribution of results. This distribution allows estimation of the range (and other statistics) of possible results, as well as the most likely result. Deterministic models are easier and faster to use; however, they give no information about any possible variation in their results. Most complex, real-world situations have intrinsically variable results. Hence, information about this variability can be very valuable.

## Complexity

More complex (i.e., complicated) models can support finer granularity. Finer granularity supports, but does not guarantee, more precision and accuracy. More complex models require more input data and are harder to examine for problems. Simple models usually run faster and take less time to set up and analyze.

Treating and modeling a system as a complex adaptive system is different from producing a complicated model. The underlying real-world system that is the subject of a PMESII model is certainly a complex adaptive system; however, the PMESII model may or may not be constructed to address typical complex adaptive systems questions. The other chapters of this book focus on these issues.

## The Nature of Agent-Based Models

Agent-Based Models (ABMs) are stochastic simulations that are built around dynamically interacting objects called agents. These agents are usually all of comparable size and scope. In some ABMs the agents are all identical copies of the same object, while in others there are two or more types of agents. The agents have preset attributes and behaviors; however, the choice of data inputs can dramatically change the model's purported real-world ideal. For example, the same agents in an ABM can represent soldiers with a mission in one model and a set of mines in the ocean in another model. Generally, ABMs contain some spatial representation, $(x, y)$ or $(x, y, z)$ coordinates, that can be used to model real-world distances.

The simulations that are most commonly referred to as ABMs have simple sets of rules; however, the simulations are dynamic. The rules mesh together during the simulation run and produce often surprising, emergent behaviors. The stochastic nature of the models and the sensitivity to small changes means that thousands to millions of runs are required to understand the range of behaviors, their frequencies, and associations with input data. The outcomes are non-linear, that is, not predictable beyond small time increments. Further, the results from $N$ iterations will likely differ from the results of $N+M$ iterations.

The emergence of complex behaviors from simple sets of rules is the principal reason for using ABMs. The claim is that just as the complexity of chemistry emerges unpredictably from the simple rules of atomic electron interactions, the best way to understand the complexity of human interactions is to investigate the emergence of complex interactions from simple simulation rules in an ABM.

The more complex ABMs incorporate variable behaviors. These ABMs support connected sequences of runs in which the results of previous runs are used to modify the behaviors in subsequent runs. These ABMs are adaptive and can generate the co-evolution of the behaviors of one or more groups of agents (sides).

The ABMs of interest here are the models that include PMESII-type attributes, such as emotions, opinions, and social grouping valences. These attributes present problems because they are labeled as PMESII-type attributes; however, it is not clear that they actually represent what their labels claim. Because our understanding of the true relationships among such variables is poor, validating the code is difficult. Further, the claim is made by some ABM proponents that the primitives that make up the relationships are what should be modeled and that by observing the emergent behaviors, users can make correlations between these and real-world behaviors with these labels, producing the desired validation.

The ideal ABM would be a completely protean, content free model in which any situation in any portion of reality could be modeled simply by changing the data and changing the human meanings attached to the objects and labels of the model. Thus the dots on the computer screen that the model uses to portray the agents can be thought of as individual soldiers or as floating mines. The rules governing movement toward or away from other agents can be thought of as social rules of liking or disliking or as representing the physical constraints of chains and wave action. Actual ABMs are not ideal, having some categorical representation built in to provide a framework for the ABM construction. Thus a particular ABM might not be flexible enough to represent both soldiers and mines. Some ABMs might include more geographical representations than others. However, in general, most of the model is contained in the data and the human interpretation that is attached to the implementation.

## Systems of Models

Systems of models are composed of multiple models. Most models can be decomposed into submodels; however, when these submodels are integral parts of the model, the model is not generally thought of as a system of models. The submodels of a system of models are generally autonomous models that could be used as standalone models.

The coupling of the models is an important characteristic of a system of models. The two basic types are tight coupling and loose coupling. Tight coupling is characterized by the direct passage of data from one model to another, with no human intervention. This data passage is considered direct even if the two models run on different computers or if there is an intervening translation program or if there is an intervening time delay (such as is determined by a time-scheduling system, e.g., the High-Level Architecture (HLA) [14]), as long as the passage is automatically controlled by a computer program, not by human intervention. Loose coupling is characterized by human intervention in the passage of data, whether for translation of outputs from one model into inputs for another or for human decision making. The Flexible Asymmetric Simulation Technologies (FAST) system {used in evaluating strategies}, described in the next section, is an example of a loosely coupled system of models and the Conflict Modeling, Planning and Outcome Experimentation (COMPOEX) system {used in evaluating strategies}, described by Starr [15], is an example of a tightly coupled system of models. Loosely coupled systems of models take on some of the characteristics of human in the loop models and must be treated as such factors in the use of PMESII models

Hartley et al. [11] describe the process that the FAST Team evolved for using its set of tools in the analysis of complex operations that required the evaluation of PMESII variables. The first issue was that no single model was capable of providing all of the capabilities needed. The ISSM could evaluate the effects of DIME interventions on PMESII variables and interpret the state space of PMESII variables on the measures of socio-political effectiveness; however, it was not a simulation and could not generate changes over time through feedback loops. Each of the simulations in the toolbox could model some DIME activities and generate changes; however, each embodied a limited (incomplete and only partially validated) model of the necessary DIME activities and PMESII factors. Thus the computer models, while providing the benefits of consistency and rapid computation of complex events impossible to humans, could not generate a useful product if used in the standard manner of end-to-end simulation. On the other hand, the human analysts could see how events in one tool could modify events in another tool, but could not effectively and consistently apply those insights, post-simulation. The humans could also mitigate some of the model validation issues.

To remedy this problem, the team created the concept of Punctuated Simulation. This concept is illustrated in Fig. 5.13. The "Road to War" was encoded in the ISSM to provide a common PMESII starting point for the simulations. Each simulation was run for a short period of time (60 simulated days to provide two 1-month data sets) to generate information for one or more of the Battles (described above). These two data sets were encoded into the ISSM, using outputs from the simulations and evaluations from the human analysts. The ISSM provided the evaluation of the impact of these inputs in socio-political terms and a new, common starting point for the simulations. Thus despite the fact that each simulation lacked important "understanding" of the entire PMESII situation, its second run, being based on this common situation (not the situation as described at the end of its own first run), incorporated the effects of the missing variables. Following the second run of the

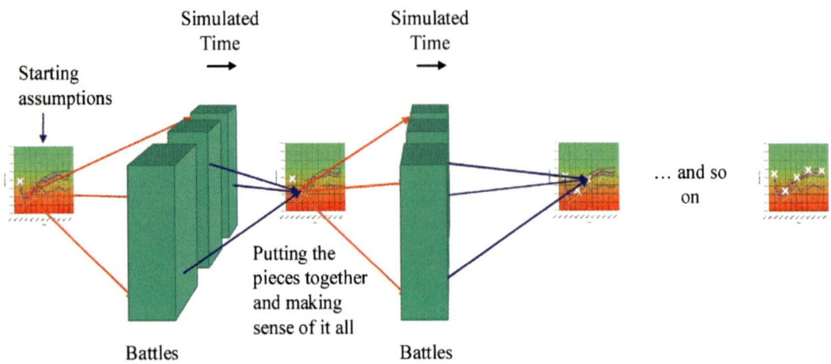


**Fig. 5.13** Punctuated simulation

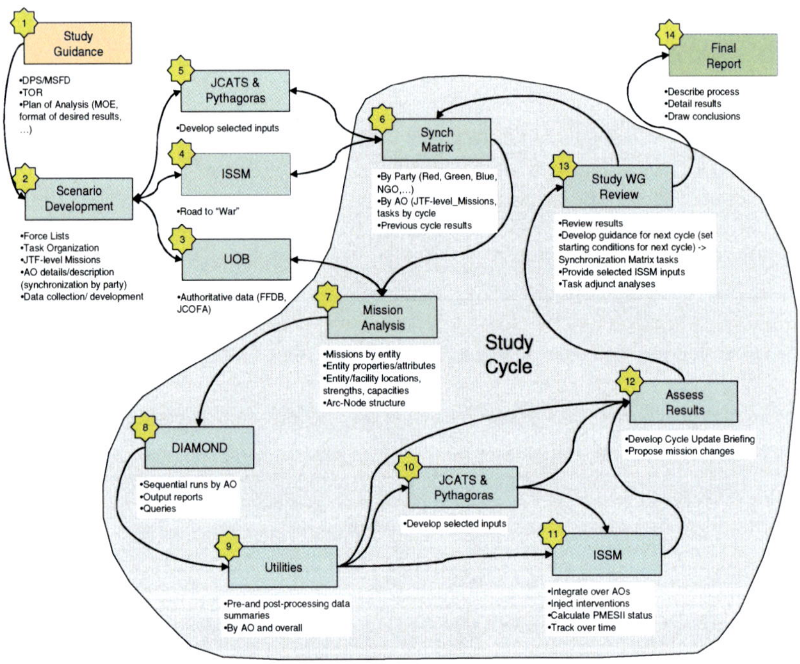


**Fig. 5.14** Overview of analysis methodology

simulations, the new outputs and analyst evaluations were again fed into the ISSM. This process was repeated until the scenario was completed.

The second problem was to embed the punctuated simulation into a structure that gave rigorous support to the analysis of the problem sets likely to be of interest. The team chose to use the standard military planning procedures as a basis. Figure 5.14 provides an overview of the adapted process.

The study began with the study guidance and supporting materials (step 1, in the figure) and a requirement for a final report (step 14). While the analysts were developing the scenario (step 2), the Unit Order of Battle Data Access Tool (UOB DAT) was used to prepare the force lists to be used (step 3) and the ISSM was used to convert the study guidance into a "road to war" that translated the text description into the PMESII variable values (step 4). The Pythagoras ABM and potentially other tools, such as the Joint Conflict and Tactical Simulation (JCATS) {used for kinetic strategy evaluation} were used to develop selected inputs (step 5).

At this point, the analysis entered into a cycle of steps. Analysts created a series of synchronization matrices (step 6) and a mission analysis (step 7) from the existing situation and the campaign concepts of the commanders of the various parties to the conflict (as articulated in the particular defense planning scenario upon which the study was based). These elements were used to define the scenario for the Diplomatic and Military Operations in a Non-warfighting Domain (DIAMOND) simulation {used for evaluating strategies}, which was run for a simulated period (step 8). The outputs of DIAMOND were transformed in a set of utilities (step 9) for input into the ISSM (step 11) (and potentially input to Pythagoras and/or JCATS (step 10), with further use as inputs to the ISSM). The outputs of the ISSM and certain outputs from DIAMOND were used to assess the situation at the end of the current cycle (step 12) from a modeling perspective. These results were used by the study working group to assess the situation from a study perspective and create the plan for the next cycle (step 13).

When sufficient cycles were completed, step 13 was followed by the creation of the final report (step 14), rather than a new cycle. In the study report, the ISSM and Pythagoras produced the principal measures of merit.

This structure fits the situation of incomplete and only partially validated model coverage in a standard analysis process. With modification, it might also be useful in an exploratory analysis situation.

## Problems in Modeling DIME/PMESII Systems

The problems specific to modeling DIME/PMESII factors can be divided into four categories: variables, relationships, invariants, and data.

- What are the relevant variables and who says so? Which variables influence each variable and who says so?
- What is the functional relationship between influencing variables and influenced variable and who says so? Which relationships are deterministic and which are probabilistic and who says so? What are the distributions for the probabilistic relationships and who says so?
- Which of these things are invariant with scenario because they describe "human nature" and who says so? How do the non-invariant things vary with scenario variations and who says so?
- What are the proper data to use as inputs and who says so? What do the "answers" mean and who says so?

These problems are real problems. They impact the creation of models and the level of believability in their results. These problems make it hard to build and use DIME/PMESII models. However, real people must-and-do make decisions in the problem space that DIME/PMESII models should address. Absent a formal model, these people use mental, conceptual DIME/PMESII models. Some points should be kept in mind:

- Some people have better mental models than other people;
- Some people apply them more consistently than other people;
- Some people know more about the available evidence than other people;
- Some people learn from their (and other's) mistakes and improve their mental models better than other people; and
- 50 % of people in any group are below average for that group.

Initially, any formal model, whether textual or programmed, is based on a mental model. However, a formal model

- May start as a composite of the mental models of many people;
- Can be applied more consistently than a mental model;
- Can be consciously created from a basis of evidence; and
- Can be explicitly modified as more evidence is obtained.

Further, a formal model can be more easily viewed as separate from the user and, thus, be used as an advisory tool.

## VV&A of PMESII Models

The U.S. Department of Defense (DoD) has defined verification, validation, and accreditation (VV&A) as follows [16]:

*Validation*: The process of determining the degree to which a model or simulation is an accurate representation of the real world from the perspective of the intended uses of the model or simulation.

*Verification*: The process of determining that a model or simulation implementation accurately represents the developer's conceptual description and specification. Verification also evaluates the extent to which the model or simulation has been developed using sound and established software engineering techniques.

*Accreditation*: An official determination that a model is acceptable for a specific purpose.

VV&A consists of processes that are difficult to perform for any model and complete verification and validation (V&V) of large simulations is virtually impossible. The IEEE Computer Society published a very good book by Knepell and Arangno [17], which give a good, readable description of V&V of standard simulations.

## VV&A of General PMESII Models

PMESII models entail additional complexities. The PMESII space is enormous and our understanding of the nature of the variables and their interactions is limited, at best. A team lead by Stuart Starr [15] generated a good, readable prescription for VV&A of PMESII models. Several observations are important.

- Because of the size of the state space, the PMESII tools of interest may consist of a family of interacting tools that employ different modeling paradigms (e.g., Bayes nets, agent-based models) and behavior paradigms/theories (e.g., lateral pressure theory of international relations; compromise theory of opinion dynamics).
- The models represent "reality" with a variety of disparate variables. Some of these variables are artificial (e.g., "political capital") while others are measured indirectly.
- Given the uncertain nature of future conflict (e.g., traditional, irregular, catastrophic, disruptive challenges), there is a need to select appropriate models for the problem of interest and to tailor them to reflect the operations of interest.
- The models may reflect behavior that is generic (e.g., an Islamic cleric) or specific (e.g., a specific Islamic cleric, such as al-Sistani).
- There is a need for a mixture of accreditation protocols to reflect potential variability in the variables and models of interest. This variability includes, inter alia, variable types (e.g., conceptual, physical), model kinds (e.g., deterministic, stochastic, manifest emergent behavior), model types (e.g., generic, specific), model coupling (e.g., independent, weakly coupled, strongly coupled), and model level detail (e.g., high, medium, low).
- There must be multiple levels of accreditation, depending on the application.
- Finally, VV&A for PMESII models must manifest three key features. They must be "continuous" (i.e., occur throughout the phases of model conceptualization, development, deployment, operation, refinement); "entrenched" to ensure that the users do not violate bounds (e.g., parameter settings, initial conditions) or model choice (e.g., employ an inappropriate model); and "people-centric" (i.e., deal with education and training of model creators, users, supporters, customers).

In addition, it is important to realize that the level of fidelity of a PMESII model to the real world cannot be expected to be as high as that of a kinetic model—we simply do not know enough about DIME/PMESII interactions to produce theories concerning these interactions that are as good as our theories about kinetic interactions. Hence, our models will suffer from inferior basal theories. See the Modeling & Simulation Coordination Office website for a discussion of Human Behavior Representation (HBR) validation [14, 18].

## VV&A of Human in the Loop Models

It is important to realize that the particular humans who are "in the loop" are part of the model. Their expertise (or lack thereof) informs their mental models which guide their actions. Some effort must be expended in identifying and taking into account the effect they have on the results of the overall model.

## VV&A of Systems of Models

The VV&A of a system of models is nominally the same as the VV&A of a single model; however, the decomposition into individual models permits a natural decomposition of the V&V process into the V&V of each individual model, the V&V of the interconnections of the models, and the V&V of the entire system. The accreditation is (generally) of the entire system, not of the individual models.

In both loosely coupled and tightly coupled systems, the V&V of the interconnections must critically examine the syntactic and semantic nature of the data passed through those interconnections and must be especially careful that the modeling assumptions that create and use the data match each other. However, loosely coupled systems of models take on some of the characteristics of human in the loop models and must be treated as such.

## VV&A of ABMs

There is very little content in a completely protean ABM and, thus, little to validate. Standard verification procedures are required to ensure the quality of such an ABM; however, there is little value in the validation, as the instantiated model (ABM + defining data) will yield a different model. However, most ABMs are not completely protean. If the model purports to have an input variable for "love" of one agent for another, an attempt should be made to determine the basis for such a label and the implementation of this basis should be checked.

Once the complete model has been instantiated, however, complete PMESII validation measures are in order. Care is required to ensure that changes in the input data from that which has been validated do not result in a new model, requiring new validation.

## The Effects of Model Purpose on Validation

There are several possible purposes of a model: prediction, education, training, exploration of possibilities, and understanding are some of the purposes. These are broad categories of purposes, with subdivisions within each. A model that is adequate for one purpose need not be adequate for another.

This point can be easily seen when considering model airplanes. The purpose of scale model airplanes is to show the airplane's configuration. The exact size is not important; the correct materials are not important; and the model airplane is not expected to fly. However, a flying model airplane is expected to fly. Its exact size is not important; the correct materials are not important; and the configuration should be recognizably similar to the original, but not exact. The purpose drives which features are abstracted as important and which are sacrificed as unimportant.

DIME/PMESII models are no different. A prediction model should yield correct predictions (within stated bounds), no matter what method is used to obtain them. A training model should train its audience correctly (in the desired domain), with no negative training. A model geared toward understanding might make no predictions at all.

Thus, the purpose must drive the tests of validity, both in what is covered and in the level of fidelity that is expected.

## Conclusion

DIME/PMESII models are models with all the benefits and problems of any other model. When the domain of interest is terrorism, with kinetic and non-kinetic effects and multiple possible causes, the technical domain of the models must include the real-world factors involved in terrorism. Measures of intervention input (tons of supplies delivered to a port, or even to the populace) are important, but not sufficient. We must also understand the impacts (usually not a single impact) of an action. Further, the decomposition ensures that we understand that we should consider more than just a single dimension of intervention, whether that dimension be legal or military or economic, and that we must consider more than just a single dimension of the situation.

The DIME/PMESII taxonomy, as a descriptive decomposition, may not be the best taxonomy. For example, some have suggested that Financial, Intelligence, and Law enforcement elements should be added to DIME, yielding DIMEFIL. Others have suggested alternatives to PMESII. However, DIME/PMESII as a general concept is entirely adequate. Letting DIME stand for all of the intervention options and letting PMESII stand for a set of descriptors of the situation allows us to organize our thinking. Thus the Social Ontology described above, which expands on the DIME/PMESII concept, contains the spirit of the DIME/PMESII taxonomy, without being constrained by all of the details.

As long as we use the term loosely, we see that the models employed to investigate terrorism generally fall into the category of DIME/PMESII models.

## References


1. Wikipedia (2009). *RAF*. http://en.wikipedia.org/wiki/Red_Army_Faction. Accessed 6 Jan 2009.
2. Allen, J. G. (2006). *Modeling PMESII systems*. Washington: DARPA.
3. Hartley, D. S., III. (1996). *Operations other than war: Requirements for analysis tools research report*. Oak Ridge: Lockheed Martin Energy Systems, Inc.. K/DSRD-2098.
4. Staniec, C., & Hartley, D. S. III. (1999). *OOTW Analysis and Modeling Techniques (OOTWAMT) Workshop Proceedings*, January 1997. Alexandria: MORS.
5. Hartley, D. S., III. (2001). *OOTW toolbox specification*. Oak Ridge: Hartley Consulting.
6. Chaturvedi, A. R. (2003). *SEAS-UV 2004: The theoretical foundation of a comprehensive PMESII/DIME agent-based synthetic environment*, June 24. Simulex, Inc., http://simulexinc.com.
7. HQ DA (2006). Systems perspective of the operational environment. *The Operations Process*, FM 5-0.1.
8. Hartley, D. S., III. (2006). *Operations other than war (MOOTW) flexible asymmetric simulation technologies (FAST) prototype toolbox: ISSM v4.00 analysts' guide*, Dynamics Research Corporation.
9. Hayes, B. C., & Sands, J. I. (1997). *Doing windows: Non-traditional military responses to complex emergencies*. Washington: CCRP.
10. Wagenhals, L. W., & Levis, A. H. (2006). *Dynamic cultural preparation of the environment: Course of action analysis using timed influence nets*, July 21. Fairfax: George Mason University.
11. Hartley, D. S., III, Holdsworth, D., & Farrell, C. (2006). *OOTW FAST prototype toolbox: Analysis process*. Orlando: DRC.
12. Gruber, T. (1992). *What is an ontology*. http://www-ksl.stanford.edu/kst/what-is-an-ontology.html. Accessed 5 July 2012.
13. Hartley, D. S., III. (2012). Ontology structures for modeling irregular warfare. *Military Operations Research, 17*(2).
14. Modeling & Simulation Coordination Office. (2012). *HLA*. http://www.msco.mil/hla.html. Accessed 5 Sept 2012.
15. Starr, S., et al. (2007). *Verification, validation, and accreditation (VV&A)*. Vienna: Evidence Based Research, Inc.
16. DoD 5000.59. (2007). http://www.dtic.mil/whs/directives/corres/pdf/500059p.pdf. Accessed 5 Sept 2012.
17. Knepell, P. L., & Arangno, D. C. (1993). *Simulation validation: A confidence assessment methodology*. Los Alamitos: IEEE Computer Society Press.
18. Modeling & Simulation Coordination Office. (2012). *Human behavior validation*. http://vva.msco.mil/Special_Topics/HBR-Validation/default.htm. Accessed 5 Sept 2012.

# Chapter 6
# Net-Centric Logistics: Complex Systems Science Aims at Moving Targets

**Thomas Ray**

## Introduction

In the second decade of the twenty first century, armed conflicts abound, and it's not your grandfather's war anymore. It's not even your grandfather's idea of what war is supposed to be.

While it's always risky to try and predict where the next war is going to be fought and how big it is going to be—there are just too many political, social, and economic variables—the current state of geopolitics and military readiness allows us to make some reasonably intelligent predictions of how war is going to be waged in the immediate future:

In the first place, definitions of "victory" are elusive. Prior to the Korean conflict, the opinion that victory means "the utter destruction of an enemy's will to fight" was likely near universal. In less complicated times, wars between nations were hardly distinguishable—except in scale—from a fight between individuals. The loser could choose only surrender (which most often resulted in slavery) or death.

That one difference—scale—turns out to be a significant factor in a world where high-powered substitutes for sticks and stones all but guarantees a Pyrrhic result for all sides in every conceivable case of large scale war. Diplomatic and political solutions, up until the end of the US–Soviet cold war, thus largely focused on mutually assured destruction (MAD). The only solution to a perfect chess game is stalemate.

Since, the steady rise of third world nations and growing economic parity among global powers (notably, the US and China) may already have been seen to contribute to a subtle shift toward strategic policy initiatives that suggest a desire for mutually assured prosperity (MAP). Evidence of increased cooperation (by both government and non-government entities) in nation building and modernization enterprises,

T. Ray (✉)
Technical Writer, US Army (ret)



P.V. Fellman et al. (eds.), *Conflict and Complexity*, Understanding Complex Systems,
DOI 10.1007/978-1-4939-1705-1_6

along with disaster relief and humanitarian aid, trade, treaties, arms reduction—even given daily reports of wars and rumors of wars, famine, and calamity from every corner of the Earth—suggests that on balance, newly enfranchised citizens of the world's exponentially growing population expect to share the exponentially growing benefits of citizenship.

We do not choose the acronym MAP capriciously. An individual's personal map to prosperity is as critical as that of a culture or nation.

In a 2007 *Science* article, Lim et al. [1] reached the counterintuitive conclusion: "Peaceful coexistence need not require complete integration." We say counterintuitive, because the assumption of integration of societies—in the context that requires assimilation of cultural values, beliefs, laws into a common well of political doctrine—has been the mainstay of liberal thought from the eighteenth century Enlightenment to the Marxist-Leninist movements of the present age. Lim et al. [1] make a case that preserves the liberal ideal while invoking the importance of multiple scales of interaction that maximize cooperation and minimize the potential for violent conflict: "Violence arises due to the structure of boundaries between groups rather than as a result of inherent conflicts between the groups themselves."

We may see a particular structural model, then, as a mode of communication by which individuals, and cultural/political organizations of individuals, can freely contribute to the common well and drink from it, without being drowned in some doctrine of forced behavior.

The motivation for Lim et al. [1] derives from Bar-Yam's extensive research in complex systems, culminating in the theory of multi-scale variety. This theory generalizes the principle that lateral, rather than hierarchical, distribution of activity and information drives system effectiveness: "In considering the requirements of multi-scale variety more generally, we can state that for a system to be effective, it must be able to coordinate the right number of components to serve each task, while allowing the independence of other sets of components to perform their respective tasks without binding the actions of one such set to another" [2].

We speak of a system in which groups, down to the least element—the individual—can be effective without sacrificing self-determination to a hierarchical order. Even though the fact remains that cultural/political groups often impose severe restrictions on the individual mobility of their members, a global system that maximizes cooperation suggests to rational societies that local liberalization is an asset to maintaining their independence among the community of nations. This is non-contradictory only if such cooperation is voluntary and non-coercive, not "… binding the actions of one … to another."

Why nations, groups, and individuals would *want* to cooperate is the theme of this chapter. Underlying is a hidden assumption we wish to make obvious: human free will transcends cultural and political boundaries; if the means to cooperate is available, the will follows.

The means is known by the generic name, *logistics*.

## Military Science Often Leads Innovation

Perhaps it is no wonder that what has always threatened to destroy civilization has also contributed so much to it.

The ancient Greeks gave us military technology and battle tactics still in use today. The ancient Romans built roads yet unrivaled in our age, most often used to supply soldiers and advance trade to the empire's frontier.

Greek city–states are long gone. The Roman Empire is a distant memory. And still, military technology ironically continues to occupy the leading edge of contributions that make the human condition better, from life-saving technology and personal mobility to food preservation and space exploration.

What the military most importantly and profoundly recognizes, is that any good thing can be turned into a weapon. Alfred Nobel would no doubt agree. And the chief good thing about running a large scale army is that it takes a remarkable amount of knowledge, discipline, organization, money, and training to keep combatants supplied with beans and bullets wherever they are called on to fight, or to restore order following disaster—yet logistics itself hasn't been widely acknowledged as a weapon. Not yet. The nature of twenty first century conflicts may change that. As we said, predicting the next war is risky business, though we will venture a few predictions based on current events:

- Twenty first century conflicts will likely continue to emphasize fewer combatant personnel with more relevant and refined skills delivering more discriminating firepower on more fronts—combined with more sustained local policing actions and more humanitarian relief missions.
- To accommodate challenges, logistics management necessarily shifts from a few centralized supply mechanisms to many small and redundant systems linked in a robust global network.
- Force projection and force protection, communication, transportation, and sustainment will demand leaner, faster, more agile methods of delivery and response to changing hot spots.

US and allied forces are already serving in far flung locations, and logistics philosophy has already started to shift from centralized warehousing to a shorter logistics tail, i.e., more widely dispersed sources and fewer levels of equipment repair and maintenance. Still, the idea of weaponizing logistics is a departure from the support role that logistics has traditionally played. The difference is between a passive "gear in the rear" mode of supply and an active combat role whose primary purpose is to suppress combat operations on the front line with a self-sustaining and indefinite armed presence.

That makes sense when it's hard to tell where the front line is, because the supply line has to move as fast, or nearly so, as warfighters do. "Rapid mobility" of supply essentials, nevertheless, is severely limited if mechanization is the only means of mobility, which in fact is a very old doctrine in fighting wars of counterinsurgency:

- Modern counterinsurgency includes examples of short-term self-sustaining units designed for rapid mobility, such as US Army cavalry units fighting frontier wars in the nineteenth century, and posted remotely.
- More recently, US Army Stryker Brigades function in much the same capacity as the cavalry.
- Total transformation of the US Army to Rapid Mobility has been a front burner topic since at least the early 1980s, though total realization of the goal is evolving.

Net-centric warfare, on the other hand, is a new concept, driven by sophisticated communications technology that facilitates network-linked resources—to identify remote targets, focus fire, and quickly shift response to new targets as battlefield conditions change.

Consider that net-centric logistics—i.e., a complex systems approach to logistics—operates on that same idea of communications technology to achieve fast response in delivering food and ordnance, equipment replacement parts, information … and to shift course as needed. It frees the "rapid mobility" concept from dependence on highly mechanized transport, replacing it with in-place resources that can be brought to bear locally, by coordination of global communication.

Because communications technology is the key, and applying Bar-Yam's model of multi-scale variety, we find:

- The main characteristic of a robust communications network is transparency among a critically sufficient number of members, with sufficient resources dedicated to a task.

That brings us to the weaponization of logistics, which like all good tactical weapons is adapted specifically to its target.

## Complex Systems Science Integrates Network with Need

A remarkable 2006 result of Dan Braha and Yaneer Bar-Yam [3] demonstrated that in a self-organized communication network, a continuously shifting hub of distributed activity causes the map to sometimes vary quickly and radically on local scales over short time intervals, even while the map itself shows little global change aggregated over long time intervals.

This abstract model would mirror complex military movements and communications, if we considered the map as a theater of operations the size of the globe. That is, each communicator in the network has at their workstation all the necessary resources to deliver a message and coordinate events, sometimes acting as the hub of activity, sometimes as the beneficiary of information, and sometimes as provider of information. Point is, the metastability of the system over time suggests that a continually shifting range of activity represented by changing hub configurations is self-limiting; as a result, the global domain is largely protected from the danger of positive feedback—i.e., a loss of system control and potential widespread self-reinforcing destruction. In the history of the world wars, one can identify such unchecked feedback of escalating hostilities. Even in the present world, one can make a good argument that the specter of damaging positive feedback, festering in individual areas of the world—among failed governments, local armed resistance to despotic regimes, piracy, organized crime and racketeering, human trafficking—is tinder for a future conflagration.

The efforts of stable nations to contain crime and hostilities are largely based on applying force in a hierarchical top–down manner. If we look at world events in the context of a self-organized system of interdependent subsystems, however, we see that such tactics are destined to be self-perpetuating; suppression of activity even if demanded to temporarily contain a situation, is anti-creative, and if Per Bak [4] is right, futile in the long run. Bak's self-organized critical, nonlinear model of change isn't just mathematics—applied to biology, it supports the Gould-Eldredge punctuated equilibrium model of evolution [5]. Most recently, mathematician Gregory Chaitin [6] has proposed "mathematical biology," further reinforcing the power to understand our physical world through abstract modeling. These ideas suggest the existence of natural ways—paths of least resistance—to aid the containment of destructive violence while promoting the creative growth of both societies and individuals.

What all these models require in common, and which extends into the evolution of all systems and subsystems whether physical or purely abstract—is an assured variety of inputs that sacrifices some efficiency in order to maximize creativity. Bar-Yam [7] recognized that cybernetics pioneer Ross Ashby's "law of requisite variety" [8] is a theorem in a world where excess resources and system redundancy are assets to the creative process. In other words, while it is counterintuitive to an individual to think of efficiency as a liability, the self-limiting process of a self-organized system can only be as efficient as the availability of resources in a particular hub-connected subsystem, to meet its needs.

So it was that Bar-Yam introduced multi-scale variety, the idea that independent subsystems allowed to organize around task coordination at different times on different scales makes the larger system effective. One can summarize: locally efficient use of resources assures global effectiveness in the creative growth of resource availability—with the caveat that local subsystems remain independent, because otherwise the drain on local resources will reduce subsystem effectiveness and cause an undesirable positive feedback loop by lack of sufficiently varied resources to sustain required tasks.

The problem of bounded rationality—the limitation of individual elements of a linked system, be they human, network node, or subsystem, to acquire sufficient information for central control decisions—had long been cast in systems thinking as a problem of hierarchy, vertical or nested. Bar-Yam's network, of multi-scale variety with distributed control, allows communications technology to be integrated laterally into the system to solve or mitigate the problem of bounded rationality.

In so doing, the system is open to self-correction; some problem might be time-limited in a particular part of the system, yet is not unresolvable in the whole, as the hub of activity shifts to a different subsystem, allowing the time-limited system the freedom to recover.

Duncan J. Watts and Steven Strogatz [9] of "Watts-Strogatz graph" fame showed a "small world effect" for fixed numbers of network nodes, laying groundwork for realistic models of self organization that allow spontaneous and orderly growth system wide.

We find:

- Complex networks mirror complex military movements & communications globally.
- "Law of Requisite Variety" demands laterally distributed decision-making power, vice centralized control, to coordinate resources by task requirements.
- Linear top–down authority is of limited effectiveness in dealing with nonlinear models of change ("avalanche model" of self-organized criticality).
- Lateral distribution of authority facilitates feedback among principals that shrinks time and space ("small world" effect).

## Enterprise Business Models Teach Effective Knowledge Management

The knowledge industry grew up in the turmoil of the information revolution. The history of the mid twentieth century saw an explosion of available information that transformed the way that commercial companies (particularly in the United States) do business.

New titles were added to boards of directors, including CIO—chief information officer—and whole corporate departments were dedicated to information technology management. Competitive pressures for better, faster, cheaper goods and services—and information—largely replaced the central production model with strategic marketing and distribution that places production closer to points of consumption. Branches of giant retail and wholesale enterprises are scattered to every corner of the US.

This transformation was largely facilitated by innovative information delivery systems including electronic data interchange (EDI) for immediate paperless transactions (purchases, billing, etc.) and transparent inventory—by which a customer can immediately view and access stocks from multiple locations, making timely decisions based on best cost, fastest availability, without holding expensive inventory on site. Just in time inventory (JIT)—made possible by innovations in information and distribution technology, give customer business strategies more flexibility and reduced inventory cost. Product delivery innovations include hub and spoke distribution—which blunt long-haul shipping costs by minimizing heavy truck and rail distance, using lighter means of transport at strategically placed local terminals.

Changes in business organization also brought about radical changes in leadership and management strategies. Quality improvement programs sprang up to emphasize the important role of individual innovation and contribution to the corporation. For the first time in corporate history, individuals—at least in more enlightened companies—were regarded as value-added assets rather than replaceable tools.

Today, knowledge management continues to come into its own:

- Timely and accurate knowledge empowers individuals and networked local organizations to make enterprising, cooperative decisions.
- Effective knowledge management leads to efficient problem solving—customer-focused, task oriented, and strategically flexible [10].
- Self-similar (scale invariant) levels of activity facilitate timely system feedback that multiplies the effectiveness of problem resolution.

In military jargon, a "force multiplier" is any weapon or circumstance that aids the effectiveness of a combat mission while reducing risk. A long logistics tail introduces considerable risk to troops at the consuming end of the chain, because any weakness in the largely linear operation—from supply shortages to ground intelligence breakdowns and delivery time lags—weakens the effectiveness of combatant commanders to respond and suppress enemy activity in the most timely manner.

It's also expensive to maintain an armed presence that depends on supplying troops from afar. The US military does have intermediate sources of supply and equipment replacement/repair based close to ground fighting forces; yet, the greater share of expense for these facilities and operations is still borne by US taxpayers. While the United States has been a historically willing supporter of foreign intervention in defense of its allies and American interests abroad, foreign bases more resemble gated outposts on a frontier that no longer exists.

What if such an outpost were a locally owned though dependable resource in the struggle to quell terrorist disruption of civilized commerce and communication? How does one ensure dependability without a fully vested interest? What keeps any organization from "going rogue" and withholding cooperation from the larger group?

There's a highly effective didactic tool by which many children learn to share equally: one is given the choice to either cut the cake, or to pick which piece to eat. We aren't children, though the principle is sound for all cooperative endeavors: equal vestment means compelling interdependence, not necessarily equal ownership.

## Logistics Transformation Echoes Industry Transformation

In peacetime, military logistics has historically mirrored commercial distribution. The state since WWII, however, has largely been that of sustained military readiness and conflict short of an all-out mobilization that would have forced modernization of buying and distribution methods.

To meet current and future requirements for armed conflict, the US military's supplier, the Defense Logistics Agency (DLA) has begun to modernize.

The DLA's transformation from a wholesale to retail philosophy is a cultural sea change; growing and maintaining a globally linked network of independently operating distributors requires a major investment in communications technology, employee training, and infrastructure.

- Challenges to implementation include:
  - Information security. Facilitating data transparency among network members (inventory, shipping, etc.) without compromising operations security and communications security.
  - Host country cooperation. Closer working ties between the U.S. State Department and the Department of Defense may be necessary to secure strategic distribution, on the same needs, and mutual benefit basis, as US military bases were negotiated in other eras.

If it seems all too idealistic to divest ownership of land and commodities and increase investment in the technology to deliver and manage services, one has to be aware that *exactly that is happening in the US industrial/corporate world right now.*

## National Security Has Become a Global Enterprise

The dedication of the United States to minimally managed, if not quite unfettered, capitalism has almost reached its logical conclusion:

- With its eroded (and still declining) industrial base and growing status as a net importer, the United States is positioned globally as a service provider economy.

This causes alarm for some, perhaps many. In the context of weaponizing logistics to combat global terrorism, however, the position is ideal. In our opinion, the greatest strength of the US can no longer be measured in stockpiles of weapons that hopefully won't be used, or in trying to restore a lost industrial base against the relentless campaign of US firms to export American jobs to cheaper labor markets. This trend is unlikely to stop, barring an unexpectedly radical change in American politics, until the world's labor market is equalized.

What the US has readily available now is an expertise in data and resource management, communications, and other high tech services that foreshadows an unrivaled leadership role in managing the world's resources, so long as the position of leader is shared with equal deference to the positions of the world's resource owners and managers. What this amounts to is a transformation of the idea of ownership from commodities to knowledge management. Potential wealth, any would-be terrorist knows, is useless without the means to trade it for real wealth. To draw again on our analogy, control of the knife is equal to owning the cake.

- The US is unlikely to be outmatched as a world leader in building and maintaining the communication and distribution systems that not only help ensure economic robustness for both US and host countries, but strengthen the military supply chain as well.

As with all military advantages, the window of opportunity to weaponize logistics is temporary and fleeting. Already, we see in certain US-owned service industries, the exporting of aspects of the knowledge industry—such as help call centers—to overseas locations. Without a continuing investment incentive to support and maintain US leadership in information and communication, the position will eventually erode and be lost; given the increased pace of technology, it could happen quickly.

Net-centric logistics has the ability to transform US foreign aid from a giveaway program to an investment program. Scandals ranging from piracy to corrupt political leadership and inept management of resources are common; we believe that reciprocal agreements that de-incentivize local political control in favor of local ownership and a management contract that includes security services, have a fair chance of working—now that widespread citizen revolts, primarily in the Mideast and Africa, have to a varying degree successfully forced broader sharing of political power.

## Endnote: Swords into Plowshares

A native of Zambia, Oxford and Harvard educated economist Dambisa Moyo quickly rose on the international stage by promoting a plan for countries of the African continent to break free of what she describes as the crippling effects of aid from the West, such as perpetual poverty and corruption [11]. While Dr. Moyo's thesis is far from universally accepted (indeed, one is challenged to find any economic philosophy that is), it has ignited a worldwide dialog over the value of investment versus aid. Whether public or private, however, we agree in principle that cyclical investment fuels sustained economic well-being while aid quickly burns out.

It has always been one of the tacit goals of US foreign policy in the latter twentieth century and into the twenty first century, to secure American interests by aiding potential adversaries before they become adversaries. Today, America's adversaries are scattered globally and linked both by a more or less effective communication network, and a common ideal—to preserve their national identity against what they perceive as a corrupt Western way of life. Referring back to the findings of Lim et al., this is the most volatile "… structure of boundaries between groups …" that is possible. That is, the boundary is of the *enemy's* choosing, enabling a war on their own terms, to inflict casualties that but for a word—"war"—would be wanton murder.

A preemptive Western strategy assumes that we, too, can choose our boundaries. A robust network of cooperative global resources surrounds potential terrorists with an irresistible incentive to abandon their plans to attack—it deprives them of a motive (which is and always was fueled by public opinion) while frustrating the means. Rational actors are expected to capitulate; what's left should be manageable by conventional crime-fighting methods.

- The benefits of an enhanced global supply network include, but are not limited to:
  - Sustained counterterrorism/counterinsurgency presence with rapid mobilization capability.
  - Natural disaster and armed conflict humanitarian relief on the spot.
  - Interlocking political interests with host country, to prevent and combat destabilizing terrorist and insurgent activity.
  - Interlocking commercial interests with host country, in using local labor and resources to supplement supply chain needs and deliver faster service to US and allied warfighters.

Whereas in the past, preemptive battle tactics meant destruction of the enemy's will to fight, we find today that violent suppression only makes stronger the roots of future conflict. When utter destruction is impossible, victory can be realized only by sustained cooperation.

**Acknowledgment** Key points were presented in outline form as a problem statement for the LOGTECH seminar program, sponsored by The Institute for Defense and Business at the Kenan-Flagler Business School, University of North Carolina, Chapel Hill, June 2009.

## References

1. Lim, M., et al. (2007). Global pattern formation and ethnic/cultural violence. *Science, 317*, 1540.
2. Bar-Yam, Y. (2004). Multiscale variety in complex systems. *Complexity, 9*(4), 37–45.
3. Braha, D., & Bar-Yam, Y. (2006). From centrality to temporary fame: Dynamic centrality in complex networks. *Complexity, 12*, 59–36.
4. Bak, P. (1996). *How nature works: The science of self-organized criticality*. New York: Copernicus.
5. Gould, S. J., & Eldredge, N. (1972). Punctuated equilibria: An alternative to phyletic gradualism. In T. J. M. Schopf (Ed.), *Models in paleobiology*. San Francisco: Freeman, Cooper and Co.
6. Chaitin, G. (2012). *Proving Darwin: Making biology mathematical*. New York: Pantheon.
7. Bar-Yam, Y. (2005). *Making things work: Solving complex problems in a complex world*. Cambridge: Knowledge Press.
8. Heylighen, F., & Joslyn, C. (1993). The law of requisite variety. In F. Heylighen, C. Joslyn, & V. Turchin (Eds.), *Principia cybernetica web*. Brussels: Principia Cybernetica. http://pespmc1.vub.ac.be/. Accessed June 2012.
9. Strogatz, S. (2003). *Sync: The emerging science of spontaneous order*. New York: Hyperion.
10. Wiig, K. (2003). *A knowledge model for situation handling*. White Paper from Knowledge Institute, Inc. http://www.krii.com. Accessed June 2012.
11. Moyo, D. (2011). *How the West was lost: Fifty years of economic folly–and the stark choices ahead*. New York: Farrar, Straus and Giroux.

# Part II
# Applications and Case Studies

# Chapter 7
# A Fractal Concept of War

**Maurice Passman**

## Introduction

Over the past 10 or so years, there has been a general recognition by the military Operational Research community of the applicability of fractal mathematics to the modelling of 'Information Age' warfare. 'Information Age' warfare, as defined by texts such as Moffat's Complexity Theory and Network Centric Warfare [1], is typified by dispersed, dynamically evolving scenarios that involve local force clustering and reclustering. This classification is contrasted with the force-on-force 'Industrial Age' warfare epitomised theoretically by Lanchester's equations. Work in the area of fractal mathematics, scaling, Self Organised Criticality and metamodels have all suggested that new, fundamental ideas are needed to understand the nature of conflict. In this paper we attempt to summarise how the Lanchester Laws were extended to include a limited fractal notion of combat (the Fractal Attrition Equation—FAE) but also to sketch a new line of development that has its foundation in the concepts first expressed by Moffat and Passman [2] in their original paper. This new work, described in more detail in a following paper, will introduce a physical theory that formulates local and non-local fractal interaction between the combat elements in terms of deterministic mechanics but also integrates randomness in the form of Boltzmann's principles of statistical mechanics. Central to this work is the notion of 'leverage' and the ever-changing dialectic between manoeuvre and attrition.

Prior to the fall of the Berlin Wall there was no real imperative, within the military Operational Research community, to consider anything except the Lanchester equations to form the mathematical basis of combat simulation or predictive models. It was only with the socio-military changes that occurred with the breakdown of the Soviet controlled Eastern Bloc, with the necessity of modelling operations other

M. Passman (✉)
Director Adaptive Risk Technology, London, UK
e-mail: adaptiverisktechnology@gmail.com

 
P.V. Fellman et al. (eds.), *Conflict and Complexity*, Understanding Complex Systems,
DOI 10.1007/978-1-4939-1705-1_7

than war, that the mood changed. Up until then, the attrition based Lanchester's equations were pretty much the sole theoretical choice of the modeller: the equations had a good history, models based on the equations could be validated, and their theoretical basis made sense in that they attempted to describe the relationship between quality and numbers of combatants.

It was at this point that Moffat and Passman had their entry. The idea that Statistical Thermodynamics could be used to examine agent-based combat models naturally led onto to investigating Percolation Theory, Critical Systems, Barenblatt's scaling and dimensional analysis methods, Renormalisation Theory and Per Bak's Self Organised Criticality (SOC). These eventually resulted in the derivation of the Fractal Attrition Equation. The main aim of this work was to develop mathematical metamodels of agent-based simulation models. Underlying this avenue of research was the realisation that other workers—outside of military Operational Research—were working on similar ideas, for example, Theodore Modis and his S-shaped growth curves that can be mathematically described by log–log/power law behaviour (Conquering Uncertainty 1998) and the understanding that the mathematics of conflict is essentially that of open, self-similar systems.

Self-similarity can be satisfied by power law scaling:

$$Q(r) = Br^b$$

Using the scaling relationship to evaluate $Q(r)$ and $Q(ar)$:

$$Q(r) = Br^b$$
$$Q(ar) = Ba^b r^b$$

If $k = a^b$ then $Q(ar) = kQ(r)$

Self-similarity can also be satisfied by more complex scaling to produce Modis-like S-curves:

$$Q(r) = Br^b f\left(\frac{\log r}{\log a}\right)$$

where $f(1+x) = f(x)$. Using the scaling relationship to evaluate $Q(a)$ and $Q(ar)$:

$$Q(ar) = Ba^b r^b f\left(\frac{\log ar}{\log a}\right) = Ba^b r^b f\left(\frac{\log a + \log r}{\log a}\right)$$
$$= Ba^b r^b f\left(1 + \frac{\log r}{\log a}\right) = Ba^b r^b f\left(\frac{\log r}{\log a}\right)$$

If $k = a^b$ then $Q(ar) = kQ(r)$.

One of the first steps in formulating a fractal metamodel approach and examining the mechanics of combat was expounded by Moffat and Passman. They created a metamodel that was used not only to describe the nature of conflict but also to encompass Lanchester concepts [3]. This methodology was based upon the evidence that the statistics conflict, e.g. in Turcotte and Roberts Fractals 6, 4, 1998 demonstrates power law behaviour.

The necessity at the time was to develop mathematical approaches to 'Information Age' conflict, i.e. low intensity conflicts, operations other than war etc. All this neatly dovetailed into Command and Control ideas and field applications of emergent phenomena that were being implemented on the ground, particularly by the US armed forces, i.e. Network Centric Warfare. The FAE can be seen as the culmination of this approach.

From one viewpoint, however, this line of research failed miserably in that it still did not adequately explain the dynamic interaction between manoeuvre and attrition warfare. Ultimately, the mathematics of the FAE describes (all be it cleverly) artificial agents moving, clustering, and reclustering, over a grid that is superimposed upon a flat, plain surface. Whilst there is some correspondence between fractal dimensions for some types of historically recorded conflict and that of cellular automata models, a coherent 'Fractal Theory of Conflict' this does not make.

On examination of Lanchester's original writings, one finds that the equations succinctly illustrate both the limitations and the advantages of his methods. For 'duels' and indirect fire Lanchester proposes a Linear Law by which you need twice as many or twice as good for a two to one chance of success. For situations where all units on one side can interact with all units on the other side, Lanchester has a 'Square Law' where you need twice as many (or four times as good) for a two to one chance of success. Lanchester's work related adequately to attritional warfare where large blocks of forces interacted (the 'Soviets are coming over the hill to the green field beyond' type scenarios).

The difficulty in finding any direct historical evidence to demonstrate where the equations consistently and uniformly describe the complicated process of combat has not seemed to harm their common usage nor did it stop workers post-Lanchester expanding the equations into areas where they were originally not envisioned; it was 'obvious' that the equations were founded upon 'common sense'. Historical analysis for the Iwo Jima campaign provided some agreement with the Lanchester formalism and although some additional evidence has been provided recently from the Ardennes campaign [4] and the battle of Kursk [5], the evidence for the rigorous comparison of Lanchester attrition type processes (i.e. examining time correlated combat strengths and casualties for both sides) remains inadequate.

Taylor [6] listed the most important limitations of Lanchester's original models:

1. Constant attrition rate coefficients
2. No force movement in space
3. Homogeneous forces
4. Battle termination not described
5. Deterministic as opposed to stochastic
6. Not verified against historical data
7. Cannot predict attrition rate coefficients

8. Tactical decision processes not modelled
9. Battlefield intelligence not considered
10. Command and control not considered

Whilst Johnson [7] has examined the history of these attempts and concluded that a Lanchester attrition type process between the square and linear laws is in operation, the key to what is actually going on lies back in Lanchester's original work. Lanchester examined the casualty evolution from the Battle of Trafalgar and viewed the battle as a series of concurrent and consecutive sub-battles separated by time and space. The correspondence between sea battles and the flat surface of a cellular automata model is good. Note also that with sea battles there are no turning effects; battle is purely attritional.

Simulations of combat at sea should therefore give a good correspondence to reality, moreover, using an FAE that has a force-on-force attrition expression incorporating battlefield clustering in time and space, but has mathematical roots in Lanchester, may give a better representation of reality, but again falls short in describing the full dynamic of land combat. That something fundamental is missing can be seen in the data used; it is curious that the historical evidence used in the attempt to support Lanchester type mechanisms seem to be of conflicts that are in terrain types that emphasise linear, attritional warfare: the Ardennes campaign where forested areas and bridges limited manoeuvre and Kursk where extensive defensive works again limited manoeuvre and created a linear, attritional battlefield. It is this issue we eventually specifically wish to remedy.

## The Manoeuvre-Attrition Dynamic

Our principal concern with attrition models of combat is that they have an incomplete view of what warfare is actually about and that the FAE is both incomplete in its description of the reality of combat but also does not encompass a mathematical formalism that is fully developed to describe the physics of combat. In a way, the attritional viewpoint is derived from the nature of the historical data that is used to validate the models built; attrition data drives modellers to create attrition models. Again, missing is the constant interaction between manoeuvre and attrition—the manoeuvre-attrition dynamic—and an understanding of this dynamic over the multitude of scales of combat. The attritional view is 'two dimensional' in the sense that there is no implication of *leverage* between opposing units in time and space. This leverage should also be contextualised by viewing the leverage generated by a turning force with the attrition effects of a fixing force. Since we are interested in the 'why' and not the 'what' of combat (i.e. the *Physics* of combat) we should use Physics terminology to briefly explain where we are. Simpkin explained this succinctly in his book Race to the Swift [8].

Momentum is defined as the product of velocity and mass. Momentum represents the resistance of a body to any change in velocity. We can view momentum as the intrinsic physical manoeuvre value of a combat force; as opposed to its counterpart, physical fighting power. Momentum can also be seen as the rate of change of leverage. The application of momentum to an opposing force as part of a flanking manoeuvre signifies the rate of change of leverage applied onto the opposing force. Greater leverage is applied by extending the level arm. Our contention therefore is that there is a continuing dialectic, within time and space, between an attrition, physical fighting power-based mechanism and that of a manoeuvre-based momentum-leverage mechanism. It is this dialectic that we wish to examine. In our next paper we consider the physical effects of this leverage and its integration into a fractal description of combat viewed from the perspective of a continuing interaction of attrition and manoeuvre warfare.

An example of the importance of this attrition-manoeuvre dynamic here can be appreciated by consideration of a historical example: the 1973 Yom Kippur ground war on the Sinai front. The single most striking feature of the 1973 Yom Kippur war was the contrasting dichotomy between the operating styles of the Egyptian and Israeli armies, i.e. of that between attrition and manoeuvre warfare respectively. Some background is required here. The equipment, organisation and function of the Israeli army reflected its strong emphasis in non-linear Deep Battle. This doctrine called for strong armoured forces to breach an opponent's line by attacking over a narrow front, moving and exploiting at high tempo at tactical, operational and strategic depths. This kind of battle was non-linear in the sense that the objective was not (directly) the destruction of enemy forces but rather that the penetrating armoured columns would disrupt communications, instil panic, destroy HQ, logistical and combat support units and make the status of enemy combat units untenable by bold manoeuvre or by turning actions.

The deep-thrusting armoured units, in this doctrine, would be operating in a hostile environment where both enemy and friendly dispositions were obscure and where headquarters would be unable to provide direct and detailed orders. The Israeli army therefore employed a very flexible command structure that relied on the intelligence and the bold initiative of its own commanders. The main reason for the Israeli army utilising this doctrine was very simple: the inability of Israel's relatively small population to absorb heavy losses. The doctrinal choice of Deep Battle places heavy responsibility on the officer core and ultimately trades off officer casualties for fewer rank-and-file losses, however, manpower quality within Israeli culture and society was never an issue. The main point we are making here is that the key element of this type of warfare *does not necessarily* rely only on the clustering or swarming behaviour of armoured forces (by clustering and swarming we mean the degradation of enemy forces by concentrating friendly forces around opposition units and then dispersing) but rather the physical and psychological effects of turning movements on enemy dispositions at all levels (tactical, operational and strategic). Note that clustering from this perspective is a function of attrition warfare, i.e. bringing together the greatest firepower most efficiently to kill the opposition. In practice, with manoeuvre warfare, the physical destruction by 'clustering' is usually undertaken by slower (i.e. infantry bound) follow-on forces. Take these elements together with other aspects of real life combat that are not usually taken into consideration in attritional combat modelling (such as overkill), then care should be taken when comparing cellular automata model results and the results of force exercise 'trials' (such as the Chinese Eye exercise) with historical analysis results.

The Egyptian army planned to win the'73 war by maximising its strengths and minimising its weaknesses. The Egyptian army lacked the tradition and command structure of independent leadership. Egypt's greater manpower numbers, particularly in infantry, were able to impose a linear form of warfare by creating an impenetrable anti-tank wall of fire. In this way, the Israeli side had to fight the kind of battle that the Egyptian army was best suited for yet found it very difficult to impose and practice the non-linear, manoeuvre warfare it was best suited for. The ground (in this case the Suez Canal) presented a barrier to large scale thrusts and the Israeli lack of infantry and artillery (so as to be able to make combined arms attacks) made it very difficult to rupture the Egyptian line. Historically, the rupture of this line was made easier by the Egyptian attacks of 14 October, however, eventually the Israeli army, by modifying doctrine at a tactical level, was able to create a bridgehead between the Third and Second Armies.

Viewed from this perspective, the Middle East conflicts post-1973 have all had very similar characteristics, with the Arab side attempting to impose, via ground and doctrine, an attritional, linear battle, and with Israel attempting to counter this by trying to impose manoeuvre/non-linear battle. Seen in this light, the 2006 Lebanon War with Hezbollah's use of bunkers, poor-going terrain, high concentrations of both infantry and anti-tank weaponry seems more like an extension of the '73 war rather than some newer, Fourth Generation level of warfare. Our task, therefore, is to represent the dynamic of attrition and manoeuvre warfare in the mathematical language of fractal mathematics. This is what we will begin to address in a following paper.

We therefore believe that the objectives for a new view of warfare modelling are three-fold:

1. To examine where Fractal Mathematics can give insight into the enormous volume of historical analysis.
2. The creation and development of mathematical expressions for the dynamic that exists in reality between attrition and manoeuvre warfare. Ultimately the goal here is to express a 'Unified Field Theory' for the Physics of combat integrating both attrition and manoeuvre warfare elements.
3. The design of simulation *tools* that can accurately model real combat whether armoured warfare on a major scale or low intensity conflicts and the effects of human elements such as morale, the will of the commander and cohesion.

## Lanchester Equations

Frederick Lanchester during the First World War proposed two systems of equations describing attritional warfare. These systems depended on whether fighting was 'collective' or not. For two opposing forces, Red and Blue of strengths $R$ and $B$ respectively, he proposed a 'Square Law' for modern collective combat where:

$$\frac{dR}{dt} = -k_B B(t), \quad R(0) = R_0$$
$$\frac{dB}{dt} = -k_R R(t), \quad B(0) = B_0$$

The equations were called a Square Law as they can be rearranged to provide an equation of state as:

$$\frac{\left(R_0^2 - R^2\right)}{\left(B_0^2 - B^2\right)} = \frac{k_B}{k_R}$$

Individual or 'ancient' combat was described by a Linear Law:

$$\frac{dR}{dt} = -k_{BR} B(t) R(t) \quad R(0) = R_0$$
$$\frac{dB}{dt} = -k_{RB} R(t) B(t) \quad B(0) = B_0$$

The equations were labelled a Linear Law as the equation of state, in comparison to the Square Law, had powers of unity:

$$\frac{\left(R_0 - R\right)}{\left(B_0 - B\right)} = \frac{k_{BR}}{k_{RB}}$$

The equations were later extended to include the heterogeneous strengths of combat units, that is, the strengths representing differing elements of each combatant (infantry, artillery etc.).

## Metamodels

Here we follow Barenblatt's formulism. Our general metamodel is of the form:

$$a = f\left(a_1, a_2, \ldots a_k, b_1, \ldots b_m\right)$$

If we change the gauge of each of our independent dimensional variables $a_1, a_2, \ldots a_k$ then their values will change accordingly. Changes are independent of each other. The variables $b_1, \ldots b_m$ have dependent dimensions so we can write their dimensions in terms of products of powers of the dimensions $a_1, a_2, \ldots a_k$, i.e.,

$$[b_j] = [a_1]^{p_j} \ldots [a_k]^{r_j}$$

where the square brackets denote 'dimension of'. It also follows that:

$$[a] = [a_1]^p \ldots [a_k]^r$$

Under a gauge transformation we have:

$$b_j' = A_1^{p_1} \ldots A_k^{r_1} b_j, \quad a' = A_1^p \ldots A_k^r a$$

The transformation group **G** generated by $\{A_1 \ldots A_k\}$ is a continuous gauge group. It therefore follows that:

$$\Pi = \frac{a}{a_1^p \ldots a_k^r}$$

$$\Pi_j = \frac{b_j}{a_1^{p_j} \ldots a_k^{r_j}}$$

are gauge invariant and are dimensionless. We can therefore write:

$$\Pi = F(a_1, a_2, \ldots, a_k, \Pi_1, \ldots \Pi_m)$$

where F represents some functional relationship. In other words:

$$\Pi = \frac{1}{a_1^p \ldots a_k^r} f(a_1, a_2, \ldots, a_k, \Pi_1 a_1^{p_1} \ldots a_k^{r_1}, \ldots \Pi_m a_1^{p_1} \ldots a_k^{r_m})$$

$\Pi$ is independent of the variables $a_1, a_2, \ldots, a_k$ therefore $\Pi$ is a function of only $\Pi_1 \cdots \Pi_m$. If we label this function $\mathbf{\Phi}$ then we can write:

$$a = a_1^p \ldots a_k^r \Phi\left(\frac{b_1}{a_1^{p_1} \ldots a_k^{r_1}}, \ldots \frac{b_m}{a_1^{p_m} \ldots a_k^{r_m}}\right)$$

This allows the consideration of three types of metamodel:

*Type 1* Where the characteristic function $\mathbf{\Phi}$ tends to a non-zero finite limit as $\Pi_j$ tends to zero or infinity and thus $f$ is a product of power monomials whose values can be determined by dimensional analysis.

*Type 2* One whose characteristic function Φ tends to power law asymptotics as $\Pi_j$ tends to zero or infinity.

$$f = \Pi_j^{\alpha} \ldots a_k^{r} f\left(\frac{\Pi_1}{\Pi_j^{\alpha_1}}, \ldots \frac{\Pi_m}{\Pi_j^{\alpha_m}}\right)$$

*Type 3* One where power law asymptotic behaviour is not observed and the characteristic function has no finite limit different from zero.

## Power Law 'Fractal' Lanchester Equations

One approach is to consider the Lanchester equations as metamodels resulting from Barenblatt's scaling equations and have been generalised to power law type equations with monomial exponents:

$$\frac{dR}{dt} = -k_B^C B(t)^D R(t)^E$$
$$\frac{dB}{dt} = -k_R^F R(t)^G R(t)^H$$

Extensive work by Hartley [9] in mapping conflict databases demonstrated that for the linear relationship for exponents $(D-G) = (H-E) = \alpha - 1$ and $(C-F) = \beta$ held with $\alpha = 1.35$:

$$Ln\left(\frac{R_0^2 - R(t)^2}{B_0^2 - B(t)^2}\right) = \alpha Ln\left(\frac{R_0}{B_0}\right) + \beta$$

$$Ln(Helmbold\ Ratio) = \alpha\ \ Ln(Force\ Ratio) + \beta$$

The derivation of this equation by Helmbold holds some noteworthy insights. Helmbold realised that obtaining suitable data was difficult and therefore resorted to an approach that only required knowledge of the duration of battle, initial and final force strengths. An empirical approach was taken as comparison with historical results for time correlated strengths and casualty data was very difficult to obtain. The line of best fit illustrates the relationship between initial and final strengths that would apply if Lanchester-like attrition processes were the sole factors producing these historical analysis results. The trend of this graph suggests that the 'common sense' attritional foundation of Lanchester's concepts is appropriate. Within a Lanchester perspective, this trend line demonstrates that different attrition coefficients cluster around a best fit relationship, however, this relationship does not explain the graph's volatility. Other processes, therefore, must be taken into

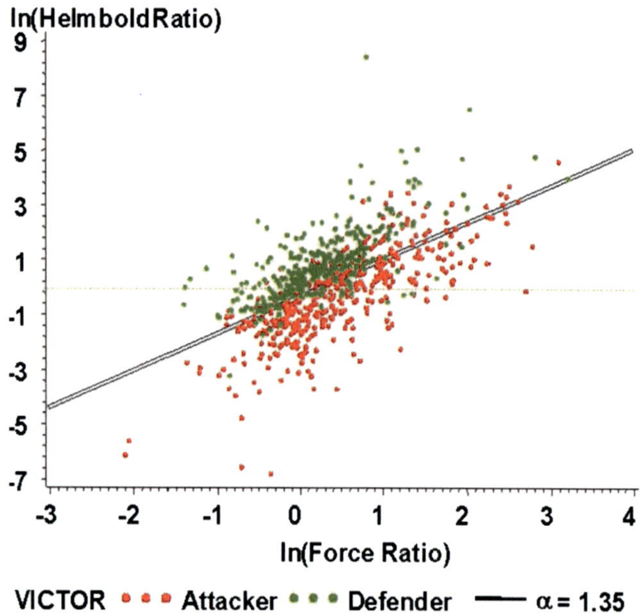


**Fig. 7.1** Helmbold relationship from Hartley [Hartley, D.S. (1991). Report K/DSRD-263/R1, Martin Marietta Centre for Modelling, Simulation & Gaming.]

consideration. To be clear: the trend demonstrates attrition but the volatility demonstrates manoeuvre (Fig. 7.1).

## Agent-Based Modelling and the Fractal Attrition Equation

Moffat and co-workers originally developed Lauren's fractal methodology to develop a fractal attrition equation for cellular automata models, such as MANA. For a time step of duration $\Delta t$ an agent travelling at velocity $v$ will travel a distance $d = v\Delta t$. The agent is then able to search a square area dimension $d$ for opposing agents during the time interval. If the opposition agents are treated as clusters and the clusters have the fractal distribution described by fractal dimension $D$ then the probability that at least one opposition agent will be encountered in that box is proportional to $d^{-D}$. If the average size of a cluster is $n_d$ then the average number of opposition agents encountered in the box is proportional to $d^{-D}n_d$. If $R(t)$ is the number of opposition agents and $\varphi(x,t)$ is the number of clusters then the expectation value for the number of agents killed by opposition agents during that time step is given by:

$$E\left(\frac{\Delta B}{\Delta t}\right) \propto k(\Delta t)^{2-D} R(t)\varphi(x,t)$$

Evoking Barenblatt's metamodelling approach we can obtain:

$$E\left(\frac{\Delta B}{\Delta t}\right)=f\left(k,(\Delta t)^{2-D},R(t)\varphi(x,t)\right)$$

And therefore

$$E_{0<t<T}\left(\frac{\Delta B}{\Delta t}\right)\propto k^{\alpha}\Delta t^{\alpha-1}R(t)$$

Where $\alpha = D/2$. When $\alpha = 1$ then it can be shown that this case corresponds to the assumption of Lanchester's Square Law. This model is generally written in the format below and labelled the Fractal Attrition Equation (FAE):

$$E_{0<t<T}\left(\frac{\Delta B}{\Delta t}\right)=E_{0<t<T}\left(-ck^{D/2}\Delta t^{-\left(1-\frac{D}{2}\right)}R(t)\right)$$

The above expression applies to all moments of a random variable if the system under consideration is assumed to be self-similar thus:

$$E_{0<t<T}\left(\left|\frac{\Delta B(t+\Delta t)-B(t)}{\Delta t}\right|^{p}\right)\propto \Delta t^{g(D,p)}$$

therefore the Fourier transform of the second moment of the random variable yields a power spectrum for the frequency distribution of a particular side's casualties:

$$\left|\Im(B)(f)^{2}\right|\propto |f|^{-(D+1)}$$

The measurement of fractal dimension sits well for a cellular automata simulation where box counting techniques may easily be used:

$$D=\lim_{d\to 0}\frac{LnN}{Ln(1/d)}$$

where $N$ is the number of boxes necessary to cover all automata.

Utilising the results from a cellular automata model such as MANA, the loss exchange ratio for both sides may be obtained:

$$\frac{B(t+\Delta t)-B(t)}{R(t+\Delta t)-R(t)}=\frac{(k_R\Delta t)^{D_R/2}\,R(t)}{(k_b\Delta t)^{D_b/2}\,B(t)}$$

For cell size $d$, assuming each cell has at most one entity and the entities have a fractal distribution, we can assume that

$$R \propto \frac{1}{d^D}$$

and that an equation of state can then be written as:

$$\frac{|\Delta B|}{|\Delta R|} \propto \left(\frac{B_0}{R_0}\right)^Q$$

The exponent formulation has been shown to be consistent with historical analysis; Thornton has shown that the number of casualties, C, has a dependence described by:

$$C \propto (\text{Number of attacking infantry/Number of defending infantry})^{0.685}$$

Note that this relationship states that attacker casualties increase as the number of attackers increases this is not the same - actually the opposite - as the Lanchester square law. The above formalism is consistent with the FAE as the clustering ability and the difficulty of having a larger force is taken into consideration via the fractal dimension.

The idea posited, in particular by Lauren [10] is that for some types of combat, the dynamics of interaction evolves towards an attractor state which is independent of initial conditions. His view is again based upon entity based automata paradigms where a straight line front between two tactical opponents buckles into a fractal shape, whose fractal dimension can be derived as a function of the force ratio of the forces involved. As a result, the combat front will buckle over time and in the limit will have fractal dimension $D = 1.685$. The value of 0.685 relates to open terrain, whereas in urban areas Moffat [11] reports that the value is closer to 0.5, i.e. a fractal dimension of 1.5 for the attractor state.

If the process is viewed as similar to a cellular automata model for inversion percolation then the fractal dimension for the interface boundary lies in the range 1.33–1.89 [12], which generally agrees with the historical fractal dimension of 1.5–1.685.

Moffat et al. in his 2005 paper also suggests that the FAE can be extended beyond the context of a metamodel for MANA simulations by considering the notion of the FAE being a two step process involving first, detection, and second combat. We therefore replace $E\left(\frac{\Delta B}{\Delta t}\right) = f\left(k, (\Delta t)^{2-D}, R(t)\varphi(x,t)\right)$ with:

$$E\left(\frac{\Delta B}{\Delta t}\right) = f\left(\left(k, d^{-D}, R(t)\varphi(x,t)\right)\right)$$

Applying Barenblatt's approach we obtain:

$$E\left(\frac{\Delta B}{\Delta t}\right) \propto k^{\alpha} d^{\beta} R(t)$$

If we wish to observe how the FAE compares with the Helmbold representation, then, if we take $d$ to be the length of the cell of the smallest area which at most contains a single Red entity and also assume that $k$ to be $d$ and $k$ are numerically identical for Blue and Red forces, then we obtain:

$$\frac{R(t)}{B(t)}(k\Delta t)^{\log_d(R_0/B_0)/2} = \frac{R(t)}{B(t)}\left(\frac{B_0}{R_0}\right)^Q$$

Therefore

$$B(t)\Delta B = \left(\frac{B_0}{R_0}\right)^Q R(t)\Delta R$$

If we divide the intervals between the ending and the beginning of an engagement into a series of time steps we can make a connection between the initial and final force strengths:

$$\sum_{\text{intervals}} B\Delta B = \left(\frac{B_0}{R_0}\right)^Q \sum_{\text{intervals}} R\Delta R$$

$$\sum_B B = \left(\frac{B_0}{R_0}\right)^Q \sum_R R$$

$$\ln\left(\frac{\left(B^2(t)+B(t)\right)-\left(B_0^2+B_0\right)}{\left(R^2(t)+R(t)\right)-\left(B_0^2+R_0\right)}\right) = Q\ln\left(\frac{B_0}{R_0}\right)$$

Relaxing the assumption that $k$ is the same for both Blue and Red forces and parametrising their values using:

$$k_R = rk_B$$

Then:

$$\ln\left(\frac{\left(B^2(t)+B(t)\right)-\left(B_0^2+B_0\right)}{\left(R^2(t)+R(t)\right)-\left(B_0^2+R_0\right)}\right) = Q\ln\left(\frac{B_0}{R_0}\right)+0.5D_R\ln r$$

The FAE therefore has a similar formalism to the Helmbold equation. Moffat et al argue that for the cases examined, the force strengths considered are sufficiently large enough that the difference between $B^2$ and $B(B+1)$ is immaterial. It is noted by the proponents of the FAE (notably Lauren and McIntosh [13]) that the fractal dimension can be considered as a representation of the degree of clustering of each force and therefore a surrogate for Command and Control, i.e., the ability of each side to manage and manoeuvre its combat elements into position and detect enemy elements. In other words, the force with the most favourable fractal dimension is likely to have the better Command and Control. Whist this may be true, note these authors are still using the syntax of attrition warfare and that fractal dimension still refers to attrition warfare combat multipliers—fractal dimension is a function of swarming/clustering ability in order to direct fire, not a way to describe the leverage ability of one unit on another.

## The Time-Series Approach

Whilst the FAE gives excellent insight into the fractal processes of cellular automata models it does not give direct tools in predicting combat casualty results from historical data. Moffat and Passman therefore proceeded in a slightly different way (see Moffat, Complexity Theory and Network Centric Warfare, Information Age Transformation Series, 2003). The ethos adopted was to build an initial practical, research instrument—a kind of developmental tool set—for predicting combat casualty rate patterns from real casualty data. Kuhn WWII data (Ground Force Casualty Patterns, Report FP703TR1, 1989) was used as being the most comprehensive and readily available. Success was to be determined not by exactly fitting (or over fitting) the real data but rather by observing if the predicted results were of the same form as the real data (i.e., gaining insight into the Physics of the process was the aim). The first part of the times series (up to day 38) was used to train a number of different time-series predictive methods. The 'fractal' predictive method used was based upon the assumption that circumstances remained sufficiently constant so that the power spectrum of the process was linear when plotted on a log–log scale. A commercially available package—the Chaos Data Analyser—produced by the American Institute of Physics was used and a number of predictive methods (included in the package) were also undertaken as a comparison. For the following Figs. 7.2, 7.3, 7.4, and 7.5 it should be noted that:

- The plots are casualties per 1,000 on the y axis and days on the $x$ axis.
- The first plot is the actual data
- The second plot is the maximum entropy prediction
- The third plot is the neural net prediction
- The final plot is the fractal prediction

What is noticeable in this preliminary experiment is that, for these data points at least, the casualty behaviour appears to be of the same form as for the real data, i.e., 'spiky'. Kuhn has suggested that the casualty rate data pattern consists of a number of 'hotspots', the number of these 'hotspots' not increasing with larger force concentrations, and second, combat is characterised by high rates of casualties lasting for short periods of time, interspersed with low casualty rates. Again, these results do not seem to suggest a modelling regime associated with attrition warfare.

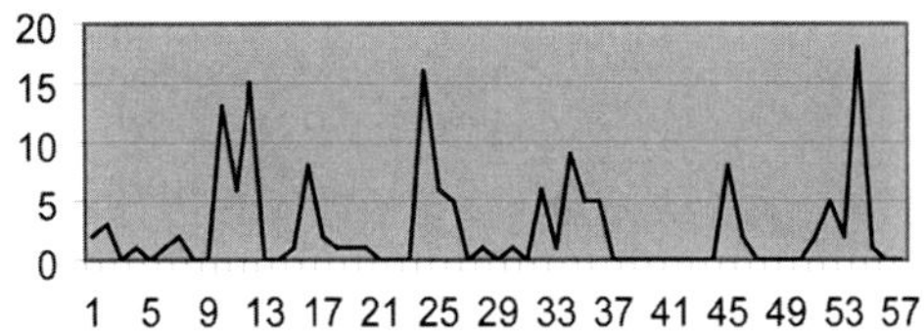


**Fig. 7.2** US second Armoured Division casualty time series for Normandy landings 1944 (From Moffat, J., Complexity Theory and Network Centric Warfare, 2003)

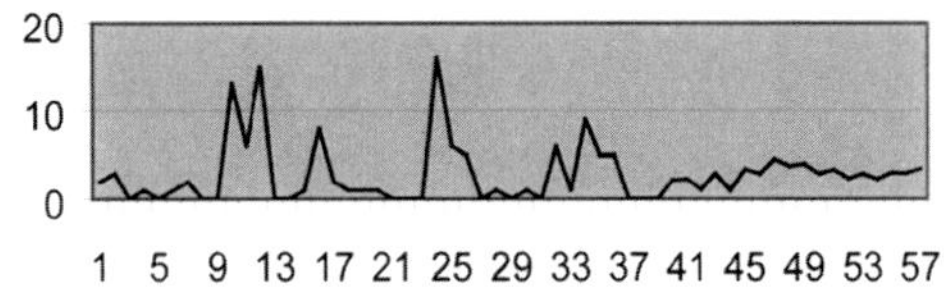


**Fig. 7.3** Maximum entropy prediction (From Moffat, J., Complexity Theory and Network Centric Warfare, 2003)

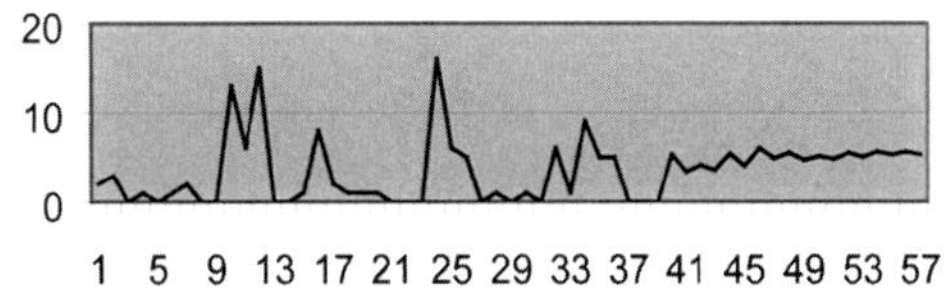


**Fig. 7.4** Neural net prediction (From Moffat, J., Complexity Theory and Network Centric Warfare, 2003)

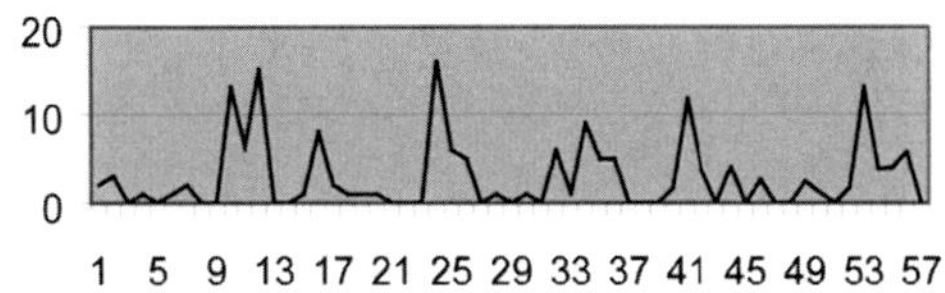


**Fig. 7.5** 'Fractal' prediction (From Moffat, J., Complexity Theory and Network Centric Warfare, 2003)

## Conclusions

In this work we have reviewed the construction of a metamodel type scheme of aggregated combat using the concepts of fractals, self-organisation, self-similarity, and scale free systems. We have presented evidence that suggest that experiments with cellular automata models such as MANA, together with mathematical techniques such as dimensional analysis, have linked historical data to fractal fingerprint behaviour. What still seems a long way off is the development of tools to examine, model and predict the wider range of combat and associated processes beyond attrition. This we hope to remedy in a future paper.

## References

1. Moffat, J. (2003). *Complexity theory and network centric warfare*. Washington: Department of Defense Command and Control Research Program.
2. Moffat, J., & Passman, M. (2002). Internal report, defence evaluation and research agency, centre for defence, analysis metamodels and emergent behaviour in models of conflict. Submitted in a cleared format to simulation in operational research in 2002, externally published on web in 2003 in hard copy form in 2004 – see reference 3.

3. Moffat, J., & Passman, M. (2004). Metamodels and emergent behaviour in models of conflict. Simulation in operational research. *Simulation Modelling Practice and Theory, 12*(7–8), EX1–EX2, 475–616.
4. Bracken, J. (1995). Lanchester models of the Ardennes campaign. In *Warfare modelling* (pp. 417–435). Alexandria: MORS.
5. Bauman, W. J. (1999). *Quantification of the battle of Kursk*. Newport News: USCAA.
6. Taylor, J. G. (1983). *Lanchester models of warfare* (Vols. 1 and 2). Washington: Operations Research Society of America.
7. Johnson, W. (1990). *Lanchester's square law in theory and practice* (pp. 484–485). Fort Leavenworth: US CGSC, AS A225.
8. Simpkin, R. E. (1985). *Race to the swift*. London: Brassey's.
9. Hartley, D. S. (1991). Report K/DSRD-263/R1. Martin Marietta Centre for Modelling, Simulation & Gaming.
10. Lauren, M. K. (2000). Modelling combat using fractals and the statistics of scaling systems. *Military Operations Research, 5*(3), 47–58.
11. Moffat, J., Smith, J., Witty, S. (2006). Emergent behaviour: theory and experimentation using the MANA model. In: Proceedings of JAMDS.
12. Hoshen, J., & Kopelman, R. (1976). Percolation and cluster distribution. I. Cluster multiple labelling technique and critical concentration algorithm. *Physics Review B, 1*(14), 3438–3445.
13. Lauren, M. K., & McIntosh, G. C. (2004). Art of war hidden in Kolmogorovs equations. *Chaos, 17*, 013121.

# Chapter 8
# Disrupting Terrorist Networks: A Dynamic Fitness Landscape Approach

**Philip Vos Fellman, Jonathan P. Clemens, Roxana Wright, Jonathan Vos Post, and Matthew Dadmun**

Over a period of approximately 5 years, Pankaj Ghemawat of Harvard Business School and Daniel Levinthal of the Wharton School have been working on a detailed simulation (producing approximately a million fitness landscape graphs) in order to determine optimal patterns of decision-making for corporations. In 2006, we adapted this study, combining it with our own work on terrorism to examine what would happen if we inverted Ghemawat and Levinthal's findings and sought to provide disinformation or otherwise interfere with the communications and decision processes of terrorist organizations in order to optimize poor decision-making and inefficiencies in organizational coordination, command, and control.

The bulk of this study was then presented at the annual meeting of the North American Association for Computation in the Social and Organizational Sciences. We present here an updated version of that study, emphasizing the rather counter-intuitive finding that "soft" targets have almost no value and that unless one can influence key factors, an effort directed at the easy to reach elements of terrorist organizations may actually be worse than mounting no effort at all. We presented a somewhat more detailed version of this research at the sixth International Conference on Complex Systems, held in Quincy, Massachusetts, also in 2006.

P.V. Fellman (✉) • J.P. Clemens • R. Wright • J.V. Post • M. Dadmun
American Military University and InterPort Police, Charles Town, WV, USA
e-mail: shirogitsune99@yahoo.com

J.P. Clemens
Intel Corporation, DuPont, Washington, USA

R. Wright
Plymouth State University, Plymouth, New Hampshire, USA

J.V. Post
Computer Futures, Altadena, California, USA

M. Dadmun
Southern New Hampshire University, Manchester, New Hampshire, USA

 
P.V. Fellman et al. (eds.), *Conflict and Complexity*, Understanding Complex Systems,
DOI 10.1007/978-1-4939-1705-1_8

This research was subsequently validated by U.S. Special Forces Major Derek Jones, in his thesis for the School of Advanced Military Studies at the United States Army Command and General Staff College at Fort Leavenworth, Kansas, "Understanding the Form, Function, and Logic of Clandestine Cellular Networks: The First Step in Effective Counternetwork Operations."[1] Major Jones argued that a major flaw in most current theoretical models of terrorism is that they only lead to the destruction of the revealed nodes, often the "disposable" or "amateur" portion of the network, while the dedicated professional cells remain unharmed [1]. In this regard, we conclude with the recommendation that some fundamental rethinking may be required if the United States is to effectively defend itself from future terrorist attacks.

## Introduction

The study of terrorist networks as well as the study of how to impede their successful functioning has been the topic of considerable attention since the odious event of the 2001 World Trade Center disaster. While serious students of terrorism were indeed engaged in the subject prior to this time, a far more general concern has arisen subsequently. Nonetheless, much of the subject remains shrouded in obscurity, not the least because of difficulties with language and the representation or translation of names, and the inherent complexity and ambiguity of the subject matter.

One of the most fruitful scientific approaches to the study of terrorism has been network analysis [2–6]. As has been argued elsewhere, this approach may be particularly useful, when properly applied, for disrupting the flow of communications ($C^4I$) between levels of terrorist organizations [4, 7–11]. In the present paper we examine a recent paper by Ghemawat and Levinthal [12] applying Stuart Kauffman's NK-Boolean fitness landscape approach to the formal mechanics of decision theory. Using their generalized NK-simulation approach, we suggest some ways in which optimal decision-making for terrorist networks might be constrained and following our earlier analysis, suggest ways in which the forced compartmentation of terrorist organizations by counter-terrorist security organizations might be more likely to impact the quality of terrorist organizations' decision-making and command execution.

## General Properties of Terrorist Networks

Without attempting to be either exhaustive or exhausting, recent research on terrorism has revealed several relevant characteristics of terrorist organizations which a prudent modeler ought to keep in mind. These networks are first and foremost, *covert,* which means that they have hidden properties, and our information about them is necessarily incomplete, hence demanding complex methodological tools for

---

[1] http://www.dtic.mil/cgi-bin/GetTRDoc?AD=ADA505161.

determining the properties of the network structure [2–5, 13, 14]. While we are primarily concerned in the present paper with formal properties of terrorist networks, it does bear keeping in mind that at the operational level they are *purposive*, which lends them not only formal characteristics, but depending upon the organization in question, a considerable ideological history [15, 16], and in some cases, rather serious (path-dependent) constraints on recruiting [11, 17] targets and methods [6]. Some other, rather interesting properties of terrorist networks include the fact that they are often separated by larger than normal degrees of distance between their participants, a condition arising from their covert nature [4, 5, 10]. Curiously, this kind of structure appears to have an emergent shape, which can be mapped as a distributed network [5, 11, 14], commonly illustrated by a social network diagram of the 9–11 Hijackers and informally referred to as "the dragon."

Carley et al. [7] have developed useful models for distinguishing cohesive vs. adhesive organizations as well as defining probably outcomes for the removal of higher visibility nodes. Formal models of network analysis can also suggest where removal of key nodes or vertices can disrupt the organization's ability to transmit commands across hierarchical levels of the organization, thus leading to command degeneration [2, 9]. The difficulty with this approach is that an important aspect of the dynamics of terrorist networks is that they are learning organizations [15, 18].

If one bears all of these features in mind, some of the complexities of dealing with terrorist organizations become immediately apparent. Terrorists are slippery foes, they are hidden, they have redundant command structures, they change their membership (not all of which changes are visible), and they learn from their mistakes. Nobody who has to deal with terrorist threats wants to see those threats, and the organizations that make them, evolve. The obviousness of this proposition is evidenced by the US reaction to 9–11. What then, are the possible approaches?

Complexity science has afforded a number of approaches to evolution in general [19–21] as well as to the evolution of organizations and the ways in which complexity science may be applied to problems of organizational behavior. In particular, Kauffman's NK-Boolean fitness landscape model appears to offer a number of fruitful heuristics [11, 22–24]. In 1999, seeking to define the formal properties of an optimal business organization decision-making process, Pankaj Ghemawat of Harvard Business School and Daniel Levinthal of the Wharton School ran an agent based simulation of decision-making in order to define the ways in which decisional interdependence and the interdependence of business units affect overall performance (fitness). In the section which follows, we will explore a number of their findings and suggest how they might be applied to inhibiting the fitness of terrorist organizations.

## The Structure of the Ghemawat–Levinthal NK Simulation

A primary goal of the simulation was to model interdependent choices. Levinthal and Ghemawat focus on this aspect of decision-making because they are attempting to understand the formal structure of decision-making in organizations with interdependent parts. While they rapidly come to focus on the same measures that we have seen used to characterize terrorist networks, hierarchy and centrality [2–4, 7–9, 13, 25, 33], plus an additional factor of randomness (which most of us are wont to deny) they come at these factors from a slightly different approach than what one might anticipate. N and K are chosen simply as (a) the number of total decisions modeled and (b) the number of decisions which depend upon other decisions. As they explain (p. 16):

> The model has two basic parameters, N, the total number of policy choices and K (<N), the number of policy choices that each choice depends upon. More specifically, each of the choices is assumed to be binary, and choice-by-choice contributions to fitness levels are drawn randomly from a uniform distribution over [0,1] for each of the 2K+1 distinct payoff-relevant combinations a choice can be part of. Total fitness is just the average of these N choice-by-choice fitness levels. Note that with K equal to its minimum value of 0, the fitness landscape is smooth and single-peaked: changes in the setting of one choice variable do not affect the fitness contributions of the remaining N-1 choice variables. At the other extreme, with K equal to N-1, a change in a single attribute of the organism or organization changes the fitness contribution of all its attributes, resulting in many local peaks rather than just one, with each peak associated with a set of policy choices that have some internal consistency. No local peak can be improved on by perturbing a single policy choice, but local peaks may vary considerably in their fitness levels.

However, a pure NK approach suffers from the disadvantage that all choices are assumed to be equal. To avoid this problem and to model a richer decisional landscape they employ an adjacency matrix, moving us into the familiar Carter Butts [2, 13, 25–27] territory of connected graphs [26, 27], and formal, axiomatically determined complex systems.[2]

[2]Replacement of the interactivity parameter, K, with an adjacency matrix is meant to let us generalize the NK approach in the directions presently of interest. A few general observations can be made about special types of and the fitness landscapes that they induce over the choices and linkages they embody. Thus, given disconnected graphs, fitness landscapes are smooth as the choices corresponding to disconnected vertices are varied—irrespective of the values of other variables. Such vertices therefore lend themselves to the notion of universal (and uncontingent) best practices. And for star graphs, in which one central choice influences the payoffs from each of $N-1$ peripheral choices but other linkages among choices are absent (corresponding to an adjacency matrix with 1's in the first column and along the principal diagonal and 0 s everywhere else), getting the first choice right is sufficient, in conjunction with a standard process of local search in an invariant environment, to lead the organization to the global optimum. But what about graphs more generally? Exhaustive enumeration of all the graphs with $N$ vertices and analysis of their fitness landscapes is unlikely to prove productive for even moderately large $N$: the number of 6-vertice graphs is 157, 7-vertice graphs 1,044, 8-vertice graphs 12,346, and so on. Restricting attention to connected graphs doesn't help much with the numbers problem since the number of connected graphs grows much more quickly than the number of disconnected graphs: with $N$ equal to 5, disconnected graphs account for about 38 % of the total, but with $N$ equal to 8, that figure is down to less than 10 %. We therefore pass up the opportunity to engage in exhaustive (and exhausting) enumeration. We begin, instead, by considering two classes of adjacency matrices that highlight two fundamental sources of asymmetry among choices, in terms of hierarchy and centrality, and comparing them with the canonical NK structure on which previous work has focused (pp. 17–18).

## Adjacency Matrices

With respect to this process, in Kauffman's original NK-Boolean dynamic fitness landscape model all of the potential choices that a firm could make were considered to be equally important and the search for higher levels of fitness was carried out through a random walk across the fitness landscape [24]. However, under these conditions the NK model could not account for the asymmetric relationship between strategic choices that decision-makers faced. In order to better represent the asymmetric nature of the choices facing a corporation (or, as in our case, a terrorist organization), Ghemawat and Levinthal replace the interactive parameter $K$, as described above, with an adjacency matrix:

> How different choices (the vertices in the graph) are linked (the lines in the graph). In such a matrix, choice variable j's effect on other variables is represented by the patters of 0 s and 1 s in column j, with a value of 1 indicating that the payoff to the variable in the row being considered is contingent on variable j, and a value of 0 denoting independence. Similarly, reading across row i in such a matrix indicates the variables the payoff of choice variable i is itself contingent upon. The principle diagonal of an adjacency matrix always consists of 1's, but the matrix itself need not be symmetric around that diagonal.

In order to simplify their examination of the relationship between asymmetric choices Ghemawat and Levinthal elected to look at two adjacency matrices that highlight the classical types of choice asymmetry: hierarchy and centrality. In the hierarchical matrix choice 1 is the most important influencing all other choices below it, choice two is the second most important, and so on to the final choice (in this case choice 10 were $N=10$) which is influences by all proceeding choices but influences only itself. For the centrality matrix choice 1 is the most central both influencing and also being influenced by all other possible choices, choice two is the second most central being influenced by all other choices and influencing all choices with exception 10 and so on. These two matrices are benchmarked against a traditional NK structure. The matrix in this case is structured such that there will be $K$ 1's in each row and column but they will be randomly distributed across the matrix.[3] In their simulation, Ghemawat and Levinthal put $K=6$ which proved the same number of peaks at the other two matrices [28].

## Hierarchy

Ghemawat and Levinthal treat hierarchical decisions as directed trees where the 1 appears to the left of the principal diagonal. In this regard as we have explained above, choice 1 is the most hierarchically important, choice 2 the second most important, etc. [12].

[3] In most cases Ghemawat and Levinthal use a Poisson distribution or another, uniform distribution, noting, in any case that the probability distribution is not likely to be a mathematically relevant factor in the overall distribution of decision outcomes (i.e., the probability distribution is not the determinative property).

## Centrality

In contrast, their treatment of centrality involves interconnected decisions and, hence produces an almost perfect 90° rotated distribution [12][4]:

## Modeling Policy Choices

Levinthal and Ghemawat then benchmark what happens in these two types of structures against the random (but symmetric) activity which is built into the canonical NK structure. As they explain:

> For all three structures, an organization's policy choices are represented by a vector of length $N$ where each element of the vector can take on a value of 0 or 1 (not to be confused with the 0 s and 1 s assigned, respectively, to denoting the absence or presence of linkages between every pair of policy elements). The overall fitness landscape will then consist of $2^N$ possible policy choices, with the overall behavior of the organization characterized by a vector $\{x_1, x_2, \ldots, x_N\}$, where each xi takes on the value of 0 or 1. If the contribution of a given element, *xi*, of the policy vector to the overall payoff is influenced by *Ki* other elements—in ways that vary across the three structures we will analyze—then it can be represented as *f*(xi, xi1, xi2, …, xiKi). Therefore, each element's payoff contribution can take on *2Ki*+1 different values, depending on the value of the attribute itself (either 0 or 1) the value of the *Ki* other elements by which it is influenced (each of these *Ki* values also taking on a value of 0 or 1) and—less commonly highlighted—the luck of the draw. Specifically, it is common to assign a random number drawn from the uniform distribution from zero to one to each possible f(xi, xi1, xi2, …, xKi) combination with the overall fitness value then being defined as
> Σi=1 to N f(xi, xi1, xi2, …, xiKi)/N. (pp. 19–20)

Their simulation structure assumes for the random benchmark that $K=6$, primarily because this value generates roughly the same number of local peaks as the hierarchical and central distributions.[5] Similarly they set $N=10$, which is sufficient

[4] (p. 19) "The particular form of hierarchy we explore in this paper has 1s as all the entries to the left of the principal diagonal (see Fig. 8.1a). Choice 1 is hierarchically the most important, choice 2 the second most important, and so on. In contrast, in a set of interaction patterns ordered by a centrality measure, policies vary in terms of their interdependence with other policy choices and this interdependence is taken to be symmetric (to distinguish it as sharply as possible from the one-way influences of hierarchy). As a result, the 1s to the left of the principal diagonal are mirrored by 1s to its right. Whether the 1s cluster centrally in the adjacency matrix, however, depends on the order in which choice variables are labeled. The particular form of centrality we explore in this paper embodies a structure and a labeling scheme that has 1s as all the entries to the left of the inferior diagonal (but distributed symmetrically to the left and the right of the principal diagonal)—see Fig. 8.1b. Thus, choice 1 is most central, choice 2 s is most central, and so on.

[5] Ghemawat and Levinthal provide five additional caveats, of which the three important for our purposes are: "A number of important assumptions, based on prior applications, are built into this specification. First of all, there is the emphasis on choice under uncertainty. In addition to its arguable descriptive realism, initial uncertainty helps explain why an organization launched over a fitness landscape may not instantly alight on the globally optimal policy vector. Second, there is the assumption that randomness takes the form of a uniform distribution. While some might argue that this distribution is too diffuse, we retain this assumption to provide at least some basis for numerical comparability with prior work, which suggests, among other things, that the structure of the fitness landscape is not sensitive to the particular probability distribution employed" [29]. Third, there is the

to generate more than a million distinct graphs, which allows them to report results averaged over a thousand independent landscapes which share the same structure. These landscapes will be either hierarchical, central or random, characterized by the particular adjacency matrix structure for each type, but with a distinct seeding (0, 1) from a uniform random distribution for the fitness of the policy variables.

## Results of the Ghemawat–Levinthal Simulation

The first question which the authors ask "what are the effects of presetting a certain number of policy choices equal to their values at the global optimum with the remaining choices determined by a process of local search?" is interesting from a complexity science point of view, but not immediately obvious in its application to terrorist organizations. The reason for this is that while answering this question allows Ghemawat and Levinthal to address issues of strategic planning and "grand strategy" in business organizations, it doesn't really provide a reliable guide for the $C^4I$ functioning of terrorist organizations. If this were all that their simulation achieved, it would have rather limited interest for us. However their second question "what happens when one of the *N* values of the policy variable is preset to a value inconsistent with the value of that variable for the global peak?" is of very substantial interest as it speaks to exactly the kind of distortions which we would wish to induce in the terrorist decision-making chain.

The first interesting result of the simulation is the difference between an optimal preset of policy configurations, the hierarchical, the central, and the random simulation arrays[6]:

Ghemawat and Levinthal describe this as:

> With a degree of match of 1, only the first, most strategic, variable is set equal to the global optimum. As more variables are matched with their settings at the global optimum, fitness rises steadily according to Fig. 8.2. However, the global optimum is not approached until nearly all policy variables are specified to equal their settings at the global optimum. Similarly, hamming distances tend to be quite large (see Fig. 8.3).
>
> The gap between the curve depicting performance under the random network structure and the other two curves indicates the power of presetting more strategic variables to their values at the global optimum. In contrast, the gap between the realized fitness level and the value of 1, indicates the loss from the not fully articulating the optimal policy array.

Admittedly, while this is an interesting result, so far it does not tell us much more than the intuitively plausible idea that an inability or failure to articulate global optima means that you probably never get there. There is, however, a slightly more interesting subtext here with respect to the complexity of rugged fitness landscapes, something originally articulated by Farmer, Packard and Kauffman ("The Structure of Rugged Fitness Landscapes" in [19]):

---

equal weighting of different choices in terms of their direct contribution (potential) to overall fitness. Again, we retain this prior assumption even though we intend to focus on asymmetries among choices. Putting different weights on the direct contributions of choice elements does not seem to us to be the best way of gaining insight into the indirect contributions that choice elements can make to overall performance by virtue of the linkages among them. (p. 21)

[6] Ibid.

**a**

| | 1 | 2 | 3 | 4 | 5 | 6 | 7 | 8 | 9 | 10 |
|---|---|---|---|---|---|---|---|---|---|---|
| 1 | 1 | 0 | 0 | 0 | 0 | 0 | 0 | 0 | 0 | 0 |
| 2 | 1 | 1 | 0 | 0 | 0 | 0 | 0 | 0 | 0 | 0 |
| 3 | 1 | 1 | 1 | 0 | 0 | 0 | 0 | 0 | 0 | 0 |
| 4 | 1 | 1 | 1 | 1 | 0 | 0 | 0 | 0 | 0 | 0 |
| 5 | 1 | 1 | 1 | 1 | 1 | 0 | 0 | 0 | 0 | 0 |
| 6 | 1 | 1 | 1 | 1 | 1 | 1 | 0 | 0 | 0 | 0 |
| 7 | 1 | 1 | 1 | 1 | 1 | 1 | 1 | 0 | 0 | 0 |
| 8 | 1 | 1 | 1 | 1 | 1 | 1 | 1 | 1 | 0 | 0 |
| 9 | 1 | 1 | 1 | 1 | 1 | 1 | 1 | 1 | 1 | 0 |
| 10 | 1 | 1 | 1 | 1 | 1 | 1 | 1 | 1 | 1 | 1 |

**b**

| | 1 | 2 | 3 | 4 | 5 | 6 | 7 | 8 | 9 | 10 |
|---|---|---|---|---|---|---|---|---|---|---|
| 1 | 1 | 1 | 1 | 1 | 1 | 1 | 1 | 1 | 1 | 1 |
| 2 | 1 | 1 | 1 | 1 | 1 | 1 | 1 | 1 | 1 | 0 |
| 3 | 1 | 1 | 1 | 1 | 1 | 1 | 1 | 1 | 0 | 0 |
| 4 | 1 | 1 | 1 | 1 | 1 | 1 | 1 | 0 | 0 | 0 |
| 5 | 1 | 1 | 1 | 1 | 1 | 1 | 0 | 0 | 0 | 0 |
| 6 | 1 | 1 | 1 | 1 | 1 | 1 | 0 | 0 | 0 | 0 |
| 7 | 1 | 1 | 1 | 1 | 0 | 0 | 1 | 0 | 0 | 0 |
| 8 | 1 | 1 | 1 | 0 | 0 | 0 | 0 | 1 | 0 | 0 |
| 9 | 1 | 1 | 0 | 0 | 0 | 0 | 0 | 0 | 1 | 0 |
| 10 | 1 | 0 | 0 | 0 | 0 | 0 | 0 | 0 | 0 | 1 |

**Fig. 8.1** (**a**) Hierarchy. (**b**) Centrality

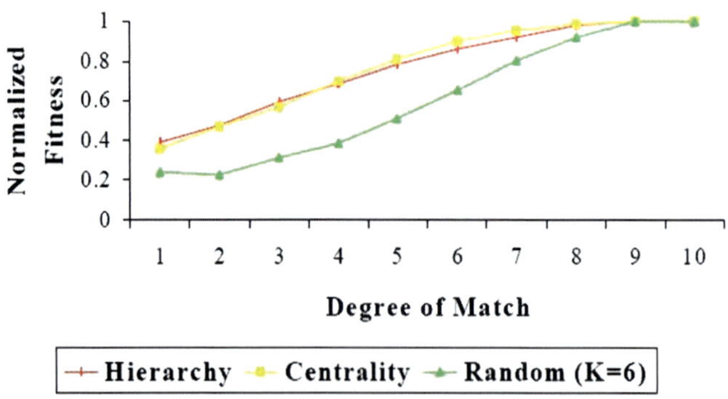


**Fig. 8.2** Value of partially articulated activity maps

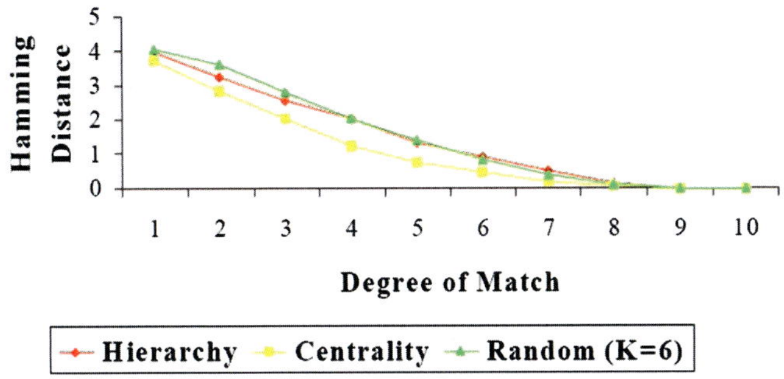


**Fig. 8.3** Partially specified activity maps and proximity to global optimum

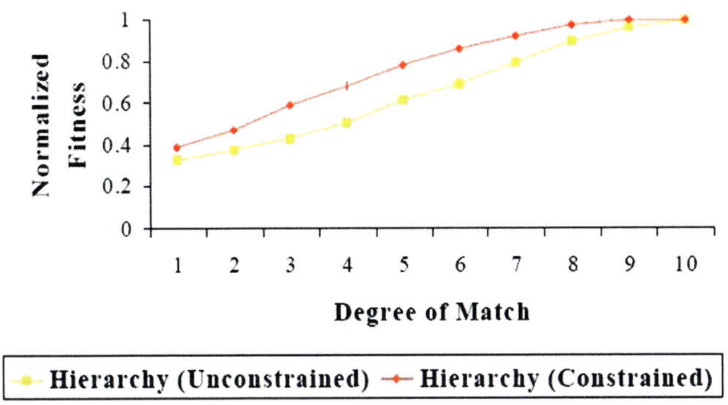


**Fig. 8.4** Value of constrained local search

> To make more sense of these patterns, it is useful to note that the fitness landscapes we are analyzing are quite complex, typically comprising over 40 local peaks. In such worlds, the powers of local search are relatively limited. Local search rapidly leads to the identification of a local peak but conveys no assurance about the local peak's global properties (i.e., its fitness value relative to the global optimum). Presetting the most strategic variables to their values at the global optimum does lead to the identification of a better-than-average local peak (recall that the normalized fitness value would have a value of zero if the average realized fitness level equaled the average value of local peaks in the fitness landscape). However, a high level of specificity is necessary to obtain the highest possible fitness levels or configurations close to the global optimum: in rugged landscapes, there are just too many positive-gradient paths that lead to local peaks other than the global one. (p. 26)

What is perhaps most interesting here are the suggestions that (1) the events of 9–11 are likely to prove the exception, rather than the rule (also giving some optimism to the long-run possibility of ultimately negating terrorist threats, something as unthinkable today as the end of the Cold War was 40 or 50 years ago) and (2) the indication that even very hierarchical terrorist organizations are unlikely to reach an optimal policy set through local search (and, following Kauffman, there simply are no other search mechanics available). Figure 8.4 provides what may be a slightly

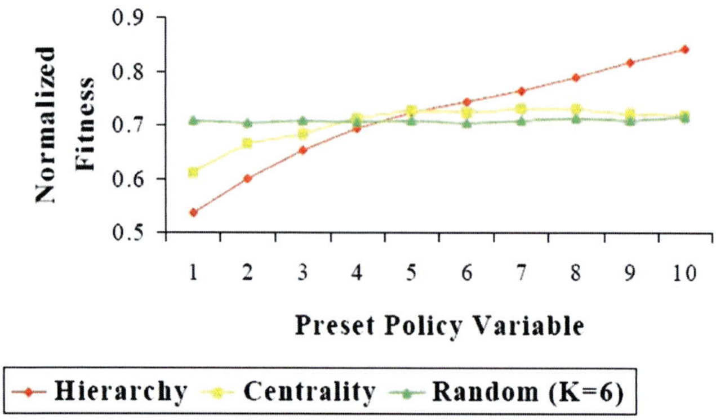


**Fig. 8.5** Constraints of history

more interesting insight. Much in the fashion of Lissack [22], Ghemawat and Levinthal use an approach similar to "patching" in order to simulate a feedback situation where initial errors in policy choice can be corrected and the corrections incorporated into subsequent searches. They note that (p. 26):

> ...while the articulation of and insistence on adherence to a single (or low-dimensional) strategic choice may not be sufficient to lead to the identification of a high-performing set of choices, a lack of such strategic discipline is likely to lead to even less attractive results. Compare the top line in Fig. 8.4, tracing the value of partially articulated activity maps in a hierarchical context in which preset choices cannot be varied (á la Fig. 8.2) with the bottom line, which looks at a hierarchical context in which the preset policy choices *can* be revised in the process of local search. It turns out that with the degree of match of 1, the latter, "unconstrained" approach underperforms the "constrained" approach, and the gap between the two widens for intermediate degrees of match prior to convergence as the degree of match hits 10. Similarly, the unconstrained approach fails to generate smaller hamming distances than the constrained approach. In that sense, strategic discipline *is* useful.

In this regard, as we have previously suggested [10, 30] forcing compartmentation of terrorist cells and interrupting communications between command echelons is likely to significantly impede the ability of those cells to function effectively. From a counter-terrorism perspective, however, the most interesting part of the simulation is the section which deals with "the constraints of history." In this case, Ghemawat and Levinthal artificially inject a variable inconsistent with the global maximum and then observe how the three different types of organization adapt to this distortion. Their Fig. 8.5 summarizes the (normalized) fitness levels achieved in the simulation when they preset a variable to a value which is inconsistent with the global optimum and this is shown below (p. 27):

The green line, representing a random pattern of interactions, not surprisingly has a fitness which is essentially "history independent." The more surprising finding is that while under conditions of hierarchy the fitness level changes quite rapidly as the mismatched variable is shifted from the most important decision to decisions of less importance, centrality conveys virtually no "recovery advantage" at all. This is a striking contrast to Porter [31] who argues that under conditions of "third order strategic fit" (strong interlinkage/strong centrality) the decision chain becomes "as strong as its strongest link." In dealing with terrorist organizations, which are primarily hierarchical in nature [6], what this finding says is that disinformation is a useful tactic (or strategy) only if it succeeds in influencing one of the key decisional variables. In other words, disinformation at the local level is unlikely to have any lasting impact on terrorist organizations. This finding also challenges the institutional wisdom of assigning case officers in the field to this type of counter-terrorism operation [32, 33]).[7] In fact, from an operational point of view, the hierarchical nature of terrorist organizations means that there may be something of a mismatch in the entire targeting process. As Ghemawat and Levinthal note, "Less central variables not only do not constrain, or substantially influence the payoff of many other choices, but they themselves are not greatly contingent upon other policy choices. Being contingent on other policy choices facilitates compensatory shifts in policy variables other than the one that is preset. As a result of the absence of such contingencies, the presetting of lower-order policy choices is more damaging to fitness levels under the centrality structure." (p. 27) The problem, however, is that decision-making in terrorist organizations is apt to be operating under conditions of hierarchy rather then centrality. Perhaps even more annoyingly, those organizations operating against terrorist cells may themselves be organized in a more modern fashion, availing themselves of divisional inter-linkages and flat management structure, so that the mismatch between operational objectives and terrorist organizations may, in fact, prove organizationally damaging to the counter-terrorist organization both in a relative and an absolute sense. In this regard, pointless, or fruitless counter-terrorism operations, particularly when conducted at the field level may do considerably more damage than good, particularly to the organization striving to combat terrorism. This is a feature of counter-terrorism that is not entirely unfamiliar to case officers who have operated in this capacity [33, 34].

In their final section, Ghemawat and Levinthal examine choice structures.[8] To do this, they set up three classes of variables: independent variables (1–3), whose pay-

[7] Ghemawat and Levinthal (2000) test this another way and come to essentially the same conclusion:"Another striking feature of this set of simulations concerns how few of the optima with presetmismatches constitute local peaks of the fitness landscape. Given the importance of configurationaleffects, one might reasonably conjecture that constraining one variable to differ from theglobal optimum would lead to the selection of a different, non-global, peak in the fitness landscape.However, Figure 9 indicates that this is relatively uncommon except as one turns to presetting theleast important variables under the hierarchy and centrality structures." (pp. 28–29).

[8] "The broader suggestion is that the 'natural' adjacency matrices we have looked at so far mix up at least three very different types of effects: influence, contingency and autonomy. Variables may be more or less influential to the extent that they affect the payoffs to other variables. In an adjacency matrix, this is represented by the prevalence of 1 s in the relevant column. Independent of influence, the payoffs from specific variables may be more or less contingent on other choices, as reflected in the number of 1 s in the relevant row of the adjacency matrix. And autonomy is characterized by variables that are neither influential nor contingent: variables that correspond, in graph-theoretic terms, to disconnected vertices. In this subsection, we look at a choice structure—distinct from the three that we have already examined—that distinguishes particularly clearly among these three effects." (p. 29)

**Fig. 8.6** Influence, contingency, and autonomy

| | 1 | 2 | 3 | 4 | 5 | 6 | 7 | 8 | 9 |
|---|---|---|---|---|---|---|---|---|---|
| 1 | 1 | 0 | 0 | 0 | 0 | 0 | 0 | 0 | 0 |
| 2 | 0 | 1 | 0 | 0 | 0 | 0 | 0 | 0 | 0 |
| 3 | 0 | 0 | 1 | 0 | 0 | 0 | 0 | 0 | 0 |
| 4 | 1 | 1 | 1 | 1 | 0 | 0 | 0 | 0 | 0 |
| 5 | 1 | 1 | 1 | 0 | 1 | 0 | 0 | 0 | 0 |
| 6 | 1 | 1 | 1 | 0 | 0 | 1 | 0 | 0 | 0 |
| 7 | 0 | 0 | 0 | 0 | 0 | 0 | 1 | 0 | 0 |
| 8 | 0 | 0 | 0 | 0 | 0 | 0 | 0 | 1 | 0 |
| 9 | 0 | 0 | 0 | 0 | 0 | 0 | 0 | 0 | 1 |

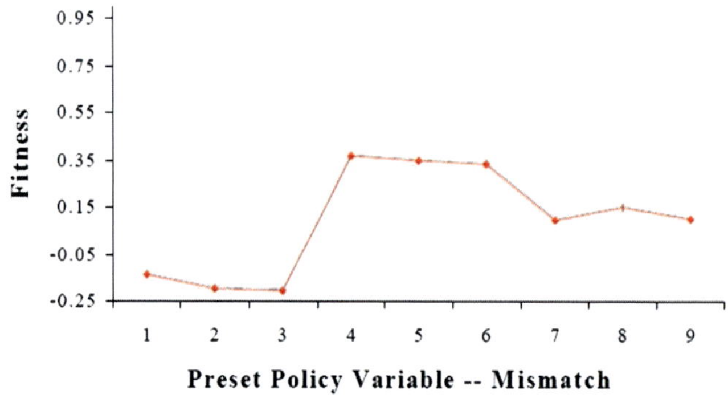


**Fig. 8.7** Constraints of history: normalized fitness levels

off is not dependent upon that of any other variables, but which influences the payoff of the dependent variables (4–6), and three variables (7–9) which are simply independent of all others (Fig. 8.6).

They illustrate their findings in Fig. 8.7:

Interestingly, Ghemawat and Levinthal are surprised that constraining the independent variables to values different from the global maximum affects overall fitness more than the constraint of contingent variables:

> Figure 8.7 indicates that constraining one of the "influential" variables to differ from the global maximum has a profound effect on the relative fitness level of the constrained optimum. Surprisingly, constraining the independent variables to differ from the global optimum has a larger impact than constraining the seemingly more important "contingent" variables. The reason for this is that the presence of contingency allows for the possibility of substituting or compensating changes in policy variables. While tightly linked interaction patterns have generally been viewed as fragile, the equifinality that high levels of interaction engender also allows for a certain robustness. In contrast, when an autonomous variable is misspecified, that doesn't create negative ramifications elsewhere in the system of policy choices; at the same time, however, there is no opportunity to compensate for the misspecification.

The most probable source of their surprise is the literature on strategic linkages (which they cite heavily at the beginning of their study). What their simulation reveals, however, is that coupled decisional structures have a compensatory feature which makes them less likely to certain kinds of decisional disruption than hierarchical decision structures. The implicit lesson for intelligence organizations is that in dealing with terrorist organizations efforts would best be directed at hitting them hard, at the highest levels of the hierarchy and leave the "cleanup" to local law enforcement organizations. In this context, less engagement may, in fact, yield more results. Conversely, if "sometimes less is more," clearly "sometimes more is a whole lot less."

## References


1. Jones, D. (2009). *Understanding the form, function, and logic of clandestine cellular networks: The first step in effective counternetwork operations*. Fort Leavenworth, KS: School of Advanced Military Studies, United States Army Command and General Staff College.
2. Butts, C. T. (2003). Network inference, error, and informant (in)accuracy: A Bayesian approach. *Social Networks, 25*(2), 103–30.
3. Carley, K. (2002a). Modeling "covert networks". *Paper prepared for the National Academy of Science Workshop on Terrorism.*
4. Carley, K., Dombroski, M., Tsvetovat, M., Reminga, J., & Kamneva, N. (2003). Destabilizing dynamic covert networks. In *Proceedings of the 8th international command and control research and technology symposium*. Washington: National Defense War College. http://www.casos.cs.cmu.edu/publications/resources_others/a2c2_carley_2003_destabilizing.pdf.
5. Krebs, V. (2001). Uncloaking terrorist networks. *First Monday*. http://www.orgnet.com/hijackers.html.
6. Sageman, M. (2004). *Understanding terror networks*. Philadelphia: University of Pennsylvania Press. https://www.google.com/search?sourceid=navclient&aq=&oq=University+of+Penn&ie=UTF-8&rlz=1T4GGLS_enUS588US589&q=university+of+pennsylvania+press&gs_l=hp..3.0l5.0.0.0.7013.0.2lkLgd2zg24
7. Carley, K., Lee, J. -S., & Krackhardt, D. (2001). Destabilizing networks. *Connections, 24*(3), 31–34, INSNA.
8. Carley, K. (2002b). Inhibiting adaptation. In *Proceedings of the 2002 command and control research and technology symposium*. Monterey: Naval Postgraduate School.
9. Carley, K., Diesner, J., Reminga, J., & Tsvetovat, M. (2004). *Toward an end-to-end approach for extracting, analyzing and visualizing network data*. ISRI, Carnegie Mellon University.
10. Fellman, P. V., & Wright, R. (2003). Modeling terrorist networks: Complex systems at the mid-range. In *Paper prepared for the Joint Complexity Conference*. London School of Economics. http://www.psych.lse.ac.uk/complexity/Conference/FellmanWright.pdf.
11. Fellman, P. V., & Strathern, M. (2004). The symmetries and redundancies of terror: Patterns in the dark. In *Proceedings of the annual meeting of the north american association for computation in the social and organizational sciences*. Carnegie Mellon University. http://casos.isri.cmu.edu/events/conferences/2004/2004_proceedings/V.Fellman,Phill.doc.
12. Ghemawat, P., & Levinthal, D. (2000). *Choice structures, business strategy and performance: A generalized NK-simulation approach*. Reginald H. Jones Center, The Wharton School, University of Pennsylvania. http://www.people.hbs.edu/pghemawat/pubs/ChoiceStructures.pdf.
13. Butts, C. T. (2001). The complexity of social networks: Theoretical and empirical findings. *Social Networks, 23*(1), 31–71.
14. Clemens, J. P., & O' Neill, L. (2004). *Discovering an optimum covert network*. Santa Fe Institute.

15. Hoffman, B. (1997). *The modern terrorist mindset: Tactics, targets and technologies*. Scotland: Centre for the Study of Terrorism and Political Violence, St. Andrews University.
16. Hoffman, B., & Carr, C. (1997). Terrorism: Who is fighting whom? *World Policy Journal*, *14*(1), 1–2.
17. Codevilla, A. (2004a). Doing it the hard way. In *Claremont review of books*. Fall. http://www.claremont.org/writings/crb/fall2004/codevilla.html.
18. Tsvetovat, M., & Carley, K. (2003). Bouncing back: Recovery mechanisms of covert networks. In *NAACSOS conference*.
19. Kauffman, S. (1993). *The origins of order.* Oxford: Oxford University.
20. Kauffman, S. (1996). *At home in the universe*. Oxford: Oxford University.
21. Kauffman, S. (2000). *Investigations*. Oxford: Oxford University Press.
22. Lissack, M. (1996). Chaos and complexity: What does that have to do with knowledge management? In J. F. Schreinemakers (Ed.), *Knowledge management: Organization, competence and methodology* (Vol. 1, pp. 62–81). Wurzburg: Ergon.
23. McKelvey, B. (1999). Avoiding complexity catastrophe in coevolutionary pockets: Strategies for rugged landscapes. *Organization Science, 10*(3), 294–321.
24. Meyer, C. (1996). What's under the hood: A Layman's guide to the real science. In *Cap Gemini, Ernst and Young, Conference on Embracing Complexity.* San Francisco: Center for Business Innovation.
25. Butts, C. T. (2003). Predictability of large-scale spatially embedded networks. In R. Breiger, K. Carley, & P. Pattison (Eds.), *Dynamic social network modeling and analysis: Workshop summary and papers* (pp. 313–323). Washington: National Academies Press.
26. Butts, C. T. (2000). An axiomatic approach to network complexity. *Journal of Mathematical Sociology, 24*(4), 273–301.
27. Carley, K. M., & Butts, C. T. (1997). An algorithmic approach to the comparison of partially labeled graphs. In *Proceedings of the 1997 international symposium on command and control research and technology*, Washington.
28. Ghemawat, P., & Levinthal, D. E. (2006). Choice Interactions and Business Strategy. (Former title: "Choice Structures, Business Strategy and Performance: A Generalized NK-Simulation Approach") The Wharton School, Working Paper, Revised February, 2006.http://knowledge.wharton.upenn.edu/wp-content/uploads/2013/09/13121.pdf
29. Weinberger, E. (1991). Local properties of Kauffman's NK Model: A tunably rugged energy landscape. *Physical Review A, 44*, 6399–6413.
30. Fellman, P. V., Sawyer, D., Wright, R. (2003). Modeling terrorist networks: Complex systems and first principles of counter-intelligence. In *Proceedings of the NATO conference on central Asia: Enlargement, civil – military relations, and security*. Kazach American University/North Atlantic Treaty Organization (NATO).
31. Porter, M. (1996). "What is Strategy?" *Harvard Business Review.*
32. Codevilla, A. (2004b). *Why U.S. intelligence is inadequate and how to fix it.* Center for Security Policy, Occasional Papers Series. http://www.centerforsecuritypolicy.org/occasional-papers/Why-US-Intelligence-Is-Inadequate.pdf.
33. Gerecht, R. M. (2001). The counterterrorist myth. *The Atlantic Monthly.*
34. Gilligan, T. (2003). *CIA life: 10,000 days with the agency*. Intelligence E-Publishing Company.

# Chapter 9
# Comparison of Approaches for Adversary Modeling Decision Support for Counterterrorism

**Barry Ezell and Gregory S. Parnell**

## Introduction

Intelligent adversaries are a fundamental component of terrorism risk analysis. Unlike natural and engineering hazards, intelligent adversaries adapt their behavior to the actions of the defender. They adapt to observed, perceived, and imputed likely future actions by those defending the system they are attempting to influence. Risk assessment models need to consider these potential adaptive behaviors to be able to provide accurate estimates of future risk from intelligent adversaries and appropriately support risk management decision making.

Guikema [1] suggests that a number of models have been proposed for modeling intelligent adversaries, with the major types of models being:

1. Probabilistic risk analysis uses event trees to model adversary actions along an incident chain as random variables, and estimate probabilities of actions based on expert elicitation. Attacker choices are modeled as defender uncertainty regarding red choices and model the sequence of steps to accomplish attack as chance nodes.
2. Decision analytic methods which treat adversary behavior as decisions made based on utility maximization, starting from the adversary's beliefs and preferences, as assessed by the defender, about the state of the world and the defender's actions.
3. Game theoretic methods take the decision analytic concept further, and treat the adversary adaptation as decisions made based on a strategic assessment and modeling of the defender's likely actions, sometimes relaxing rationality assumptions.

B. Ezell (✉)
Old Dominion University's, Virginia Modeling, Analysis and Simulation Center,
1030 University Blvd., Suffolk, VA 23435, USA
e-mail: bezell@odu.edu

G.S. Parnell
Department of Industrial Engineering, University of Arkansas, Fayetteville, AR 72701, USA



P.V. Fellman et al. (eds.), *Conflict and Complexity*, Understanding Complex Systems,
DOI 10.1007/978-1-4939-1705-1_9

4. Statistical and machine-learning methods. Finally, statistical and machine-learning methods develop models of adversary behavior, such as future attack patterns, based on past observations.

In this chapter, we summarize several methods associated with this classification. First, we discuss the need to integrate terrorism risk analysis into the intelligence cycle. Next, we provide a brief description of several methods, and then benefits and limitations are discussed. Then we compare an event tree and a decision tree for the same problem. We conclude with some observations.

## Integrating Risk Analysis into the Intelligence Cycle

Intelligence and risk analyses are closely linked. Intelligence analysis provides information for risk analyses and risk analysis can be used to guide intelligence collection and analysis. Willis [2] suggests a process that integrates risk analysis into the intelligence cycle such that terrorism threat assessments and warnings may be provided to decision makers. This approach includes identification of scenarios of greatest concern. The author argues that this approach can help overcome risk analysis limitations in terms of: information availability, matching resolution of results to the problem, and reflecting risk as the social construct. This includes considering the possibility of surprise.

The integration process is described as follows. First, the intelligence cycle is considered an input–output process in which raw intelligence is the input and intelligence products are the outputs. This can be described in five steps that include [3]:

1. Direction of intelligence collection,
2. Collection of new information,
3. Processing,
4. Analysis,
5. Dissemination.

Willis et al. [4] summarize the integration of the intelligence cycle process. The authors believe that the intelligence community and the homeland security community must intimately interact and at many stages throughout the process of collection, processing, and analysis.

Intelligence that may be employed to evaluate the magnitude and nature of terrorism threats is generated by collection activities. Thus, information about terrorist threats and adaptability can be obtained. As indicated by Willis et al. [5], there is risk from terrorism only when there is a credible threat of harmful attack. An approach that fits well in considering these potential threats, associated vulnerability, and consequences while developing strategies to manage is risk analysis. The product of integration leads to the identification of terrorism scenarios that present the greatest risk.

In order to prevent terrorist attacks, information of potential locations to conduct further surveillance and identifying potential terrorists is required. This also entails an understanding of timeline and resources associated with the planning an attack. The integration between the results of risk analysis with an exhaustive assessment of the planning stages of terrorist attacks, the intelligence process can provide directed intelligence products and refine future planning and direction of intelligence collection.

# Methods

## *Logic Trees*

Logic trees provide a suitable approach for capturing and processing large complex information while assessing this information into a series of smaller and less complex elements [6]. Sequential assessments must be performed to be able to analyze and address uncertainties in a sequential fashion. Logic trees are divided into two types that include:

1. Probability, event, and decision trees
2. Fault, attack, and success trees

Details that describe these methods are as follows.

### Probability, Event, and Decision Trees

Description

1. Probability trees determine the probabilities of events in the outcome space by characterizing a sequence of uncertain events. Typically, they are a succession of nodes based on probabilistic variables that branch out from each node that show the model of the potential mutually exclusive and collectively exhaustive (ME&CE) states of nature and their associated probabilities. Probability trees are used to calculate the probability of each of the possible outcomes and the expected value.
2. Event trees are used to model the sequences of uncertain events that cause or result in certain consequences. They are extensions of probability trees since they add an initiating event, mitigating events, and consequences. Thus, for each probability path, consequences are included. Event trees are used to calculate the probabilities of the consequences and the expected consequences.
3. Decision trees are logic trees that include decision nodes in addition to events. Branches of decision nodes characterize decision alternatives. Chance nodes have branches representing ME&CE states of nature and assigned probabilities. Endpoints are the leaves of each path and characterize the final consequences of a path [7]. In addition, the consequence of each path can be modeled as value/utility and the expected value/utility can be calculated [8].

Probability trees, event trees, and decision trees have been used to model very large complex systems.

Benefits

Probability trees are useful to graphically employ probability theory and characterize probabilistic relationships between two or more events. In addition, they serve as the mathematical means for characterizing more advanced tree structures, e.g., event or decision trees [9]. According to Ezell et al. [6] event trees have been used for large systems, e.g., nuclear reactor safety studies and to employed event trees to analyze cyber risks [10]. In decision analysis, decision trees are used as a decision support tool to determine the alternative with the best expected value/utility or the best risk profile in many important decisions [8].

Decision trees can be used to model the attacker's or the defender's decisions. Attacker decision trees enable characterizing actions of attackers as decisions while characterizing defenders as chance nodes in the terrorism context. Defender trees can be used to model the actions of defenders as decisions while the actions of attackers as chance nodes. Parnell et al. [11] modeled attacker and defender decisions in the same decision tree by modifying the decision tree algorithm to have the attacker maximize consequences and the defender minimize consequences. Game theory models (next section) analyze the decisions of both the attacker and the defender.

Probability, event, and decision trees assign probabilities to branches to model the certainty about the outcome of each branch for a chance node. At each node, probability assessments may be conditioned on one or more of the predecessor nodes. The presumption is that all the branches connected to that node portray the states of the previous events. These conditional probabilities assume mutually exclusive and collectively exhaustive events. Benefits are as follows:

1. Simple to understand and explain
2. Tree structure and probabilities can be generated based on subject matter experts' information
3. Easily replicable
4. Flexible. They may be integrated with other tools.

Limitations

Logic trees have the following limitations:

1. Human judgment is required to identify all possible scenarios.
2. Probability assignments are usually based significantly on subjective judgments mainly due to:

    a. Human judgment usually determines the ME&CE outcomes.
    b. Availability of numeric data is limited for many potential terrorist attacks.

3. In most cases, terrorist decisions are not random events.
4. Terrorists' preferences are dynamic and might change over time.

5. Models can become very complex for complex problems with many possible events and decisions.
6. The overall success of the process depends on the skill of the analyst involved.

Intelligent Adversary Perspective on Logic Trees

Advocates of intelligent adversary modeling argue that terrorists do not roll dice but instead they seek to maximize the potential consequences. Although event trees have been applied to many applications to terrorist risk, Ezell et al. [6] point out that event tree have limitations in how adversary choices are modeled. Event tree models typically describe "the intelligence community's uncertainty in the intelligent adversary's true utility." The assessment for terrorist preferences, potential terrorist actions, and the assignment of probabilities to terrorist actions are reliant upon the available information in the intelligence community. Finally, the authors suggest decision trees in this context presuppose:

1. The intelligent adversary's true utility for the choices in the tree is known
2. The adversary has the knowledge and rationality to actually maximize that utility which is argued not to be the case.

Using decision trees to model terrorism from the perspective of terrorists requires two important assumptions. First, it assumes ideal adversary intelligence and rationality—that the adversary *knows* which branch choice at a particular decision node best maximizes consequences. The second critical assumption, also related to the adversary's objective function in the decision mode, is that the intelligence community (or anyone else for that matter) knows the objectives the adversary is trying to maximize in the first place.

Advocates of using decision trees argue that we should not rely on probability estimates of terrorist decisions unless we have actionable intelligence and, for strategic planning, we should assume that the terrorist has sufficient knowledge to attack create the worst consequences and plan our risk management actions to reduce the worst potential consequences Parnell et al. [11].

## Fault, Attack, and Success Trees

Description

1. Top down approach deductively analyzes failure/success in which an undesired state of a system is processed using Boolean logic, reliability theory, and probability theory. Boolean logic is employed to consider a series of lower-level events. Mostly used in the field of safety engineering and reliability engineering, this method is employed to calculate the probability of a safety accident or a particular system level (functional) failure.

2. For intelligent adversary analysis fault trees are defender models and success trees are attacker models. Fault trees characterize the combinations of system failures and human errors. Success trees describe the arrangement of events that cause to success. The tasks of an intelligent adversary directed to defeat a defensive system can be captured by attack trees models.
3. Goals:
   a. Providing an understanding of why mitigating events in an event tree may fail is the goal of fault trees.
   b. Classifying the different means in which a system can be defeated is the purpose of attack trees. Attack trees can be considered a special case of fault trees. They are used to represent an adversary's successful defeat of a countermeasure.

Details

As stated before, events in a fault tree are connected with probabilities. Consider that component failures typically occur at some constant failure rate $\lambda$. The failure rate $\lambda$ is assumed to be a constant hazard function. Failure probability is reliant upon the rate $\lambda$ and the exposure time $t$:

$$P = 1 - e^{(-\lambda t)} \tag{9.1}$$

$$P \approx \lambda t,\ \lambda t < 0.1 \tag{9.2}$$

Events are modeled with gates that are AND or OR gates. A fault trees are normalized to a given time interval, such as average time. The relationship between the event hazard function to the time interval conditions the set of event probabilities. The associated uncertainty to a gate's output event relies on the input event probabilities. AND gates symbolize an arrangement of independent events. In other words, in the same gate, the uncertainty of any input event to an AND gate is unchanged by any other input event. Mathematically the probability of the AND gate output can be expressed as follows (9.3).

$$P(A \text{ and } B) = P(A \cap B) = P(A)P(B) \tag{9.3}$$

In addition, an OR gate can be represented by the set union as follows (9.4).

$$P(A \text{ or } B) = P(A \cup B) = P(A) + P(B) - P(A \cap B) \tag{9.4}$$

Failure probabilities on fault trees are usually small, therefore, the intersection $P(A \cap B)$ is a very small error term as well. The output of an OR gate can be approximated arguing that the inputs are mutually exclusive events as follows.

$$P(A \text{ or } B) = P(A) + P(B) - P(A \cap B) \approx 0 \tag{9.5}$$

An OR gate with two inputs symbolizes the probability that one or the other input, but not both. This can be represented by (9.6). Because of $P(A \cap B)$ typically is a very small term, the exclusive OR gate has limited value in a fault tree.

$$P(A \operatorname{xor} B) = P(A) + P(B) - 2P(A \cap B) \tag{9.6}$$

Benefits

1. Produces a reliability estimate which allows evaluation and improvement of the overall reliability of the system.
2. It also evaluates the effectiveness of and the need for redundancy.
3. Can be used to calculate probabilities used for other models.

Limitations

1. A limitation of the fault/success tree is that the undesired event evaluated must be foreseen by SMEs and all significant contributors to the failure must be anticipated.
2. This effort may be very time consuming and expensive.
3. The overall success of the process depends on the skill of the analyst involved.

## *Influence Diagrams*

An influence diagram (ID) is a graphical representation of a decision situation [8]. It is very similar to a Bayesian network (see next section). Probabilistic inference problems and decision making problems can be formulated and solved using influence diagrams. Formulation that follows maximum expected value/utility criterion can be solved. The relationships between decisions, events, and outcomes are symbolized as nodes and arcs in a directed acyclic graph. An ID uses an equivalent solution algorithm to the event tree for probabilities and the decision tree algorithm for decision trees.

### Description

Influence diagram conventions use nodes and arcs. There are four types of nodes and four types of arcs used in influence diagrams.

Nodes

1. Decision node that corresponds to each decision to be made—rectangle.
2. Chance or uncertainty node—oval.
3. Calculated or determined uncertainty—double oval.
4. Value nodes which are symbolized as diamonds (alternatively, or hexagons). Each influence diagram must have one value/utility node used to solve the diagram.

Arcs

1. Arrow between two uncertainties means probabilistic dependence. The direction of the arrow indicates the order of conditionality.
2. An arrow from an uncertainty to a decision means that the outcome of the uncertainty is known when the decision is made.
3. An arrow between two decisions means that when the decision at the head of the arrow is made, the choice taken in the other decision is remembered perfectly.
4. Finally, an arrow from a decision to an uncertainty means that the probabilities of the outcomes depend on the choice made in the decision.

Formally, the semantic of influence diagram is based on sequential construction of nodes and arcs, which implies a specification of all conditional independencies in the diagram. The specification is defined by the d-separation criterion of Bayesian network. According to this semantic, every node is probabilistically independent on its non-successor nodes given the outcome of its $X$ immediate predecessor nodes. Likewise, a missing arc between non-value node $X$ and non-value node $Y$ implies that there exists a set of non-value nodes $Z$, e.g., the parents of $Y$, that renders $Y$ independent of $X$ given the outcome of the nodes in $Z$.

Every influence diagram can be represented by a symmetric decision tree, in which all possible combinations of decision, events, and outcomes are represented. Conversely, all decision trees can be enriched to symmetric trees and converted to an influence diagram. Their solution algorithm is equivalent.

In the context of intelligent adversary an ID can represent a terrorist group's decision to choose a mode and a target of attack as indicated by [12]. Decision nodes can be characterized as the possible countermeasures; outcome nodes may represent the consequences portrayed as a combination of the type of weapon, target, and delivery means [6]. Information available to the US government can be symbolized as other nodes are event nodes which can describe terrorist states. Thus, IDs may be used to calculate the probability of alternative types of terrorist attacks.

## Benefits

IDs' benefits can be summarized as follows [13].

1. Represent the probabilistic structure of complex problems compactly,

2. Simplify modeling by allowing the analyst to specify single nodes that represent entire probability distributions over nearly arbitrary relationships with other variables,
3. Provide a picture of the model relationships not possible to visualize with trees,
4. In order to express conditional independence relationships that are natural in an ID, one would need to resort to the use of complicated information sets among siblings in a tree. This may help computationally but not in terms of constructing models,
5. The number of variables in the problem grows, the size of a symmetric tree grows exponentially, which is not true in the case of influence diagrams since they need to only represent the probabilities associated with their parents in the graph.
6. Facilitate communication between analysts and decision makers, and
7. Form the basis for efficient and easy-to-use computer-based tools.

#### Limitations

1. Smaller, highly asymmetric problems may be easier to understand when represented a decision trees.
2. The timing of events may be easier to identify in a decision tree, although the same information is implicit in the influence diagram.
3. To represent fully the decision alternatives, strategies, alternative events, and value of outcomes, the analyst must complete the probability and value tables in the influence diagram, a process similar to placing the value of variables in the decision tree.
4. As with an event tree and a decision tree, an influence diagram for a complex problem may require a large number of probability assessments.
5. The overall success of the process depends on the skill of the analyst involved.

### *Causal Loop Diagrams and Systems Dynamic Models*

#### Description

Human systems are complex that are difficult to represent in computational models. System dynamics (SD) is a modeling paradigm that lends itself to capture the many factors that influence complex system behavior in a dynamic model. SD is a way to model and depict factors contributing to the behavior of a system and the causal relationships that exist among those factors. To employ this modeling method one begins by developing a causal loop diagram that shows these variables and their causal linkages.

The SD framework is based on using a system science framework to be able to gain knowledge and understanding of complex systems such as the adaptive adversary problem. It is employed to represent and simulate the behavior of the system over time. Knowledge obtained from the observation of the behavior of the system is used to design and test strategies such that an accurate representation of change in behavior as well as consequences on the system constituents can be quantified and predicted. Finally, an understanding of patterns on such behavior as well as measuring of such consequences enables the intelligence community to determine potential preventive policies as well as mechanisms to minimize likelihood of terrorist events.

The SD framework acknowledges the complex interactions among many feedback loops, considers linear and nonlinear cause-and-effect, and requires an analytical perspective that considers the potential impact of effects on causes. Thus, it allows revealing and quantifying unseen dynamics that might have a relevant effect on the interaction among considered subsystems.

Empirical analysis can be incorporated to explain the relationships between individual factors and subsystems in the overall intelligent adversary context. By understanding the behavior of this complex system, where extrinsic and intrinsic factors, e.g., targets, weapons, terrorist group, intelligence, etc., interrelate, one is able to identify the degree to which factors and potential polices, e.g., awareness interventions, affect the behavior of the intelligence adversary system.

A general modeling methodology [14] to be followed includes: (1) developing a system model including actors, relationships, potential targets, etc; (2) formulating a theory that underlies this model; (3) formulating a simulation model to test the dynamic hypotheses generated from the functioning of model; (4) determining and analyzing the impact of factors and policies that influence the model; and (5) investigating, modeling, and testing mechanisms that might generate policy resistance.

### Benefits

1. Causal diagram provides a visualization of the relationships
2. Ability to assess the impact of casual relationships over time that can cause surprising nonlinear behavior
3. Integration of many viewpoints on causal relationship
4. Ability to capture feedback loops
5. SD models are precise and objective in their representation relationships laid open and thus open for intersubjective assessment.

### Limitations

1. The identification of the casual relationships and their magnitude in each time increment requires extensive SME knowledge
2. Accuracy of decision rules or policies
3. Little data may be available for soft variables and intangibles
4. Model boundary left undefined
5. The overall success of the process depends on the skill of the analyst involved.

## *Bayesian Network Analysis*

### Description

A Bayesian network (BN) is a directed acyclic graph or belief network where nodes correspond to probabilistic variables and the directed arcs specify probabilistic dependence. Ezell et al. [6] summarize the main elements of BNs as follows. The probabilistic dependence between pairs of random variables is determined by arcs in the network; the direction of the arc indicates which of two possible conditional probability distributions has been captured. The relationships between nodes can be explained by employing tree and family metaphors. Nodes that point to children are called Parents. All nodes that point to it directly or indirectly through its parents are known as Ancestors or Children. Those nodes to which a node points directly or indirectly are known as descendants.

The Bayesian approach involves experts giving information about certain events or quantities to a decision maker (sometimes called a supra-Bayesian) who then updates a prior distribution using Bayes' Theorem. Clemen and Winkler [15] point out the difficulties with obtaining the likelihood function required by the Bayesian methods. Thus, they describe four classical methods for combining probabilities and some methods of combining probability distributions by Lindley [16, 17], Clemen and Winkler [18], Winkler [19], and Mendel and Sheridan [20].

Ezell et al. [6] note that BN have been extensively employed in the modeling anti-terrorism issue, e.g., predicting distribution for lethal exposure to Sarin, a chemical nerve agent. In this case, the model can be characterized as indicates. First, the conditional probability defined for each node: Sarin Attack, Exposure Type, Detected Early, and Lethal Exposure. Sarin Attack would represent the root node and parent to Exposure Type and Detected Early. The inference algorithm applies Bayes's rule and update the probability distribution for Lethal Exposure by altering the evidence for a given variable. BN are very versatile, since this situation might change, and they are easy to update, e.g., as the evidence changes, the posterior probability changes. Netica™ or Genie™ is a well-known Software able to interact and exchange data with other models. O'Hagan [21] summarizes benefits and limitations as follows.

### Benefits

1. Bayesian statistics provides more intuitive and meaningful inferences. A Bayesian analysis gives the more direct, intuitive, and meaningful statement of the probability that the hypothesis is true.
2. Bayesian methods can answer complex questions cleanly and exactly. The computing tools now available for Bayesian statistics allow us to tackle enormously more complex problems.
3. Bayesian methods make use of all available information. This is simply a reference to the fact that the Bayesian approach includes the prior information. Since

the prior information should represent all the available knowledge apart from the data themselves, no relevant information is omitted in a Bayesian analysis.
4. Bayesian techniques are particularly well suited for decision making. What makes decisions hard is uncertainty. There is uncertainty about the consequences of any given decision, due to lack of knowledge about some relevant facts or parameters.
5. Bayesian methods can quantify those uncertainties using personal probability. Indeed, very often one explicitly derives a posterior distribution for unknown parameters based on the available evidence.

#### Limitations

1. BN involve elements of subjectivity to identify the variables, the relationships, and the probabilities,
2. Prior distribution is difficult to specify reliably, thus, Bayesian methods are more complex than frequentist methods,
3. There are few BN software packages, and
4. The overall success of the process depends on the skill of the analyst involved.

### *Game Theoretic Models*

Game theory (GT) models may be characterized as multi-agent decision problems. Von Neumann and Morgenstern, and Nash, are classical references of this modeling approach. In game-theoretic models is assumed that all the possible utilities of the different consequences for each player must be derivable and usable within the model. This assumption has important implications related to the knowledge of the possible goals and aspirations of the different players. Conversely, classical game theory considers one set of utilities for each player only.

Hurwitz and Marwala [22], summarize benefits and limitations of agent-based model as follows.

#### Benefits

1. Agent behavior is simpler to model than the overall system they comprise, and hence the system becomes easier to model.
2. The emergent behavior resulting from the agent interactions implies that systems too complex to be traditionally modeled can now be tackled, since the complexity of the system need not be explicitly modeled.
3. Large systems with heterogeneous agents can be easily handled within a multi-agent system, while this is incredibly difficult to cater for using traditional mathematics, which would make the often unrealistic demand that the components be homogenous.

**Limitations**

1. Since the emergent behavior is arrived at empirically, and is not deterministic, it is difficult to state with any degree of certainty as to why a certain outcome has been arrived at.
2. Since emergent behavior is often unexpected, it can be difficult to ascertain whether the multi-agent system (MAS) is incorrectly modeling the system in question. Thus, validation of the model becomes an important aspect of any MAS.
3. The overall success of the process depends on the skill of the analyst involved.

## *Canonical Intelligent Adversary Risk Management Model*

Parnell et al. [11] argue that there is a difference in the modeling of intelligent adversary decisions that event trees do not capture. The attacker makes decisions to achieve his objectives. The defender makes resource allocation decisions before and after an attack to try to mitigate vulnerabilities and consequences of the attacker's actions. This dynamic sequence of decisions made by first defender, then an attacker, then again by the defender should not be modeled by assessing probabilities of the attacker's decisions.

Parnell et al. [11] suggest that risk management model for bioterrorism homeland security. They suggest that it must be considered six components: the initial actions of the defender to acquire defensive capabilities, the attacker's uncertain acquisition of the agents (e.g., A, B, and C), the attacker's target selection and method of attack(s) given agent acquisition, the defender's risk mitigation actions given attack detection, the uncertain consequences, and the cost of the defender actions. In general the defender decisions can provide offensive, defensive, or informational capabilities. They do not consider offensive decisions such as preemption before an attack. They consider decisions that will increase their defensive capability (e.g., buy vaccine reserves) or provide earlier or more complete information for warning of an attack. Parnell et al. [11] summarize benefits and limitations of this approach as follows.

**Benefits**

1. It provides a more accurate risk assessment.
2. It provides information for risk-informed decision making.
3. Using COTS software the existing sensitivity analysis tools can be considered.
4. The risk analysis can be conducted by one risk analyst with an understanding of decision trees and optimization.

### Limitations

The authors assert that some of the limitations are actually the same as using event trees.

1. There are limitations on the number of agents used in the models.
2. There are challenges in assessing the probability of the uncertain events, e.g., the probability that the attacker acquires Agent A.
3. There is a limitation in modeling of the multiple consequences.
4. Another limitation may be that in order to get more realistic results, one may have to develop "response surface" models of more complex consequence models.

There are some decision tree limitations that are different from event trees.

1. There are a limited number of risk management decisions that can realistically be modeled. Care must be used to choose the most appropriate set of potential decisions.
2. There may be an upper bound in the number of decisions or events the COTS software can model.
3. Successful model operation and interpretation requires trained analysts who understand decision analysis and defender-attacker defender optimization.
4. Toward Comparison of Risk Analysis Techniques.

Merrick and Parnell [23] recommend the comparison of risk analysis techniques on the same problem to better enable insights. As an example of this approach we compare event tree and decision trees on an illustrative Vehicle Borne Improvised Explosive Devices (VBIEDs) homeland security study of two tunnels in a large urban area. The table identifies the actors, the events, probabilities (used triangular distributions), node type, consequences, event tree assumptions, and decision tree assumptions. The data in Table 9.1 is illustrative.

### Event Tree Analysis with Most Likely Values

First, we perform an event tree model using the most likely probabilities and consequences of the triangular distributions. All nodes in the event tree are chance nodes. The expected consequences are about $37M (Fig. 9.1).

### Decision Tree Analysis with Most Likely Values

Next, we perform a decision tree model using the same most likely probabilities and consequences. The first two nodes in the event three have been converted to Red decision nodes. It is important to note that both event and decision trees include exactly the same scenarios. The expected consequences are about $79M. Red would select Tunnel 1 due to the higher expected consequences (Fig. 9.2).

**Table 9.1** Illustrative VBIED tunnel attack data

| Actor | Event | Probability | Node type | Consequence ($) | Event tree | Decision tree |
|---|---|---|---|---|---|---|
| Red | Initiate VBIED attack | Tri~ (0.3, 0.5, 0.8) | Red choice | – | Red choice modeled as blue | Red decision |
| Red | Select target | – | Red choice | – | Probability split between two | Red decision |
| Red | – Bridge tunnel 1 | Tri~ (0.3, 0.6, 0.8) | Red choice | Tri~(50M, 80M, 120M) | Red choice modeled as blue uncertainty. | Red decision |
| Red | – Bridge tunnel 2 | – | Red choice | Tri~(40M, 70M, 150M) | 1-probability for tunnel 1 | Red decision |
| Blue | Detection | Tri~ (0.01, 0.05, 0.1) | Blue | – | Blue systems | Blue systems |
| Blue | Interdiction | Tri~ (0.1, 0.2, 0.3) | Blue response | – | Blue systems respond | Blue systems respond |

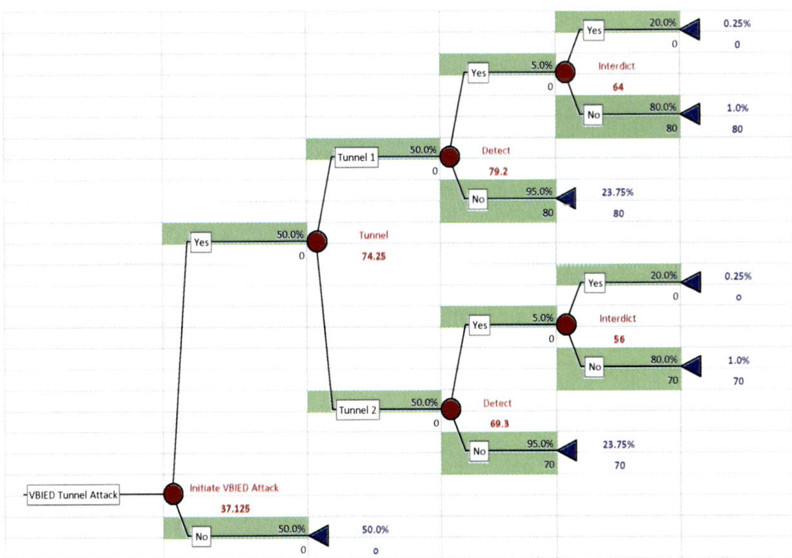


**Fig. 9.1** Event tree with most likely values

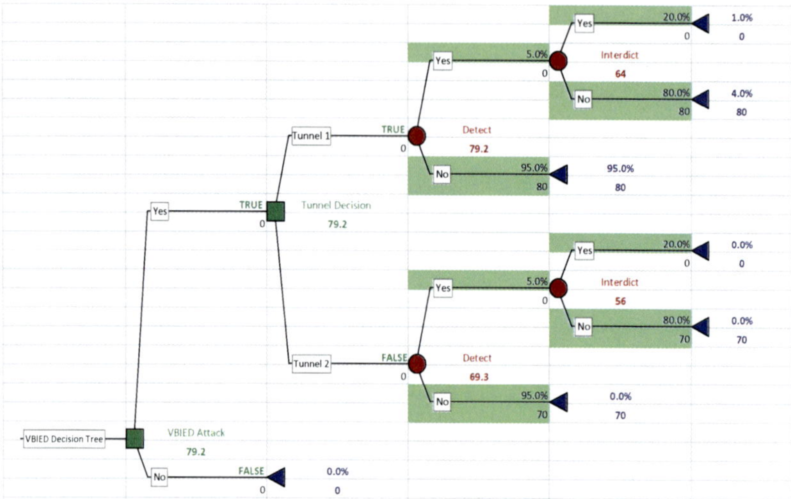


**Fig. 9.2** Decision tree with most likely values

### Comparison of Event Tree and Decision Tree with Most Likely Values

Figure 9.3 provides the event tree and decision tree risk profiles. The risk profiles are very different. The decision tree risk profile model shows the 1 % effectiveness of detection and interdiction (0.05*0.2=0.01). In the event tree model the 1 % effectiveness is lumped with the probability of no attack.

### Comparison of Monte Carlo Event Tree and Decision Tree Simulations

Figure 9.4 provides the Monte Carlo event tree and decision tree risk profiles. The risk profiles are similar to the event tree and decision tree with most likely values; however, the expected consequences ($43.44M event tree and $82.26M decision tree) are higher. The decision tree risk profile model again shows the 1 % effectiveness of detection and interdiction (0.05*0.2=0.01). In the decision tree model, Red attacks Tunnel 1 and has a lower max consequence. In the event tree model, Red selects Tunnel 2 50 % of the time and, therefore, has a higher max consequence.

## Conclusions

Intelligent adversaries are a fundamental component of terrorism risk analysis. The modeling of potential adversary attacks on our homeland security is an important opportunity to support national decision makers responsible for homeland security

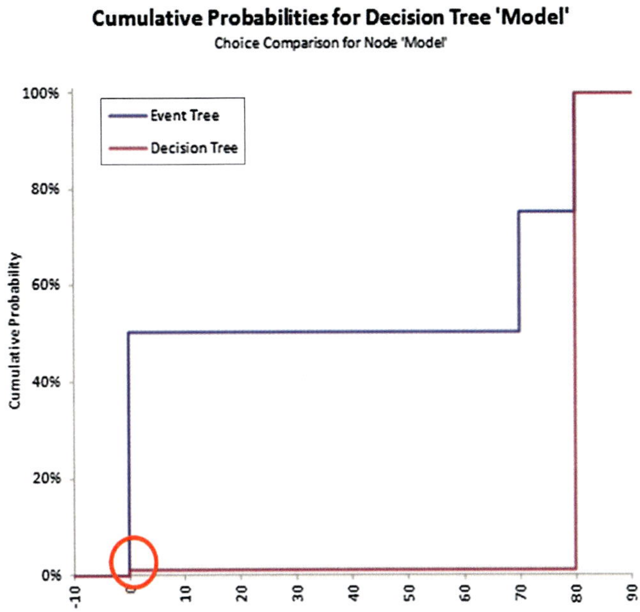


**Fig. 9.3** Cumulative risk profiles

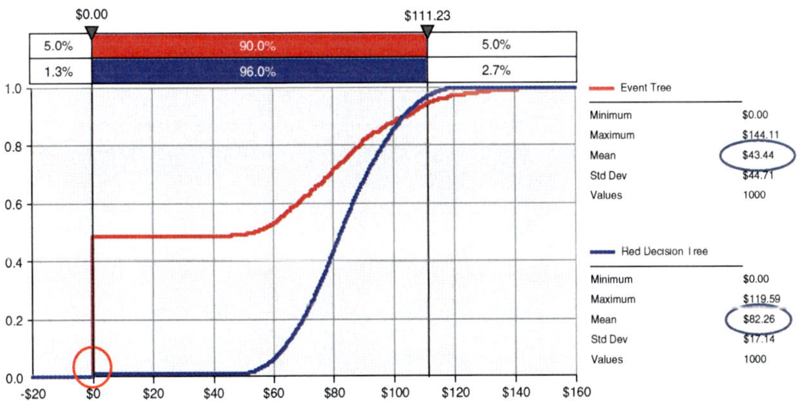


**Fig. 9.4** Comparison of Monte Carlo simulations

decisions. This chapter has reviewed some of the most common techniques: logic trees, influence diagrams, Bayesian networks, systems dynamics, and game theory. We provided technical descriptions; references; and the benefits and limitations of each technique.

We also compared an event tree and a decision tree model for a VBIED attack on two tunnels in an urban area. We found that the techniques both consider the same scenarios but resulted in very different expected consequences, risk profiles, and insights about defensive capabilities.

We recommend further comparative studies to obtain insights into the impact of modeling assumptions on risk assessments and risk management alternatives.

## References

1. Guikema, S. (2012). Modeling intelligent adversaries for terrorism risk assessment: Some necessary conditions for adversary models. *Risk Analysis, 32*(7), 1117–1121.
2. Willis, H. H. (2008). *Using risk analysis to inform intelligence analysis*. CREATE Research.
3. Krizan, L. (1999). *Intelligence essentials for everyone*. Washington: GPO.
4. Willis, H. H., LaTourrette, T., et al. (2007). *Terrorism risk modeling for intelligence analysis and infrastructure protection*. Santa Monica: RAND Corporation.
5. Willis, H. H., Morral, A. R., et al. (2005). *Estimating terrorism risk*. Santa Monica: RAND Corporation.
6. Ezell, B. C., Bennett, S. P., et al. (2010). Probabilistic risk analysis and terrorism risk. *Risk Analysis, 30*(4), 575–589.
7. Goodwin, D. (2005). *The military and negotiation: The role of the soldier-diplomat*. London: Taylor & Francis.
8. Parnell, G. S., Bresnick, T. A., Tani, S. N., & Johnson, E. R. (2013). *Handbook of decision analysis*. Hoboken: Wiley.
9. Kumamoto, H., & Henley, E. (1996). *Probabilistic risk assessment and management for engineers and scientists*. New York: IEEE Press.
10. Ezell, B., Haimes, Y., et al. (2001). Risks of cyber attack to water utility supervisory control and data acquisition systems. *Military Operations Research, 6*(2), 30–46.
11. Parnell, G. S., Smith, C. M., et al. (2009). Intelligent adversary risk analysis: A bioterrorism risk management model. *Risk Analysis, 30*(1), 32–48.
12. Pate-Cornell, M. (2007). The engineering risk analysis method and some applications. In W. Edwards, R. J. Miles, & D. von Winterfeldt (Eds.), *Advances in decision analysis*. Cambridge: Cambridge University Press.
13. Crowley, M. (2004). *Evaluating influence diagrams*. Department of Computer Science, University of British Columbia.
14. Sterman, J. D. (2000). *Business dynamics – system dynamics for a complex world*. New York: McGraw Hill Higher Education.
15. Clemen, R. T., & Winkler, R. L. (1999). Combining probability distributions from experts in risk analysis. *Risk Analysis, 19*, 187–203.
16. Lindley, D. V. (1983). Reconciliation of probability distributions. *Operations Research, 31*, 866–880.
17. Lindley, D. V. (1985). Reconciliation of discrete probability distributions. In J. M. Bernardo, M. H. DeGroot, D. V. Lindley, & A. F. M. Smith (Eds.), *Bayesian statistics 2* (pp. 375–390). Amsterdam: North-Holland.
18. Clemen, R. T., & Winkler, R. L. (1990). Unanimity and compromise among probability forecasters. *Management Science, 36*, 767–779.
19. Winkler, R. L. (1981). Combining probability distributions from dependent information sources. *Management Science, 27*(4), 479–488.
20. Mendel, M. B., & Sheridan, T. B. (1989). Filtering information from human experts. *IEEE Transactions on Systems, Man, and Cybernetics, 36*, 6–16.

21. O'Hagan, A. (2003). HSSS model criticism (with discussion). In P. J. Green, N. L. Hjort, & S. T. Richardson (Eds.), *Highly structured stochastic systems* (pp. 423–453). Oxford: Oxford University Press.
22. Hurwitz, E., & Marwala, T. (2007). Multi-agent modeling of interaction-based card games. In *Proceedings of the 3rd international North American conference on intelligent games and simulation* (pp. 23–28), University of Florida, USA.
23. Merrick, J. J., & Parnell, G. S. (2011). A comparative analysis of PRA and intelligent adversary methods for counterterrorism risk management. *Risk Analysis, 31*(9), 1488–1510.

# Chapter 10
# The Landscape of Maritime Piracy and the Limits of Statistical Prediction

**Philip Vos Fellman, Dinorah Frutos, Thanarerk "Nathan" Thanakijsombat, Pard Teekasap, and Britten Schear**

## Introduction

In 2009, maritime piracy reached its highest level since the International Maritime Bureau's (IMB's) Piracy Reporting Center (PRC) began tracking piracy incidents in 1992, surpassing levels from 2008, which was the previous record year. Of the 49 successful hijackings in 2008, 42 occurred off the coast of Somalia, including the capture of an oil supertanker, the Sirius Star. In late, 2008 and early 2009, we began a 2-year study of maritime piracy patterns, using advanced tools of non-linear dynamical systems modeling and data visualization based upon a database of 11 years of IMB Data. In 2011, Britten Schear developed a qualitative, historical analysis of the Somali pirate problem which serves as the appendix to this article, illustrating how qualitative research remains an important aspect in combating maritime piracy.

P.V. Fellman (✉)
American Military University and InterPort Police, Charles Town, WV, USA
e-mail: shirogitsune99@yahoo.com

D. Frutos
School of Business, Southern New Hampshire University,
2500 N River Rd, Hooksett, NH 03106, USA

T.N. Thanakijsombat
Southern New Hampshire University, 2500 N River Rd, Hooksett, NH 03106, USA

P. Teekasap
Rattanakosin College for Sustainable Energy and Environment, Rajamangala University of Technology, 39 Moo 1, Rangsit-Nakhonnayok Road, Thanyaburi, Pathum Thani 12110, Thailand

B. Schear
American Public University System, N George St, Charles Town, WV 25414, USA



P.V. Fellman et al. (eds.), *Conflict and Complexity*, Understanding Complex Systems,
DOI 10.1007/978-1-4939-1705-1_10

With respect to the quantitative research, our initial guide as to what we might expect to find from the IMB data was drawn from the research of Aaron Clauset and Kristian Skrede Gleditsch at the Santa Fe Institute on land-based terrorist organizations [1]. In their study, Clauset and Gleditsch found a pattern of "learning" in terrorist organizations characterized by a steady increase in the frequency of attacks, with a definable learning curve, although no learning curve effect was observed with respect to the magnitude of attacks. The increase in overall damage caused by terrorist attacks was simply the result of a larger cumulative number of attacks.

We hypothesized that given the trends of increasing maritime piracy attacks, there might or might not be a learning curve effect with respect to the number of attacks and that it would be reasonable to expect a learning curve effect with respect to the size of maritime piracy attacks. Particularly, when following piracy events in the news, it appeared in 2009 and early 2010 that each major attack was increasing in the level of the scale of targets. After a thorough analysis of the data, this trend was determined to be an artifact of news reporting. However, in confirming our null hypothesis, we were able to draw a variety of useful conclusions about maritime piracy and to draw a series of useful conclusions about the nature of this activity. After the qualitative study, we have also been able to place the quantitative results in context, hopefully contributing to the successful diminution of Somali piracy.

## The Methodology and Basic Results of the Clauset and Gleditsch Study

Clauset and Gleditsch studied 3,143 fatal terrorist attacks between 1968 and 2008 from 381 terrorist groups. Their findings indicated that "the frequency of a group's attacks accelerates along a universal trajectory, in which the time between attacks decreases according to a power law in the group's total experience; in contrast, attack severity is independent of organizational experience and organizational size." They further argued that "the frequency of a group's attacks accelerates along a universal trajectory, in which the time between attacks decreases according to a power law in the group's total experience; in contrast, attack severity is independent of organizational experience and organizational size" [1].

Their study is particularly important to those modeling terrorism, especially those using open source data as they indicate a more fruitful way of analyzing the usually sparse or sparsely detailed data on terrorist events:

> High quality empirical data on terrorist groups, such as their recruitment, fundraising, decision making, and organizational structure, are scarce, and the available sources are not typically amenable to scientific analysis. However, good quality data on the frequency and severity of their attacks do exist, and their systematic analysis can shed new light on which facets of terrorist group behavior are predictable and which are inherently contingent. Each record in our worldwide database of 3,143 fatal attacks includes its calendar date $t$, its severity $x$, and the name of the associated organization if known.
>
> For each group we quantify the changes in the frequency and severity of a group's attacks over its lifetime using a *development curve*. This curve maps a group's behavior onto a common quantitative scale and facilitates direct comparison of different groups at similar points in

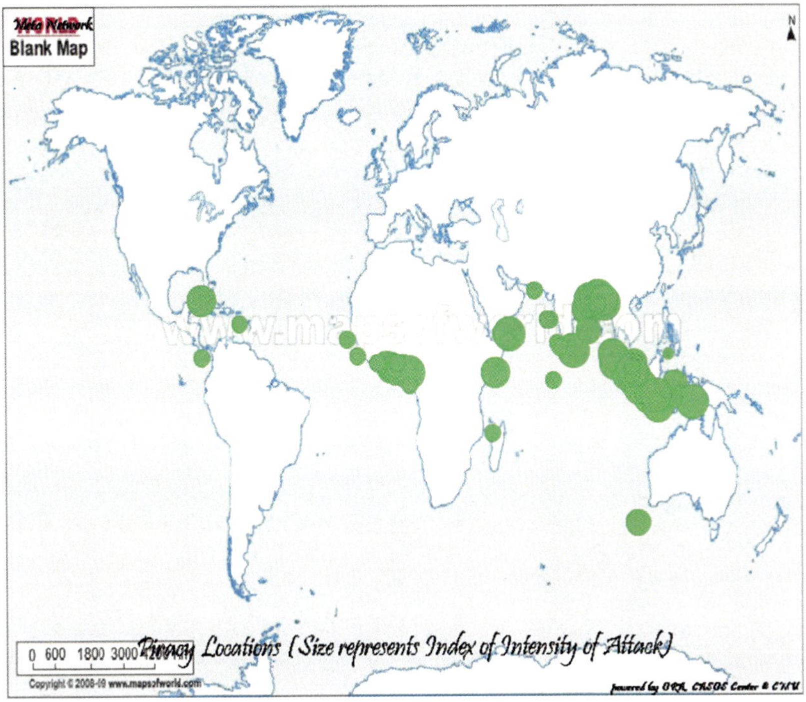


**Fig. 10.1** Network diagram representing Location and Intensity of Attack

> their histories. To construct this we plot the behavioral variable, such as the time (days) between consecutive attacks $\Delta t$ or the severity of an attack $x$, as a function of the groups maturity or experience k, indexed here by the cumulative number of fatal attacks (Fig. 10.1). Combining the developmental curves of many groups produces an aggregate picture of their behavioral dynamics, and allows us to extract the typical development trajectory of a terrorist group.

Clauset and Gleditsch are then able to model group experience curves with a relatively high degree of precision, noting that "the envelope or distribution of delays $p$ ($\Delta t$, $k$) can be characterized as a truncated log-normal distribution with constant variance $\sigma^2$ and a characteristic delay between attacks $\mu$ that decreases systematically with experience $k$." Mathematically, they represent this function as:

$$p\left(\log \Delta t, \log k\right) \propto \exp\left[\frac{-(\log \Delta t + \beta\left(\log k - u\right)^2}{2\sigma^2}\right]$$

where $\beta$ controls the trajectory of the distribution toward the natural cutoff at $t=1$ day. For small $k$, i.e., during a group's early development, this model predicts a mean delay between attacks that decays like a power law $\Delta t \approx \mu k^{-\beta}$; however, as $k$ increases, this trend is attenuated as the mean delay asymptotes to $\Delta t = 1$… Under this model, $\beta = 1$ would indicate a simple linear feedback between a group's attack rate and its experience. However, we find $\beta = 1.10 \pm 0.02$ indicating a faster-than-linear feedback between the accumulation of experience and the rate of future attacks.

This model successfully predicts that the distributions of normalized delays $\Delta tk^{\beta}$ will collapse onto a single log-normal distribution with parameters $\mu$ and $\sigma$. However, individual attacks cannot be considered fully independent ($p=0.00\pm0.03$) indicating that significant temporal or inter-group correlations may exist in the timing of a group's future attacks.

In contrast, Clauset and Gleditsch (similar to the results obtained from our data) did not find a correlation between learning and scope or intensity of damage:

In contrast, the severity of attack $x$ is independent of group experience, $k$ (Fig. 10.2c; $r=-0.024$, t test, $p=0.17$), as illustrated by the collapse of the severity distributions $p(x|k)$ onto a single invariant heavy-tailed distribution…

Clauset and Gleditsch go on to discuss a number of additional characteristics of the distribution as well as explanations for the results. Key to the interpretation of our data was the visualization of the distributions as depicted below:

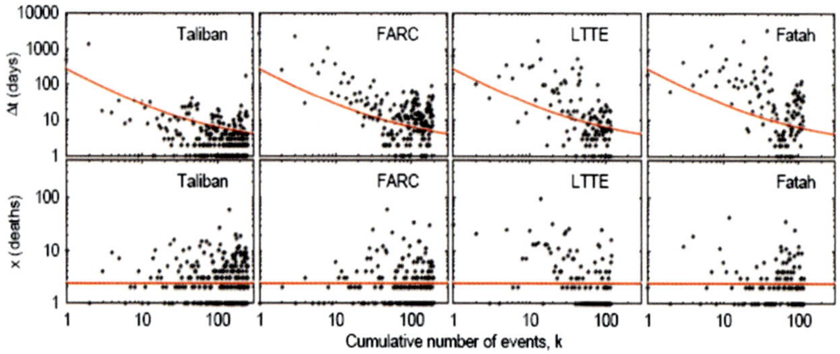


The charts above show individual development curves, giving the time until the next fatal event $\Delta t$ and the severity of the event $x$ as a function of individual group experience $k$ for the four most experienced groups (Taliban, FARC, Tamil Tigers, and Al Fatah) along with population level curves derived from analyzing all groups together. Similar results hold for less experienced groups

(Clauset and Gleditsch).

## Formatting Data in the Maritime Piracy Study and Quantitative Methodology

Maritime piracy data is supplied by the IMB of the International Maritime Organization.[1] However, this data is a very raw compilation and includes virtually any incident requiring the intervention of local law enforcement, including theft

[1] The International Maritime Bureau's Piracy Reporting Center (IMB-PRC), a private organization that monitors piracy incidents all over the world, defines "piracy" as "an act of boarding or attempting to board any ship with the intent to commit theft or any other crime and with the intent or capability to use force in the furtherance of that act" [2].

and burglary at berth, which cannot properly be considered piracy in any case. One of our first tasks was to properly format the data, eliminating those reports which were not, properly speaking, piracy, and then, because we were working with inherently different data than Clauset and Gleditsch, it was necessary to construct a scale for the incidents in order to categorize piracy incidents. In distinction to land-based terrorist attacks, piracy attacks have a profoundly property-based component, so that simply measuring fatal casualties would not be a sufficiently detailed method of categorizing maritime piracy attacks. The scale which we used to organize our data is detailed in the following chart:

Given the paucity of organizational data, we were forced to analyze piracy data by country. Because there were commonalities within regions, we also tabulated geographical data by region. For each location, we quantify the changes in the frequency and severity of the attacks over a 10-year period (1998–2008) using a development curve. The locations we have selected based on the number of attacks are: Indonesia, Malacca Straits, Somalia, and Nigeria. The developmental curve maps the "behavior" of the attacks in those locations from which we can infer the behavior of the pirate groups in the particular location. They also provide a common quantitative scale that facilitates the direct comparison of different locations at similar points in their life histories.

To construct this curve, we plot the behavioral variable, such as the time (in days) between consecutive attacks or $\Delta t$ or the severity of an attack $x$, as a function of the experience of the piracy group in the selected location $k$, indexed by the cumulative number of attempted attacks on ships underway. Combining the developmental curves of the selected locations produces an aggregate picture of the behavioral dynamics of each location, and allows us to extract the typical developmental trajectory of a pirate group.

## Results of the Study

The dynamics revealed by this study suggest several interesting features of maritime piracy and parallel many of the results discovered by Clauset et al., for land-based terrorist organizations. With respect to the frequency and severity of attacks over time, these results are independent of many commonly studied social and political factors, and suggest that there may be "new" fundamental scientific laws (other than the social and cultural trends typically studied in this field) underlying the dynamics of both terrorist and piracy attacks. As recent studies on combat force structure, resilience, and breakthrough levels suggest [3], emergent processes and fractal structure may play important roles in processes previously thought to be governed by simple power laws.

If one can replace the data smoothing typically implied by linear and multiple regression analysis with a more complete explanation of dynamical processes, and can supplement cultural and political analysis with a more complete picture of process dynamics (including invariants), then one may achieve a better understanding of both land-based terrorism and maritime-based piracy (and terrorism) than is currently available through single-mode analysis. In particular, a richer set of analytical tools may be helpful in predicting terrorist or piracy behavior when dealing with unknown, unidentified, or unidentifiable groups and may help to develop a new approach for dealing with particularly difficult to categorize or difficult to define terrorist and piracy events.

## Data Visualization

In the present study, we began by applying Social Network Analysis to our data, using Kathleen Carley's ORA software available from CASOS at Carnegie Mellon University. SNA Data visualization was employed in order to elucidate patterns of behavior among different maritime piracy groups. Network data in its purest from consists of a square array of measurements which focus on the actors and their relational properties (particularly network "nodes" and "edges"). The rows of the array are the cases, or subjects, or observations. The columns of the array are the same set of cases, subjects, or observations. Each cell of the array describes a relationship between the actors. By comparing the rows of an array, we can see which actors are similar to the other actors with whom they are connected by the first actors' choice (as in a directed graph). By looking at the columns, we can see the relationship between actors on the basis of who is chosen by other actors. These are useful ways to look at the data because they help to show which actors have similar positions in the network. A major emphasis of network analysis is seeing how the whole pattern of individual choices gives rise to more holistic, or emergent patterns [4]. Figure 10.2 shows the network diagram of Locations and Intensity of Attacks (based on the Index on Table 10.1) for the year 2003. The network diagram includes all the attacks that occurred in the year and all the locations in which attacks took place.

Figures 10.3 and 10.4 represent the location and type of weapon used in the attack and the location and number of pirates involved in the attack, respectively. As compared to the raw data from the IMB, these network diagrams provide an improved visualization of the 2003 piracy attacks since the size of the nodes gives us an indication of the intensity of the attack, or the number of pirates involved in each attack for each location. We also applied ORA software to data visualization in a two node network for time and location of maritime piracy attacks. Representative visualizations for 1998, 2003, and 2008 are provided, illustrating geographical shifts in the location of the most numerous and severe piracy attacks (Figs. 10.5, 10.6, and 10.7).

## Conclusion: The Mathematical Landscape of Maritime Piracy

Unlike Clauset and Gleditsch, we did not find the anticipated learning curve, which we had initially posited might lie along the severity axis rather than the frequency axis. Nor did we find a learning curve along the frequency axis, despite the overall

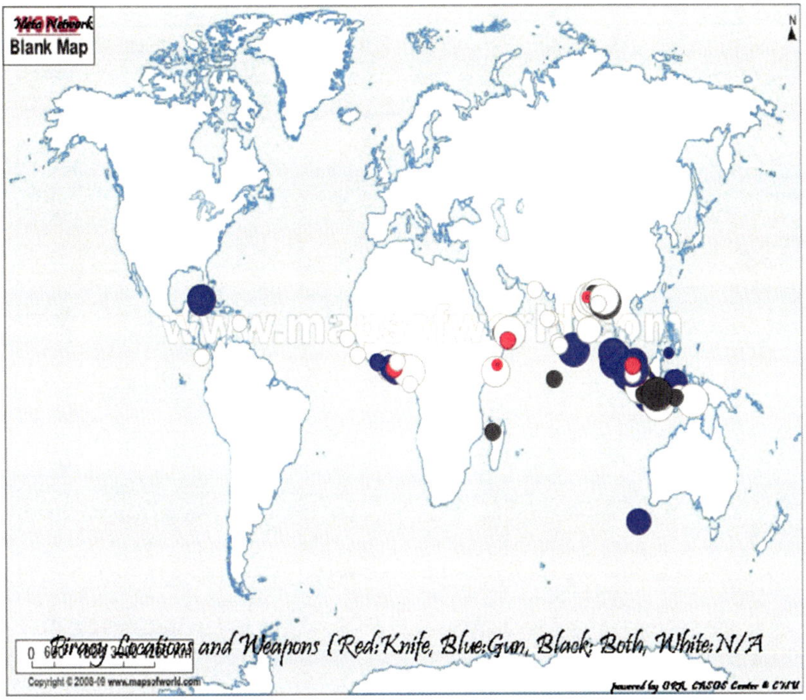


**Fig. 10.2** Network Diagram representing Location and the type of weapon used in the attacks

**Table 10.1** Description of attack

| Description | Index |
|---|---|
| No goods stolen/no injuries to crew | 1 |
| Stolen goods (any)/knives–guns–RPG | 2 |
| Stolen goods (any)/injured crew | 3 |
| Stolen goods (any)/crew killed or missing | 4 |
| Hijacked/crew recovered (or crew hostage for ransom) | 5 |
| Hijacked/crew killed/missing and ship missing | 6 |
| Ship rammed with explosives | 7 |

increase in maritime piracy incidents. One possible explanation to which we give substantial weight is that unlike terrorist networks [5, 6] maritime piracy organizations tend to be staffed by largely "temporary" workers rather than being characterized by strongly cohesive or adhesive cellular organizations. Recent data from Somalia suggests that the actual pirates undertaking the attacks are frequently

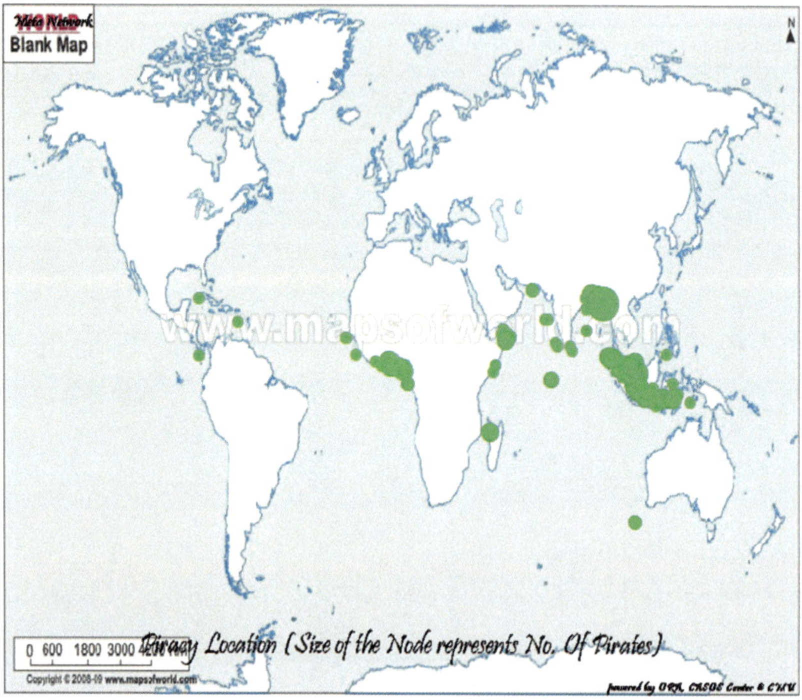


**Fig. 10.3** Network Diagram representing location and the number of pirates involved in the attacks

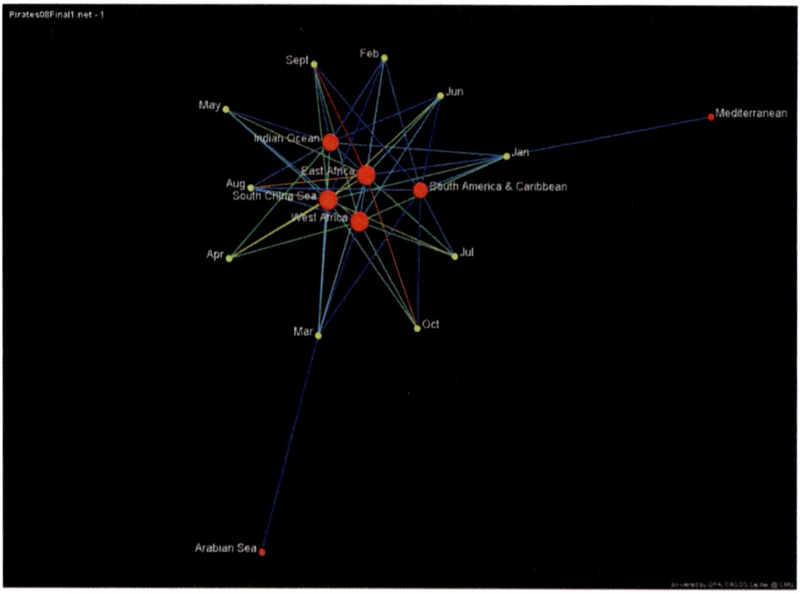


**Fig. 10.4** Two node network—time and location for maritime piracy attacks, 1998

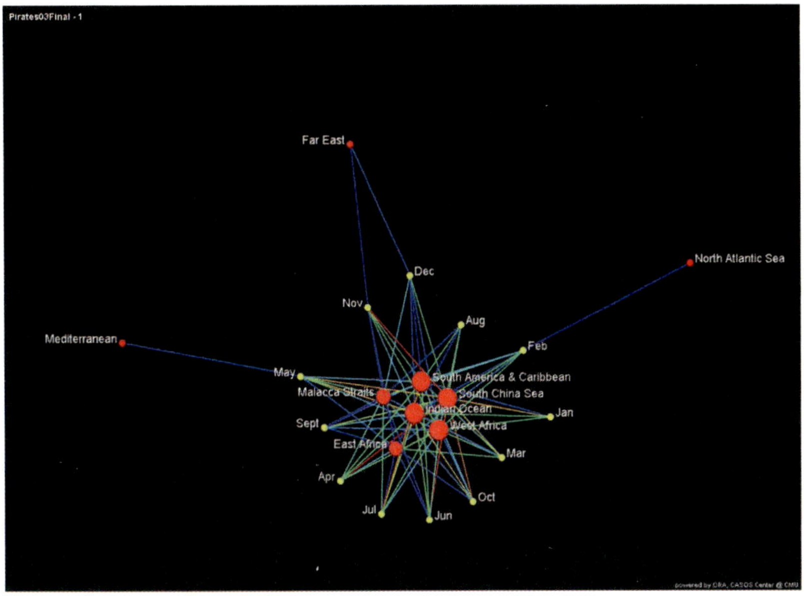


**Fig. 10.5** Two node network—time and location for maritime piracy attacks, 2003

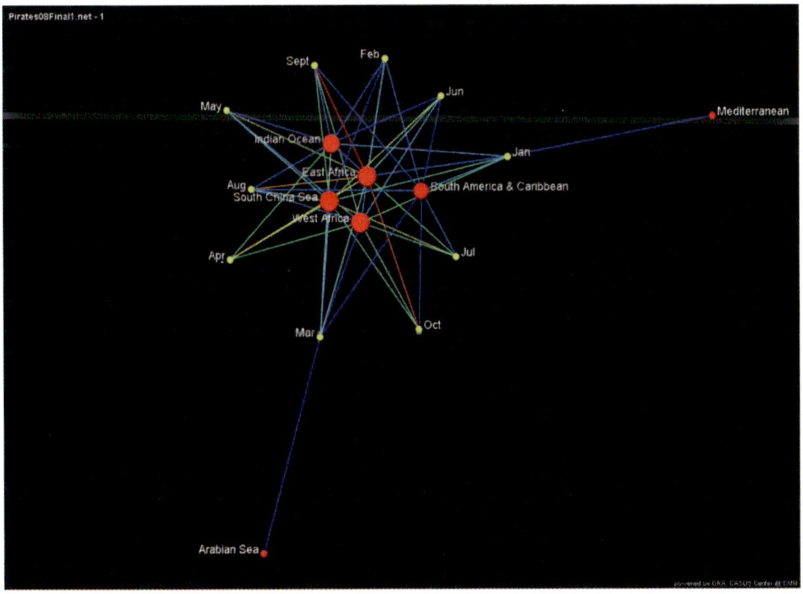


**Fig. 10.6** Two node network—time and location for maritime piracy attacks, 2008

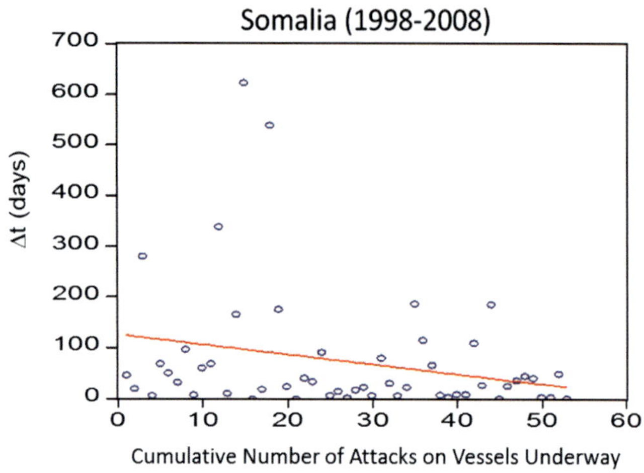


**Fig. 10.7** Somalia

forced to flee after 1–5 attacks [7]. With additional data there might be some trends which could be modeled regarding the accretion of cash and the planning of operations with respect to particular warlords or organized crime groups, however, this is beyond the scope of the present study. The following charts show the distributions for each of the areas which we studied.

In closing, we should like to emphasize that while maritime piracy is a serious and growing problem and while being attacked, detailed, robbed, or killed is every bit as terrifying as experiencing any conventional terrorist attack, maritime piracy has a different mathematical landscape than the kind of traditional, land-based terrorist groups studied by Clauset and Gleditsch. Similarly, the criminal organizations which are responsible for the bulk of maritime piracy appear to have a rather different structure than that typically associated with terrorist organizations and terrorist networks [5]. We don't, in this case, mean to minimize either the seriousness or the intensity of the piracy threat, but simply should like to demonstrate that it has a rather different structure than land-based terrorism and may require substantially different approaches to successfully reduce the level of piracy incidents [7].

## Historical Background

When Thomas Jefferson and John Adams traveled to London in 1786 to meet with the ambassador from Tripoli, the objective was to negotiate an end to the pirate attacks on American merchant ships traversing the coasts of North and West Africa.

The Barbary pirates, as they were collectively known, were outlaw sailors from Morocco, Algiers, Tunis, and Tripoli. Their often violent attacks on American ships were sponsored in part by the Arab leaders of their home nations, and the recently independent US had lost the protection of the British and French navies which had previously served to prevent such attacks. Jefferson and Adams asked Tripoli Ambassador Sidi Haji Abdrahaman why it was that his countrymen were targeting American vessels, unprovoked, and the answer that they received could easily be drawn from a modern news story: The Americans are infidels, and as such, Muslims quite simply have the right to rob their ships, and proof of this is found in the Koran [8]. The memory of the Crusades was fresh and easily reignited in the minds of Muslims, and the Barbary pirates thought of themselves as nothing less than sacred warriors, defending Islam and securing restitution for their people ([9], p. 32).

The common practice of the early 1800s for dealing with the Barbary pirates was for international ships to pay off the pirates in exchange for not being taken hostage. The practice of paying "tributes"—the more dignified term for "bribes"—was widespread, and the Spanish advised the young American government that money was the easiest means of gaining safe passage for their vessels. Thomas Jefferson was opposed. Upon being elected president, Jefferson ended the practice of tributes, instead sending the US Navy to combat the pirates at their home-ports in two Barbary Wars. By 1815, as the US had defeated the navy of Tripoli and even brought the battle ashore, piracy against US ships sailing around Africa was effectively ended.

The recent resurgence of piracy off the Horn of Africa is as alarming as it is surprising after a nearly 200-year lull. It is not uncommon for adversarial nations to face each other in conflict repeatedly over long stretches of time, but it is unusual when old tactics are not only brought back to use, but prove to be successful. Somali pirates have established a lucrative business of boarding much larger shipping vessels from small, basic skiffs and have succeeded in securing large ransoms in exchange for hostages. Modern history between the US and Somalia (the epicenter of modern piracy) is marked by mistrust and violent conflict. American overt and covert operations have failed tragically, and the threat posed by Somali pirates to US and allied ships is increasing. What is more, the power vacuum left by the disintegration of Somalia's government in 1991 is in grave danger of being filled by Islamic extremists who have access to boats, American and Soviet-era weapons, and funding via the ransoms paid to the pirates. The US will likely increase covert operations in Somalia, as has been the case in adversarial nations since 9/11/01; further, the US is not expected to risk an overt military intervention following the failed "Black Hawk Down" operation in 1993. By examining past failures, and the rare successes that have marked American dealings with Somalia, it becomes clear that both the U.S. Intelligence Community and Congress must develop new tactics to confront the threat now posed by Somali piracy and extremism. Taking a cue from Thomas Jefferson, the way forward may be through a confrontational, albeit clandestine approach.

## Somalia's Problems

Somalis have a devotion to historical memory that is comparatively unmatched in the Muslim world. Based on a clan structure that stretches back through many generations, Somali society is strictly adherent to a sort of cast system under which relationships, treaties, and even elections are dictated by clan alliance. Even the customary Somali greeting, "Yaa Tahay?" is translated as "What clan are you?" ([10], p. 27). General Mohamed Siad Barre attempted to get rid of clan-defined loyalties when he came to power in a coup in 1969, trying to force Somali loyalty only to his government. Ironically, not only was this practice unsuccessful, but when Barre and his failed administration were being chased out of Mogadishu in 1991, Barre was forced to turn to his own clan alliances for protection ([10], p. 27). Barre did not understand the sway that clan identity holds for Somalis, evidenced by the fact that many Somali children can, from a very young age, recite their clan history back as far as it stretches, predating modern borders and governments.

Just as closely as Somalia is linked to clan heritage, it is also linked to the repeated tragedy of natural disasters. The 1990s and 2000s have been marked by ruthless droughts that have deprived Somali farmers and herders of the already paltry rainfall that they receive. Fishing used to be the one guaranteed industry, as Somalia has the longest coastline on the continent, but the 2004 tsunami in Indonesia reverberated to the Somali shores, wiping out 75 % of the local's fishing gear, including boats ([10], p. 40). Somalis reported that as the waves took over the fishing villages, barrels of toxic waste dumped by foreign merchant vessels were washed in from the sea bed, sickening the populace. One major source of income for Somalis was quickly wiped out.

Throughout these natural disasters, Somalis were sadly assured that their government would not provide a means of stability. When Barre was removed from power, the notorious warlord Mohamed Farah Aideed took his place. Aideed and his clan-backed militia began killing the international aid workers that had arrived to restore order and bring food and medical supplies. Specifically, Aideed's Somalia National Alliance killed 24 Pakistani soldiers who were part of the UN relief program ([11], p. 9). When US Special Forces returned to track down Aideed, they were caught in a vicious fire fight with men armed with Rocket Propelled Grenades and AK-47s. Eighteen Army Rangers died on October 3, 1993, as the Somali militias successfully shot two Black Hawk helicopters (one to the ground) and trapped the Rangers in a maze-like street battle. As Dr. Richard W. Stewart noted in his report, "In a country where the United States, perhaps naively, expected some measure of gratitude for its help, its forces received increasing hostility as they became more deeply embroiled into trying to establish a stable government" ([11], p. 14). This pre-9/11, land-based action against an unpredictable, non-uniformed enemy marked one of the last times before the war in Afghanistan that the US chose to insert visible troops into a country in chaos, and it firmly shook the belief that greater firepower and technical sophistication could overwhelm a bare-bones insurgency.

In this case, the US forces overlooked the importance to the Somalis of demonstrating clan power. Despite the food aid, and the millions of US dollars invested, the attraction of having waged and won a battle against the much stronger Americans was greater than even personal survival. This is a theme that has haunted US operations in Somalia since, and indeed still finds its way into current discussions on Somalia in which the idea of providing for an "alternative" to militarism and violence for the Somalis is heralded as the best way to ensure for better relations between the US and Somalia. While the Somalis desperately need jobs and a stable government, the ideology of clan power cannot be overlooked. The Transitional Federal Government formed in 2004 is the temporary, internationally recognized government structure that now governs Somalia; elections are set to be held in August 2012 to establish a permanent government, and Somalia will likely vote along clan lines. Only by recognizing the importance of clan loyalty can the US hope to avoid disastrous interventions such as the 1993 operation.

## Piracy

Much like the Barbary pirates of the eighteenth and nineteenth centuries, Somali pirates today aim to capture humans, not goods. Likewise, just as was advised to Thomas Jefferson, most shipping companies and other smaller vessels prefer to simply pay their captors rather than escalate the standoff by openly arming their ships. For this reason, from 2007 to 2010 Somali pirates were able to inflict between 7 and 12 billion dollars of damage on the international economy which was achieved by attacking 450 ships and taking 2,400 hostages [12]. Most of the time, the pirates get their ransoms paid by hijacking a vessel, sailing it back into a Somali harbor, and holding the hostages on land until their company, families, or home governments pay the decided amount. The prices are not fixed, as the Somali pirates are well aware that American and European hostages will bring a much higher ransom than will sailors from third world countries. This was evidenced by the $1.5 million that was turned down by the Somalis holding American Jessica Buchanan and Dane Poul Hagen Thisted hostage in January 2012, as they were sure that they could negotiate a higher pay out.

Along with adapting to international economics, Somali pirates have also evolved rapidly to apply new tactics, and just within the last 5 years, piracy off the Horn of Africa has become exponentially more aggressive. Captain Richard Phillips, a merchant mariner who piloted the shipping container *Maersk Alabama* in April 2009 parallel to Somalia's coast, was taken captive when he exchanged himself for his crew and the ship. In the days leading up to the pirate attack on the *Alabama*, Phillips was taunted by pirate boats who seemed to be testing his reaction to their approaches, learning which evasive maneuvers would be taken, and assessing weaknesses. On the open water, tactics are not easily hidden, which is at least one cover that can be practiced by land-based troops. Phillips was shocked when a pirate skiff finally caught up to his ship, and he noticed a ladder being propped against the side of the *Alabama* so that the pirates could easily climb aboard. The ladder seemed to be fitted exactly to latch onto the metal bars around the deck; previous pirate attacks that Phillips had heard of involved grappling hooks and ropes, which slowed the process of a pirate climbing on board ([13], p. 111). What is more, Phillip's captors spoke English and were easily able to communicate their demands once on board. In a country where just over one third of the population is literate, fluency in several languages is not uncommon for Somali pirates.

One further strange adaptation that Somali pirates have made is the use of psychological tricks to diminish the resolve or will of a hostage. Capt. Phillips notes that the pirates that took him hostage told him after 2 days at sea that they were working with the US Navy, and that his detention was all part of a drill. Despite the fact that Phillips [13, p. 225] had been repeatedly beaten, and that the pirates had first come aboard his boat shooting at him, he claims to have started to doubt the reality of his situation, as well as the intentions of the navy ship that was negotiating his release. This is an ironic twist, as part of the humanitarian relief that the US sent to Somalia in the 1990s was a Psychological Operations team that set up a local newspaper and radio station that propagandized against the armed militias ([11], p. 7). While there is an obvious difference between both the intentions and tactics of the two, it does seem as if the Somali pirates have turned some of the intelligence tactics of the US back against their captives.

Pirates have also begun to target ships not only for the likelihood of being able to hijack them, but also for the value of the cargo to the owners. Again, the actual cargo is not the goal, but a large pay out from those to whom the cargo is important. Larger ships carry more cargo, but they also sail in deeper waters. This means that little skiffs used for attacks will not make it out the two or three hundred miles where the shipping containers sail. The solution for pirates has been to hijack and steal medium sized personal boats that can make the further distance and can carry the smaller skiffs; these are referred to as "mother ships," and have greatly expanded the pirates' reach into the Gulf of Aden. Noah Shachtman of Wired Magazine interviewed a Somali pirate in the summer after Capt. Phillips was taken hostage, and the pirate explained, "Often we know about a ship's cargo, owners and port of origin before we even board it. That way we can price our demands based on its load. For those with very valuable cargo on board, then we contact the media and publicize the capture and put pressure on the companies to negotiate for its release" (2009). An advanced system has even been put in place once the hostages are taken, in which a third party outside of Somalia negotiates the ransom and organizes the actual money transfer, which often involves airlifting bags of cash onto the deck of the captured ship ([12], p. 7).

The reasons that Somalis give for the rapid expansion of piracy in their waters is that over fishing and pollution by foreign vessels has depleted the fish populations in their waters, and thus destroyed their livelihood. The pirates think of the hijackings and hostage takings as reparations for the international community's thievery and desecration of Somali waters. This claim rings hollow, however, as Somali pirates do not target only fishing vessels. Shipping containers that held food being delivered by the World Food Programme were routinely targeted until 2007, when armed French ships began escorting them ([10], p. 34). Pirates were holding up delivery of the very relief that was meant for their countrymen. What is more, Capt. Phillips [13, p. 65] noted that on each of his trips along the Horn of Africa, he had noticed many schools of fish which the Somalis were not going after. While the Somalis have certainly suffered from drought and a tsunami, the reason for pursuing piracy may be that it is simply easier and more cost-effective than fishing.

## The Islamist Threat

Somalia has many of the main ingredients necessary for the spread of radical Islam throughout its society. The northwestern part of the country referred to as Somaliland claims independence but is not internationally recognized as separate from Somalia. The northeast is called Puntland, and exists, like the southern part of Somalia, as a semi-autonomous region only vaguely under government control. As with all other parts of Somali society, the borders are more clearly drawn by clan alliance than by geography or political enforcement. Despite the need to ally themselves first with clan, however, Somalis do want stability and peace within their cities. It is this desire that has allowed the Islamic Courts Union to creep into power, and by extension, the al-Shabaab militant group that is aligned with al Qaeda. In his Congressional testimony, Dr. Peter Pham of the Institute for International and Public Affairs, stated that "...the now defunct Somali Democratic Republic creates the conditions for the advent of the Islamists of the [Islamic Courts Union] sort in the same way that the Taliban in Afghanistan arose out of that country's anarchy as a force of order amid factious leaders rapacious militias" [14].

This creates a difficult situation for the US, as any overt assistance that Somalia is seen to be offered by Americans can be twisted by Islamist ideology to appear to be the work of infidels attempting to coopt Muslim lands. By the time that Somalis—who are nearly all Sunni Muslim—realize that they have fallen into the trap laid by extremist Islamic sects of offering security by way of brutal domination, an Iran-like government could already be in place. With a radical Islamic government in control in Somalia, it is not difficult for the U.S. Intelligence Community to make the leap to worrying about the potential for suicide attacks on American ships, caring no longer for the ransom, but rather for the propaganda tool of having destroyed an American symbol of strength and commerce. The US avoidance of putting troops on the ground in Somalia would be forcibly altered by an outright Somali attack on a US ship. As Jonathan Stevenson points out in his article, "The fear is that if... Islamists, on the pattern of Hizbullah and Hamas, offer antidotes to social problems that secular entities cannot or will not supply, Islamism will take deeper root in Somalia. In this scenario, it could acquire a degree of legitimacy that could rival or even defeat that of clan identity and religious moderation and create more promising conditions for the establishment of a functioning republic – this time, an Islamic one" (2007, pp. 5–20).

## American Responses

The IMB, a division of the International Chamber of Commerce, is the organization that tracks reports of piracy and works as a dispatch office for distress calls from hijacked ships; the Combined Maritime Forces is a multinational coalition that is led by the US Navy (CENTCOM), under which the Combined Task Force 151 operates, shifting command between 20 coalition nations [12]. These organizations along with NATO and the navies of several countries now operate along the Horn of Africa as a protective force that is readily mobilized when a ship reports pirate activity. As the IMB states as its mission, "The main function of the PRC is twofold: (1) To be the single point of contact for ship Masters anywhere in the world who are under piratical or armed robbery attack. The information received from the Masters is immediately relayed to the local law enforcement agencies requesting assistance. (2) The information received from the ship Masters is immediately broadcast to all vessels in the Ocean region—thus highlighting the threat to a Master en route into the area of risk" [15]. Despite this, incidences of pirate attacks have increased, and the pirates themselves have become more daring in their attacks.

It seems now that the best deterrent to pirate attacks is the offensive method of rescuing hostages. The two most recent successes at overcoming the pirates are the case of the rescue of Jessica Buchanan and Poul Hagen Thisten from Galcayo, Somalia, and the operation to rescue Capt. Richard Phillips from his captured life boat at sea. These operations, though covert only in planning and execution—both cases were widely broadcast via the media—involved the use of US Special Operations forces, and specifically the Navy SEALs. Both incidents relied heavily on the expertise of covert teams, rather than overt displays of force. It may be for this reason that both operations ended in the successful rescue of the hostages, and they should serve as the guide for future US operations in Somalia.

Legislation is slowly catching up to the needs of the anti-piracy tactics around the Horn of Africa. President Obama signed Executive Order 13536 in April 2010 stating that the threat posed by both violence in Somalia and the threats at sea necessitated a robust response not only to the pirates, but to the nationals that provide them support [12]. A further recommendation of the Subcommittee on Coast Guard and Maritime Transportation is that the "…US and international partners…interdict pirate vessels and intervene in pirate attacks. The Plan also supports disrupting pirate bases in Somalia and depriving pirates of illicit revenues" (2011). This demonstrates that Congress is moving towards a more aggressive stance on piracy, and indicates that pirates should be tracked back onto their own soil, and that their operations should be shut down at the source; this vague language hints at the use of covert actions to do so.

The European Union has, in fact, already begun the practice of attacking the pirates and their bases before they set sail. On May 15, 2012, an EU naval force helicopter fired upon pirate skiffs that were dry-docked onshore in a city north of Mogadishu, disabling several boats and inflicting no human casualties [16]. This action was in line with a new EU mandate that encourages a tougher stance against pirates, including their bases of operations on land, but prohibits any troops on the ground; this is assuredly a lesson learned from the catastrophic fate of the American Rangers in 1993.

## Conclusion

Maintaining a strong defense against piracy is important, and this means working to defeat the low-tech methods of pirates. At present there are companies working on radars that will pick up very small craft, down to the size of a dingy, solving one of the problems large shipping containers have in spotting incoming pirate boats [17]. Stronger coalitions must be built with African countries willing to prosecute Somali pirates captured at sea. And American companies need to be protected against the liability that can be incurred for having paid a ransom, which is, in the eyes of US law, supporting terrorism. The future of security and peace in Somalia and in the waters off the Horn of Africa lie, however, in an undercover war.

Covert actions in Somalia and in Somali waters are difficult to uncover, as they are still ongoing. However, the threat posed by Somali piracy and the possibility of the Islamic extremism taking hold in Somalia point towards an expanded use of covert operations that will disrupt adversarial elements of Somali society while preserving an unnoticed American presence. There are rumors that the US may have created a secret prison in Mogadishu for holding pirate suspects, and there are possibly aerial and submarine reconnaissance units gathering intelligence throughout the Gulf of Aden. In the days leading up to the EU attack on the pirate base, locals reported seeing spy planes flying overhead and unfamiliar boats sailing offshore that did not come onto land [16]. These are steps towards a more effective combating of piracy and extremism. While it may take generations to undo the damage that decades of war and clan rivalry have exacted upon the Somalis, the US can best protect its assets and the security of the region by offensively and covertly breaking down Somalia's ruthless and corrupt elements.

## Appendix I: Geographic Charts (Figs. 10.8, 10.9, 10.10, and 10.11)

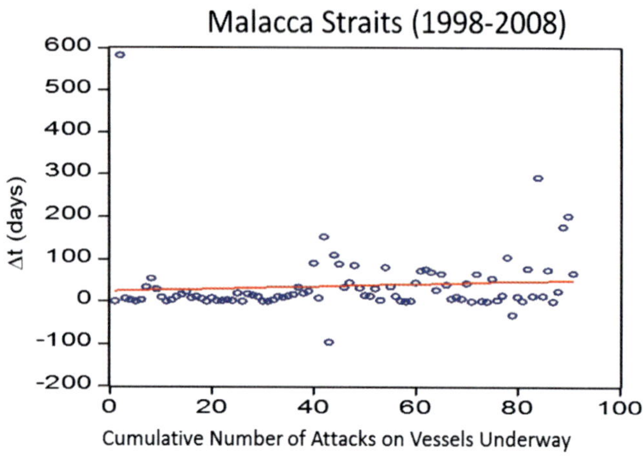


**Fig. 10.8** Malacca Straits

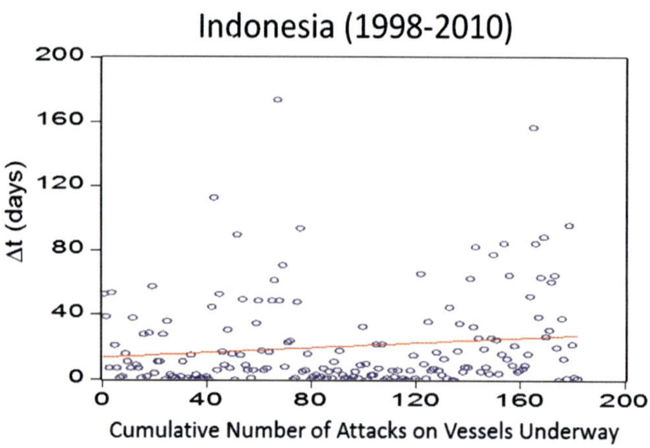


**Fig. 10.9** Indonesia

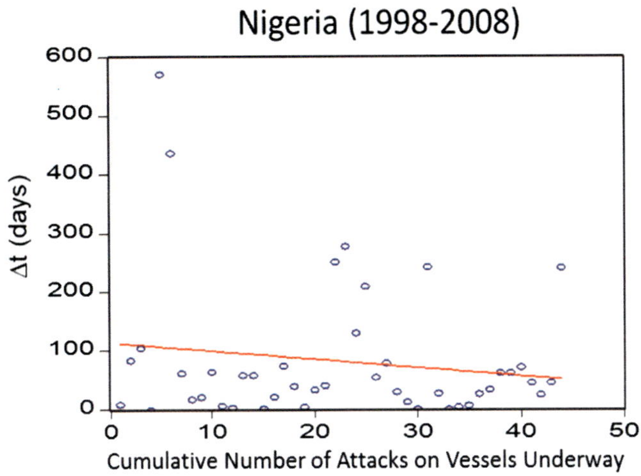


Fig. 10.10 Nigeria

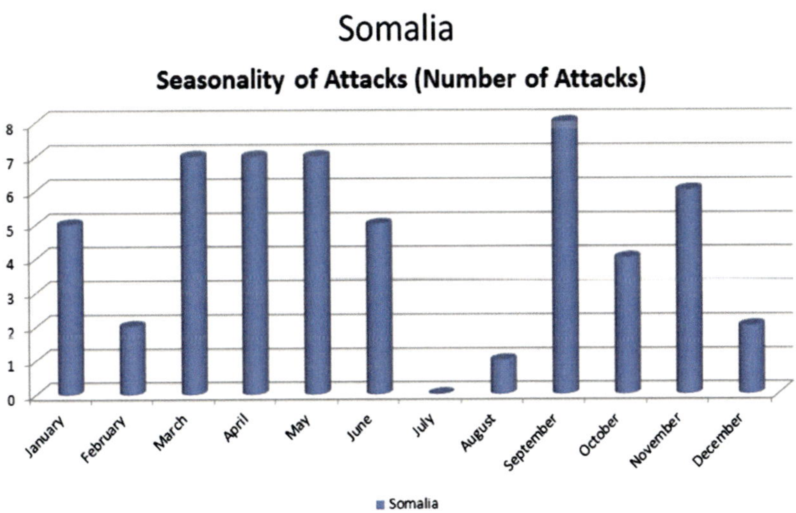


Fig. 10.11 Somalia

## Appendix II: Seasonality of Attacks by Region (Figs. 10.12, 10.13, 10.14)

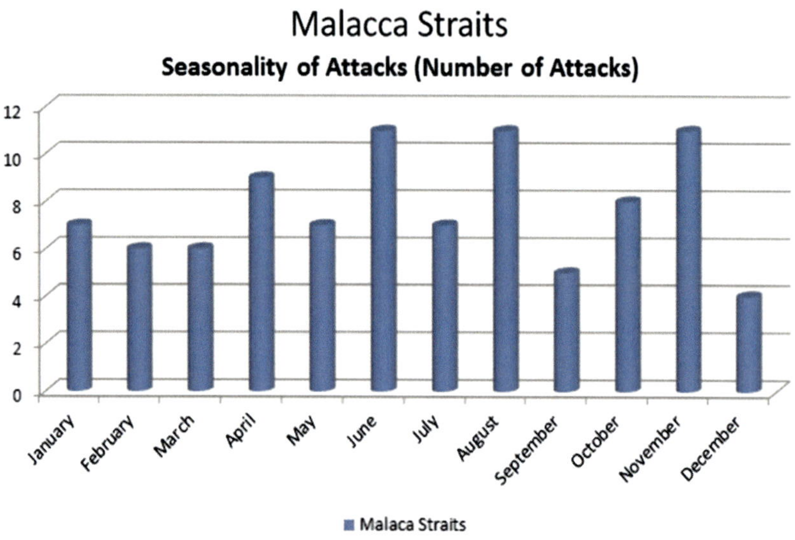


**Fig. 10.12** Malacca Straits

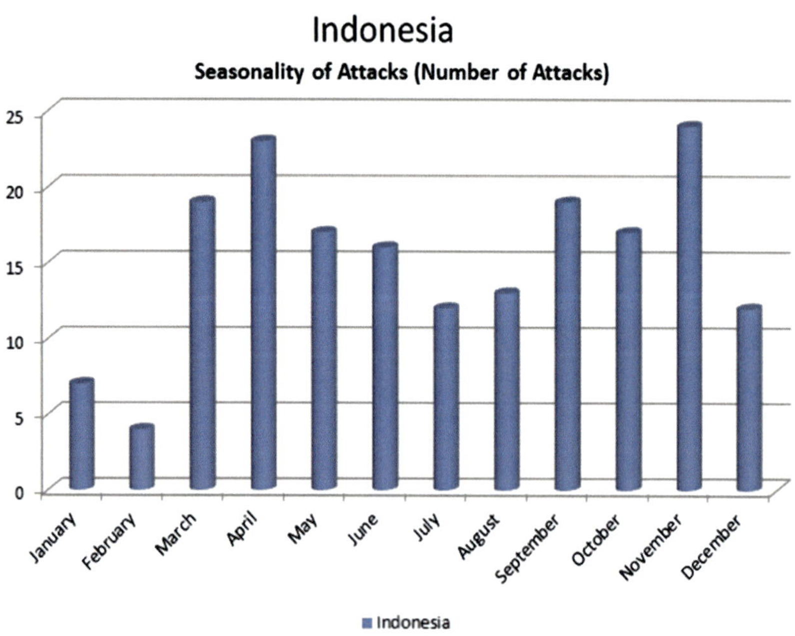


**Fig. 10.13** Indonesia

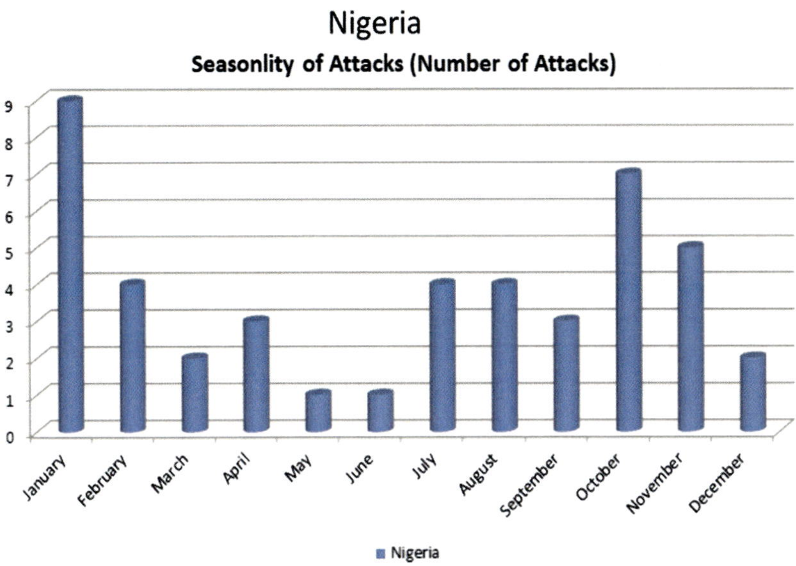


**Fig. 10.14** Nigeria

## Appendix III: Understanding the Somali Threat[2]

> Once you have a ship, it's a win–win situation. We attack many ships every day, but only a few are ever profitable. No one will come to the rescue of a third-world ship with an Indian or African crew, so we release them immediately. But if the ship is from Western country or with valuable cargo like oil, weapons or … then it's like winning a lottery jackpot. We begin asking a high price and then go down until we agree on a price.
>
> Somali pirate, interviewed July 28, 2009
> (Shachtman, Wired Magazine)

[2] By Britten Schear, American Public University System, School of Global and Security Studies.

## References


1. Clauset, A., & Gleditsch, K. S. (2009). *The developmental dynamics of terrorist organizations*. Retrieved from arXiv:0906.3287v1 [physics.soc-ph]: http://arxiv.org/abs/0906.3287
2. Abhyankar, J. (2001). Piracy and ship robbery: A growing menace. In H. Ahmad & A. Ogawa (Eds.), *Combating piracy* (p. 11). Tokyo: The Okazaki Institute.
3. Moffat, J., & Passman, M. (2004). Metamodels and emergent behaviour in models of conflict. *Simulation Modelling Practice and Theory, 12*(7–8), 579–590.
4. Hanneman, R. A., & Riddle, M. (2005). *Introduction to social network methods*. Riverside: University of California Riverside.
5. Fellman, P. V. (2011). The complexity of terrorist networks. *International Journal of Networking in Virtual Organizations, Inderscience, 8*(1–2), 4–14.
6. Carley, K., Dombroski, M., Tsvetovat, M., Reminga, J., & Kamneva, N. (2003). Destabilizing dynamic covert networks. In *Proceedings of the 8th international command and control research and technology symposium*, National Defense War College, Washington. http://www.casos.cs.cmu.edu/publications/resources_others/a2c2_carley_2003_destabilizing.pdf
7. Chalk, P. (2008). *The maritime dimension of international security: Terrorism, piracy, and challenges for the United States*. The RAND Corporation Monograph Series. http://www.rand.org
8. McNamara, R. (2012). The United States fought wars against North African pirates in the early 1800s: Thomas Jefferson wouldn't give in to the Barbary pirates. 19th Century History. http://history1800s.about.com/od/americanwars/tp/barbarywars.htm.
9. Lunsford, V. (2008). Historical piracy provides lessons for dealing with today's Somali Pirates. In N. Berlatsky (Ed.), From piracy on the high seas. Farmington Hills: Greenhaven Press, 2010.
10. Bahadur, J. (2011). The pirates of Somalia: Inside their hidden world. New York: Pantheon Books.
11. United States Forces, Somalia. (2003). After Action Report and historic overview: The United States Army in Somalia, 1992–1994. Washington, D.C.: Center of Military History.
12. Subcommittee on Coast Guard and Maritime Transportation. (2011). Assuring the Freedom of Americans on the High Seas: The United States Response to Piracy. U.S. House of Representatives, 15 March 2011.
13. 13. Phillips, R. (2010). A captain's duty. New York: Hyperion.
14. Somalia: Expanding Crisis in the Horn of Africa. (2006). Congressional hearing. Joint hearing before the subcommittee on Africa, global human rights and international operations and the subcommittee on international terrorism and nonproliferation. 109th Congress. 29 June.
15. IMB Piracy Reporting Center. (2012). http://www.icc-ccs.org/piracy-reporting-centre.
16. Gettleman, J. (2010). "The Pirates are Winning", New York review of books. New York Times, 14 October 2010.
17. Brown, N. (2010). Fighting the hydra: Multinational piracy operations move inshore. International Defense Review, 11 August 2010.

# Chapter 11
# Identities, Anonymity and Information Warfare

**Stuart Jacobs, Lou Chitkushev, and Tanya Zlateva**

## Introduction

Modern information warfare embodies many different forms and types, using various computer devices and applications as weapons. However, several common characteristics, based on fundamental intentions of the contemporary cyber warrior have been identified: (1) neutralize an adversary's (target's) ability to defend itself (2) reduce an adversary's ability to launch a counter attack, and (3) increase the adversary's willingness to submit to the demands of the attacking entity/organization.

The primary focus of information warriors is usually on rendering the target's information and communications infrastructure unusable via interfering with infrastructure service delivery capabilities. Many denial-of-service type attacks are based on flaws in the current networking protocols and mechanisms and represent one of the most common tools in information warfare.

Another intention of information warriors may be causing the target to no longer "trust" the reliability or integrity of its infrastructure. These types of attacks are detrimental for industries that rely on reliability of the communication infrastructure, such as the financial industry.

Causing the target to no longer "trust" the information processed or communicated by the infrastructure may be another goal of information warriors. It usually results in reluctance of target to use the networking resources and overall decrease in communication activities.

Finally, preventing the target from use of assets or destruction of infrastructure components and assets is a form of information warfare that can have devastating consequences in sensitive areas such as military applications.

S. Jacobs (✉) • L. Chitkushev • T. Zlateva
Metropolitan College, Boston University, 755 Commonwealth Ave, Boston, MA 02215, USA
e-mail: sjjacobs@bu.edu



P.V. Fellman et al. (eds.), *Conflict and Complexity*, Understanding Complex Systems,
DOI 10.1007/978-1-4939-1705-1_11

In all these instances the information warrior will target the adversary's critical infrastructure. Recent examples of this type of activity occurred in 2007 against Estonia [1, 2] and the spring and summer of 2008 prior to Russia moving troops into Georgia [3, 4].

A question worth considering is how information warfare differs from cyber-terrorism and cyber-crime. The goals of the cyber-terrorists are to cause fear of death and injury and increase the willingness to submit to extortion demands for financial or political gain. The cyber-terrorist will likely view the target's information and communications infrastructure as assets worth subverting either to directly create fear and terror or as a "force multiplier" in conjunction with some other form of attack (likely physical) [5].The cyber-criminal is concerned with financial gain typically from the resale of stolen personally identifiable information (PII) or theft of commercially valuable information [6, 7]. Frequently, the cyber-criminal will: (1) interfere with infrastructure service delivery capabilities to perpetrate electronic blackmail, (2) penetrate an organization's infrastructure to steal organizational information, and (3) steal identities based on stolen PII.

Based on these observations we can conclude that motivation is the main difference between cyber-warriors, cyber-terrorists, and cyber-criminals, although all three groups use the same sort of tactics and techniques. The report "Botnets, Cybercrime, and Cyberterrorism: Vulnerabilities and Policy Issues for Congress" [8] by the congressional Research Office clearly notes that these groups use similar, if not common, methodologies. The difference is simply motivation, goals, and objectives (i.e., furthering a cause, an ideology or profit) not methods and techniques.

## Problem Statement

What the computer and communications industry, and the world economies, need to recognize is that identities in the cyber world are critical to the safe and proper operation of our networked world. An asserted identity is a critical factor in all forms of electronic communication and even represents an asset to its owner.

## Identities Are Valuable Assets

A subject's identity is supposed to be a unique identifier of a subject, regardless if the subject is a human or a computing device (e.g., switch, router, server, workstation, mobile device, etc.). There are many forms of identities used in electronic communications and commerce (Ecommerce). Some of the more common identities are: data link media access control (MAC) addresses, internetworking (IPv4/

IPv6) addresses, application addresses (transmission control protocol [TCP]/user datagram protocol [UDP]/stream control transmission protocol [SCTP] port numbers), web server addresses (unified resource locator [URL]), email account identifiers (email uniform resource identifiers [URIs]), Voice over IP sources and destinations (session identification protocol [SIP] URIs). Control of MAC and IP addresses can become quite a problem in large complex intranets and backbone networks. However the use of IP addresses as identities for access control decisions is invalid as these IP addresses cannot be verified and are routinely "spoofed" as part of numerous forms of network-based attacks. Many modern applications no longer rely of internet assigned numbers authority (IANA) controlled transport protocol port numbers and instead dynamically negotiate these port numbers during application session establishment.

On the other hand, control of web URLs, email URIs and SIP URIs actually represents a level of control of the subjects or systems to which these identities are associated. URLs are bought and sold on a routine basis and have to be searchable within the Internet DNS of servers. A customer who changes their email service provider (whoever operates the email server that services the customer's email account) is forced to use a different email URI as these URIs are service provider specific. The situation with email URIs is the same with SIP URIs when considering service provider based multimedia services (i.e., VoIP, IPTV, etc.).

Service consumers need to rely on the validity of URLs and URIs for most of the Ecommerce and general communications services that society relies on in our modern inter-networked world. Rarely are mechanisms deployed that allow these service consumers to verify (authenticate) the very URIs and URLs relied upon daily. The lack of routine URL and URI authentication provides an opportunity for malicious activities to occur. When a web browser is unable to authenticate the legitimacy of a URL, the browser user is at risk of being attacked by the server in a number of ways, such as by a "man-in-the-middle" technique or downloading malware (i.e., Spyware, adware, viruses, root-kits, etc.) into the system running the browser. The same types of attacks can be used against URIs. Since the DNS infrastructure is used for mapping URLs and URIs into routable IP addresses, DNS queries should be trustable, yet there are a growing number of attacks on the DNS every day [9–12].

In the physical world authentication of identities is a common activity. For example, international travelers routinely authenticate their identity by presenting a government issued passport as an authentication credential. Also, motor vehicle operators on public roads and highways are required to carry a government issued drivers license and present it to law enforcement personnel when requested. Authenticating one's identity is routine when cashing a bank check or making a payment using a credit card, and almost everywhere a voter has to prove their identity before being allowed to vote in an election. Moreover, in almost all commercial financial activities and interacting with government entities, one must be able to provide an authenticable identity.

## Issue with Anonymity

Anonymity is the ability to conceal, hide, or withhold one's identity. Over the decades anonymity has evolved into an accepted concept in computing/networking. In the early days of networking many engineers did not consider that the technology they developed would be used for inappropriate activities. Even though RFC 791 [13] included the specification of a security field within the IPv4 header, this capability was never expanded upon. Other core protocols, such as TCP, UDP, DNS, ARP, DHCP (dynamic host configuration protocol), HTTP (hypertext transfer protocol) and SNMP (simple network management protocol), either ignored the need for security, especially the authentication of identities, or provided such simple authentication mechanisms that these mechanisms are/were useless. UDP has no authentication mechanisms. TCP has an optional mechanism based on a shared secret key and MD5 message digest but there is, even today, no networked approach for distributing these keys. DNS continues to be deployed in an unauthenticated manner. HTTP includes a simple radix-64 encoded password for authentication and a more robust message digest authentication approach (as of the publication of [14]) that relies on a server supplied secret key. ARP has never included any form of authentication; it has just been considered a "trust worthy" protocol. SNMP first relied on clear-text authentication via a password called a community name and evolved with the release of SNMPv2 to a mechanism based on a shared secret key and MD5 message digest but there is, even today, no networked approach for distributing these keys. Consequently virtually all deployments of SNMP continue to rely on the use of community names.

Societies today consider anonymity appropriate in only few situations, such as:

1. Elections where secret balloting is considered a prerequisite for maintaining fair and honest voting, although the identity of a voter needs to be verifiable prior to submitting an actual ballot;
2. Reporting crimes, or information about crimes or illegal behavior, such as with "tip-lines" and "whistle blowing"; and
3. Conducting surveys where anonymous input enhances the likelihood of collecting accurate data.

In most forms of non-cash based economic activities anonymity is not considered acceptable and the same is true for many forms of social interaction.

When the use of authenticated identities is commonly enforced in non-electronic interactions, why should anonymity be condoned over the Internet? Within the United States 46 of the 50 states now have laws concerning the protection of PII that requires the ability to authenticate identities of subjects prior to granting access to PII. Yet the majority of applications and network protocols either do not directly support authentication in a cryptographically strong manner or are not used in conjunction with authentication technologies that have been available for over a decade (i.e., IPsec, Public Key Infrastructures [PKIs], smartcards, security tokens, etc.)

## Issues with Common Networking Protocols

Most of the networking protocols in use today lack the ability of proper user authentication, or their authentication mechanism is primitive or not implemented. There are many reasons for this anomaly, but the main ones are the legacy issues with protocols that were developed during the infancy of the Internet when security threats were not imminent. At the same time, this is surprising when we know that the original Internet development was supported by the military research organizations. In the next section we will present some examples of authentication issues in networking protocols at different layers of the networks.

ARP is used for mapping network IP address to physical MAC addresses as a consequence of addressing problems that the Internet has had since the very beginning. ARP lacks authentication and uses broadcasting at the data link layer. In addition, ARP's limited cache can be overpopulated and flushed which additionally complicates the problem. ARP does not have an integral mechanism to distinguish between legitimate and illegitimate messages, and ARP replies are processed (accepted and cached) without authentication of their origin. Since an ARP request is broadcasted over the local network, any host can reply with a request to be mapped to a specific IP address. Moreover, gratuitous ARP replies can be sent without requests, resulting in diverting traffic from a host or network under attack. A large number of such replies can result in flushing of the ARP cache. All of these attacks are enabled by the lack of proper authentication within the APR protocol.

Authentication problems exist within the dynamic host configuration protocol (DHCP). Theoretically, any entity can request an IP address to be assigned to it. Moreover, anyone can respond to that request. This can result in denial of service, or communication with a compromised host. DHCP option 90 was intended to solve DHCP's lack of authentication; however, option 90 is virtually unusable due to the lack of a secret key distribution mechanism.

Although IP addresses are the main way of determining routing over Internet, they cannot be considered identities, but only attributes for packet filtering access control. Source IP address are generated by the originating host and they can be easily modified or misrepresented. Actually, the first step of many attacks over the network is spoofing of source IP addresses of IP packet.

The lack of authentication at the IP level has a major consequence for the security of routing protocols, both Interior Gateway Protocols (IGPs) (routing information protocol [RIP], open shortest path first [OSPF]) and border gateway protocol (BGP). Even for recommended IGPs, like OSPF, the default state for authentication is often off.

The transport layer security (TLS) protocol, based on public key cryptography, has been developed to provide secure communication over the Internet. However, even with this protocol there are some major authentication issues. For example, during the renegotiation phase, the protocols do not necessarily ensure continuity before and after negotiation.

DNS provides a mechanism to resolve host names into IP addresses. Lack of authentication of DNS has been a major issue. Cache poisoning occurs when a DNS server passes a query within its cache, for which it doesn't know the answer, onto another DNS server that has inaccurate information.

## Issues with IP Version 6

There is now growing discussion over IPv4 address exhaustion and the need to seriously transition to IPv6. Some individuals and groups further advocate such a transition as IPv6 [15] mandates support for IPsec [16] whereas IPsec availability on IPv4 networks is totally optional. However usage of IPsec is not mandatory and, given current trends, not likely to be utilized for authenticating network activities.

## Issues with Federated Identities

Work is currently under way on the concept of Federated Identities [17, 18]. A key component in this concept is establishment of "trust relationships" between different organizations, namely:

> A company must trust its partners to vouch for their users. Each participant must rely on each partner to say, in effect, "This user is OK; let them access this application" [17].

Establishing this form of trust relationship between governments for international travel (e.g., electronic passports) or different government entities (i.e., police, elections boards, tax collectors accepting common electronic credentials) should be quite achievable. In fact the United States already has a law that starts to provide a foundation for this approach by stating that:

> … any transaction in or affecting interstate or foreign commerce—
>
> (1) a signature, contract, or other record relating to such transaction may not be denied legal effect, validity, or enforceability solely because it is in electronic form; and
> (2) a contract relating to such transaction may not be denied legal effect, validity, or enforceability solely because an electronic signature or electronic record was used in its formation [19].

However the application of Federated Identities in the commercial world has to overcome a number of issues, namely:

- Business entities need to relinquish direct control of customer identities when federating with other business entities;
- Businesses routinely strive to maintain some degree of control over their customers (account) and the loss of direct control of identities will likely be perceived as a business disadvantage;

- Identity federation will mostly be viewed as reasonable for commercial peering/joint activities when covered under contracts; however, customer approval (with either an opt-in or opt-out process) will undoubtedly be necessary; and
- Commercial identity federation will have to comply with laws governing PII, which currently has not been given much consideration.

## Solution

As noted in our problem statement, there are numerous network vulnerabilities due to lack of authenticated identities that cyber-warriors, cyber-terrorists, and cyber-criminals leverage. However there are also a number of technologies available where some are being deployed and others being largely ignored. These technologies include the following.

### *IEEE 802.1X Network Authenticated Access Controls*

IEEE 802.1X is a robust mechanism that can prevent use of a network prior to authentication and authorization. Not only does 802.1X support simple extensible authentication protocol (EAP) methods based on passwords, it also will integrate with organization PKIs. ISPs, or any other organization, can establish a PKI and issue credentials to all of its Internet service customers or employees/users and thereby identify and control who has access. IEEE 802.1X, if deployed routinely, would likely reduce the occurrence of proxy/gratuitous ARP based attacks on hosts and subnets.

### *IP Security Authenticated Network and Host Access Controls*

A highly effective approach for eliminating anonymity and providing both peer-entity and data-origin authentication of communications between systems is the use of IP security (IPsec, [20]). While there is a delay due to security association (SA) establishment, once an SA is established, the per-packet IPsec processing overhead can be quite low. The most appropriate IPsec transform dealing just with authentication is encapsulating security payload without encryption (ESP_NULL, [21]) in transport mode. ESP_NULL provides exactly the same data-origin authentication as a message digest (MD5, [35]) or secure hash algorithm (SHA-1, [23]) digest used with a shared secret key with one major difference; ESP_NULL utilizes the internet key exchange (IKE) protocol ([24]/RFC 4306) for key management, that many other protocols ignore, and at about the same cost for per packet processing overhead. Furthermore, both network and host packet filtering access control functions ("firewalls") can easily be configured to forward or block packets based on ESP (encapsulation security payload) header values.

### TLS Authentication

TLS [25] can provide effective elimination of anonymity. If mutual authentication is based on digital signatures exchanged within the TLS Handshake Protocol, then both peer-entity and data-origin authentication of communications between applications is provided. If only server authentication occurs within the TLS Handshake Protocol, then only data-origin authentication can be assured so long as the client system performs some form of authentication over the TLS encrypted session.

### DNS Protection via DNS Security

Usage of DNS Security (DNSSEC) [26] to combat the problems associated with the "spoofing" of legitimate DNS servers or tampering with DNS query responses is very effective. The DNS protocol extensions defined in RFC 4035 [27], when coupled with the digitally signed resource records specified in RFC 4034 [28], provide to DNS clients (resolvers) with both data-origin authentication and data integrity of DNS data returned in query responses. These two security services fully allow a resolver to know that a query response was sent by an authenticated server and was not modified in transit; thus making anonymity based attacks via DNS unsuccessful.

### Defense Against Web Application Abuse

Routine use of TLS, with at least server authentication, for web application communications would eliminate many forms of attack from malicious web servers. In a mature electronic commerce world insisting that Internet accessible web servers have authenticated identities should not be considered a hardship. Legitimate web sites/businesses will want verifiable identities as this supports "name recognition" and serves to ensure their customers of the business's integrity. TLS based mutual authentication would allow web servers to strongly authenticate client systems and thereby be able to make sound authorization decisions regarding site access and uploading of information to the site. This approach does entail a web server maintaining state on its requester/clients, however this state maintenance activity has already become routine with many Ecommerce web sites. An alternative approach to TLS, especially to avoid the processing overhead of mandatory TLS encryption, would be the use of IPsec ESP_NULL for interactions not requiring confidentiality and ESP triple digital encryption standard (3DES, [29]) or ESP advanced encryption standard (AES, [30]) transforms only when confidentiality is required.

## *Defense Against Electronic Mail Abuse*

Textual, and even multipart internet mail extensions (MIME) enhanced, email was never designed with security capabilities, leading to the development of Secure MIME (S/MIME, [31]) that provides both authentication and confidentiality capabilities. If email client applications required S/MIME authentication on all received emails, and their associated attachments, then the majority of email based attacks would be effectively blocked. Receipt of "phishing" emails, attempting to fool the recipient to log into spoofed servers, would be identifiable in almost all situations. Email attachments containing "malware" are highly unlikely to be sent from authenticated sources and what entity would want to send us "spam" emails if they had to authenticate their identity.

## *Identity Management that Scales*

A method to manage authenticable identities, that bypasses many of the problems associated with the concept of federated identities, is the use of PKIs based on [32] digital certificates. A historic argument against PKIs was based on the assumption that there had to be a single PKI likely operated by some government entity that all other PKIs would be subordinate to. This argument is not valid due to the concept of cross-CA (certificate authority) certification where CAs in different PKI hierarchies are issued credentials by CAs in other PKI hierarchies and thus creating a traversable trust hierarchy tree. Cross-CA certification provides an organization the ability of verifying the validity of identity authentication credentials issued by a different organization's PKI without having to relinquish any control over identities.

Different governments can deploy national PKIs for authenticating population identities for governmental and international activities (i.e., electronic passports, shipping documents, etc.) that other governments can rely upon. This also applies between national governmental bodies and regional governmental (i.e., state, province, city and town) entities. These regional entities can deploy their own PKIs for regional or local official activities (i.e., motor vehicle operator licenses, voter authentication, etc.). Commercial and non-commercial enterprises can operate PKIs for private or commercial activities that allow customer, or user, identities to be validated by other organizations while allowing the organization that issues the identity authentication credentials to retain primary control over these identities.

However, there are a number of capabilities that are not currently present which need to be incorporated in many protocols and applications, for example:

- Most web browser and email applications support digital certificates [32] but have no ability to perform trust hierarchy traversal; these applications frequently can only validate the CA that directly signed a certificate and not validate other CAs within a PKI or across PKIs. This situation should be addressed in a timely manner;

- The technical requirements and processes for cross-CA certification need to be standardized thereby making CA trust hierarchy traversal well defined;
- Few protocols or applications include support for the on-line certificate status protocol (OCSP, [33]). OCSP support would make certificate revocation checking significantly more efficient that having to retrieve and search certificate revocation lists (CRLs);
- Standardize the use of PKCS #10 [34] as the de jure mechanism for submitting X.509v3 certificate requests from any PKI; and
- Standardize the use of PKCS #12 [35] as the de jure mechanism for secure storage and transfer of pass-phrase protected private keys and X.509v3 certificates.

OCSP services/functions, that support multiple PKIs simultaneously, could be provided as services to PKI operators on a "for fee" basis thereby making the use of OCSP economically advantageous. To further encourage PKI based identity verification, a certified secure user application for generating public/private keys and creation of PKCS #10 [34] certificate requests being sent to RAs could be developed and made freely available to all parties.

## Conclusions

We believe that modern electronic societies can no longer endorse, nor allow, anonymity to be generally accepted. Industries and societies need to insist that electronic infrastructures and services more closely reflect how non-electronic activities insist on the use of authenticated identities. Requiring authentication of network access and network-based activities (application services) and removing the cover of anonymity will increase the difficulty of perpetrating attacks (including spam, phishing, etc.) by cyber-criminals, terrorists, and those involved information warfare.

## References

1. N.Y. Times (2007). *Estonia says cyber-assault may involve the Kremlin [online]*. The New York Times. http://www.nytimes.com/2007/05/17/world/europe/17iht-estonia.4.5758556.html
2. Krebs, B. (2007). *Shadowy Russian firm seen as conduit for cybercrime [online]*. The Washington Post. http://www.washingtonpost.com/wp-dyn/content/article/2007/10/12/AR2007101202461.html
3. Schactman, N. (2009). *Top Georgian official: Moscow cyber attacked us – we just can't prove it [online]*. Condé Nast Digital. http://www.wired.com/dangerroom/2009/03/georgia-blames/
4. Schactman, N. (2008). *Russian coder: I hacked Georgia's sites in cyberwar [online]*. Condé Nast Digital. http://www.wired.com/dangerroom/2008/10/government-and/
5. Mocana, (2010). *Expert Warns of Industrial Control Security Risks [online]*. Mocana DeviceLine Blog. http://mocana.com/blog/2010/01/10/expert-warns-of-industrial-control-security-risks/

6. Vijayan, J. (2010). *Military contractors targeted in Chinese attacks, says F-secure [online]*. Computerworld. http://www.computerworld.com/s/article/9146318/Military_contractors_targeted_in_Chinese_attacks_says_F_Secure
7. Zetter, K. (2010). *Google hack attack was ultra sophisticated, new details show [online]*. Condé Nast Digital. http://www.wired.com/threatlevel/2010/01/operation-aurora/
8. United States. Congressional Research Service. (2008). *CRS Report for Congress: Botnets, Cybercrime, and Cyberterrorism: Vulnerabilities and Policy Issues for Congress, January 29, 2008*. Retrieved on 21 Jan 2010 from http://www.fas.org/sgp/crs/terror/RL32114.pdf
9. Leyden, J. (2010). *Baidu sues registrar over DNS records hack: Legal salvo lands in New York [online]*. Enterprise Security. http://www.theregister.co.uk/2010/01/20/baidu_dns_hack_lawsuit/
10. Leyden, J. (2009). *DNS attack hijacks Twitter [online]*. Enterprise Security. http://www.theregister.co.uk/2009/12/18/dns_twitter_hijack/
11. McPherson, D. (2009). *Multi-Stage BGP & DNS Attack Vector [online]*. Arbor Networks Security Blog. http://asert.arbornetworks.com/2009/01/multi-stage-bgp-dns-attack-vector/
12. Keizer, G. (2008). *Researchers unleash DNS attack code: HD Moore unveils two exploits for Dan Kaminsky's critical Internet routing bug [online]*. Computerworld. http://www.computerworld.com/s/article/9110622/Researchers_unleash_DNS_attack_code
13. RFC 791 (1981). *Internet protocol DARPA internet program protocol specification*, September 1981, Defense Advanced Research Projects Agency Information Processing Techniques Office.
14. RFC 2617 (1999). *HTTP authentication: Basic and digest access authentication*, June 1999, Internet Engineering Task Force.
15. RFC 1883 (1995). *Internet protocol, version 6 (IPv6) specification*, December 1995, Internet Engineering Task Force.
16. RFC 1825 (1995). *Security architecture for the internet protocol*, August 1995, Internet Engineering Task Force.
17. Carr, D. F. (2003). *What's Federated Identity Management? [online]*. Ziff Davis Enterprise Holdings Inc. http://www.eweek.com/c/a/Channel/Whats-Federated-Identity-Management/
18. Huntington, G. (2006). *Federated Authentication [online]*. Huntington Ventures Ltd. http://www.authenticationworld.com/Authentication-Federation/
19. United States Public Law 106-229, 114 Stat. 464 (2000). *Electronic Signatures in Global and National Commerce Act*, 30 June 2000
20. RFC 2401 (1998). *Security architecture for the internet protocol*, November 1998, Internet Engineering Task Force.
21. RFC 2410 (1998). *The NULL encryption algorithm and its use with IPsec*, November 1998, Internet Engineering Task Force.
22. RFC 2403 (1998). *The use of HMAC-MD5-96 within ESP and AH*, November 1998, Internet Engineering Task Force.
23. RFC 2404 (1998). *The use of HMAC-SHA-1-96 within ESP and AH*, November 1998, Internet Engineering Task Force.
24. RFC 2409 (1998). *The internet key exchange (IKE)*, November 1998, Internet Engineering Task Force.
25. RFC 2246 (1999). *The TLS protocol version 1.0*, January 1999, Internet Engineering Task Force.
26. RFC 4033 (2005). *DNS security introduction and requirements*, March 2005, Internet Engineering Task Force.
27. RFC 4035 (2005). *Protocol modifications for the DNS security extensions*, March 2005, Internet Engineering Task Force.
28. RFC 4034 (2005). *Resource records for the DNS security extensions*, March 2005, Internet Engineering Task Force.
29. RFC 2451 (1998). *The ESP CBC-mode cipher algorithms*, November 1998, Internet Engineering Task Force.

30. RFC 3602 (2003). *The AES-CBC cipher algorithm and its use with IPsec*, September 2003, Internet Engineering Task Force.
31. RFC 5751 (2010). *Secure/multipurpose internet mail extensions (S/MIME) version 3.2 message specification*, January 2010, Internet Engineering Task Force.
32. ITU-T X.509v3 (1997). *ITU-T Recommendation X.509: Information technology – Open systems interconnection – the directory: Authentication framework*. International Telecommunications Union Telecommunication Standardization Sector.
33. RFC 2560, X.509 (1999). *Internet public key infrastructure online certificate status protocol – OCSP*, June 1999, Internet Engineering Task Force.
34. PKCS #10 v1.7 (2000). *Certification request syntax standard*, May 26, 2000, RSA Laboratories.
35. PKCS #12 v1.0 (1999). *Personal information exchange syntax*, June 24, 1999, RSA Laboratories.

# Part III
# Broader Horizons

# Chapter 12
# The Geography of Ethnic Violence

**Alex Rutherford, May Lim, Richard Metzler, Dion Harmon, Justin Werfel, Shlomiya Bar-Yam, Alexander Gard-Murray, Andreas Gros, and Yaneer Bar-Yam**

**Overview** We consider the conditions of peace and violence among ethnic groups, testing a theory designed to predict the locations of violence and interventions that can promote peace. Violence arises at boundaries between regions that are not sufficiently well defined.We model cultural differentiation as a separation of groups whose members prefer similar neighbors with a characteristic group size at which violence occurs. Application of this model to the area of the former Yugoslavia and to India accurately predicts the locations of reported conflict. Characterizing the model's success in predicting peace requires examples where peace prevails despite diversity. Switzerland is recognized as a country of peace, stability, and prosperity. This is surprising because of its linguistic and religious diversity that in other parts of the world lead to conflict and violence. Here we analyze how peaceful stability is maintained. Our analysis shows that peace does not depend on integrated coexistence, but rather on well-defined topographical and political boundaries separating linguistic and religious groups, respectively. In exactly one region, a porous mountain range does not adequately separate linguistic groups and violent conflict has led to the recent creation of the canton of Jura. Our analysis supports the hypothesis that violence between groups can be inhibited by both physical and political boundaries. A similar analysis of the area of the former Yugoslavia shows that during widespread ethnic violence existing political boundaries did not coincide with the boundaries of distinct groups, but peace prevailed in specific areas where they did coincide.

---

A. Rutherford • M. Lim • R. Metzler • D. Harmon • J. Werfel
S. Bar-Yam • A. Gard-Murray • A. Gros • Y. Bar-Yam (✉)
New England Complex Systems Institute, Cambridge, MA, USA
e-mail: yaneer@necsi.edu



P.V. Fellman et al. (eds.), *Conflict and Complexity*, Understanding Complex Systems,
DOI 10.1007/978-1-4939-1705-1_12

Achieving peace requires a vision of what it looks like. How we imagine peace affects the steps we take and our ability to implement it in diverse locations around the world. Does peace in one place look the same as in another? Is knowledge of the specifics of each conflict necessary to negotiate peace between ethnic groups in conflict? Even if specifics are important, there are broad frameworks that guide our thinking. Recently, we introduced a complex systems theory of ethnic conflict that describes the conflicts in areas of the former Yugoslavia and India with high accuracy [1], as well as the absence of conflict in Switzerland, despite its linguistic and religious diversity [2]. Other approaches [3–14] generally consider (a) the process by which ethno-religious identity is established and if interventions could diminish its importance relative to more inclusive identities, and (b) control mechanisms of the state and of organizations of ethnic groups and if interventions could strengthen the state while subsuming or accommodating ethnic groups within state authority. More specific social and economic factors identified in the literature as contributing to violence include oppression of minorities, economic grievances, historical precedents, competition for resources, favoritism, availability of resources for violence, security fears, mobilization by elites, weak social ties, national ethnic diversity, territorial claims, religious or political polarization, incendiary media, and international influences. While most consider national conditions, a few consider local violence to identify the role of local socio-economic or geographic factors [8–10]. However, in our model specific details of history, social and economic conditions are not the primary conditions for peace or conflict. Instead the geographic arrangement of populations is key. Significantly, it points to two distinct conditions that are conducive to peace—well mixed and well separated. The first corresponds to the most commonly striven for framework of an integrated society [15]. The second corresponds to spatial separation, partition, and self-determination—a historically used but often reviled approach [16]. We also identified a more subtle third approach, that of within-state boundaries in which cooperation and separation are both necessary [2]. The success of this approach is of particular importance as the world becomes more connected. As illustrated by the European Union, the role of borders as boundaries is changing.

To model violence we assume that highly mixed regions do not engage in violence, and neither do well-segregated groups, an intuitive hypothesis with empirical support [8]. The analysis is applicable to communal violence and not to criminal activity or interstate warfare. In highly mixed regions, groups of the same type are not large enough to develop strong collective identities, or to identify public spaces as associated with one or another cultural group. They are neither imposed upon nor impose upon other groups, and are not perceived as a threat to the cultural values or social/political self-determination of other groups. Partial separation with poorly defined boundaries fosters conflict. Violence arises when groups are of a size that they are able to impose cultural norms on public spaces, but where there are still intermittent violations of these rules due to the overlap of cultural domains. When groups are larger than the critical size, they typically form self-sufficient entities that enjoy local sovereignty. Hence we expect violence to arise when groups of a certain characteristic size are formed, and not when groups are significantly smaller or larger than this size. The model of violence depends on the distribution of the population and not on the specific mechanism by which the population achieves this structure, which may include internally or externally directed migrations. By focusing on the geographic distribution of the population, the model seeks a predictor of conflict that can be easily determined by census. This may work well because geography is an important aspect of the dimensions of social space, the dynamic coarsening process is universal and other aspects of social behavior (e.g., isolationism, conformity, as well as violence) are correlated to it.

The geographical distribution theory [1, 2] considers type separation into geographical domains independent of the specification of the individual types—a universality of type behavior in collective violence. Violence arises due to the structure of boundaries between groups rather than as a result of inherent conflicts between the groups themselves. In this approach, diverse social and economic causal factors trigger violence when the spatial population structure creates a propensity to conflict, so that spatial heterogeneity itself is predictive of local violence. The local ethnic patch size serves as an "order parameter," a measure of the degree of order of collective action, to which other aspects of behavior are coupled. The importance of collective behavior implies that ethnic violence can be studied in the universal context of collective dynamics, where models can identify how individual and collective behavior are related.

The predictor we identify based on spatial census data need not describe the immediate social or institutional triggers of violence, just the conditions under which violence becomes likely. Previous research striving to characterize ethnic conflict by census data has focused on measures of ethnic or religious "fragmentation" [17–21]. Such measures characterize the diversity of a country without reference to its spatial structure, i.e. the overall proportions of ethnically distinct groups in a country. They are therefore distinct from the spatial characterization of our study. The literature is divided about whether or which correlations exist with measures of national ethnic composition. We find however that the spatial distribution of ethnic groups is a strong predictor of locations of violence.

Mathematically, evaluation of the model begins by mapping census data onto a spatial grid. The expected violence is determined by detecting patches consisting of islands or peninsulas of one type surrounded by populations of other types. These features are detected by pattern recognition using the correlation of the population for each population type with a template that has a positive center and a negative surround. The template used is based on a wavelet filter [1, 2, 22, 23]. Wavelets are designed to obtain a local measure of the degree to which a certain scale of variation (wavelength) is present. Outcomes are highly robust, and other templates give similar results. The diameter of the positive region of the wavelet, i.e., the size of the local population patches that are likely to experience violence, is the only essential parameter of the model. The parameter is to be determined by agreement of the model with reports of violence, and results were robust to varying the parameter across a wide range of values. For Yugoslavia and Switzerland, we model the effect of internal topographical and political boundaries. We assume that separate autonomous regions can be analyzed by including only the populations within each of the autonomous areas to determine the expected violence. Where boundaries are incomplete, as might be the case for mountains, lakes and convoluted political boundaries, we include only the populations that are in line of sight through gaps or past ends of boundaries to determine the expected violence within a region. An effective map of populations at each site is constructed, determined by the orientation of any boundaries relative to that site. Populations past boundaries of the line of sight are replaced by neutral populations. The result of the correlation of population with the wavelet filter is a single value at every location, the theoretical "propensity to violence," and the locations of expected violence are obtained by applying a threshold to that value. The location of groups of a certain size is indicative of a violence-prone group, but the precise location of violence is not determined. The proximity of these violence-prone groups to actual violence is tested by constructing proximity maps. The proximity to reported violence is correlated to the proximity to violence prone groups.

## Yugoslavia

To test the predictive ability of the model we performed simulations based upon census data for the former Yugoslavia and India. We assigned areas of pixelated geographic maps pixel by pixel to ethnic groups at random but in proportion to their relative population census in the region. While this does not reflect the physical geography or local mixing of groups in buildings and villages, over an area of multiple pixels it captures the regional composition of the census. For Yugoslavia, census data from the early 1990s prior to the outbreak of conflict [24, 25], as shown in Fig. 12.1a, was used to create an ethnic map (Fig. 12.1b), which was used to obtain the regions of expected violence shown in (Fig. 12.1c).

We then obtained from books [3], newspapers, and Internet sources (see supporting online text) the locations of reported violence for the area of the former Yugoslavia. Multiple independent sources were used to provide validation for each location of violence [1, 2]. We consider these reports as indicators of areas of actual violence, keeping in mind possible bias and incompleteness and that areas of widespread violence are identified only by local urban centers. In comparing such reports with model predictions, we note that the model identifies locations of groups of a particular size, but the location of the actual violence should occur somewhere in the area between adjacent groups. Despite these caveats, overlaying the locations of reported and predicted violence in Fig. 12.1d demonstrates a significant ability of our simple model to identify regions of reported violence. We performed statistical analyses comparing the predicted to the reported violence, evaluating the ability of the model to determine both where violence occurs and where violence does not occur. For comparison we randomized the locations of reported violence. We defined conflict proximity as the distance between a given position and the nearest location of violence (predicted, reported, or randomized). We calculated Pearson's correlation and other statistical measures between the proximities of predicted and reported violence, and compared them with the same measures in relation to randomized reports. We found that the model has a correlation of 0.9 with reports (0.89 to two significant digits), a level of agreement not reached in any of 100,000 randomized trials. Moreover, the predicted results are highly robust to parameter

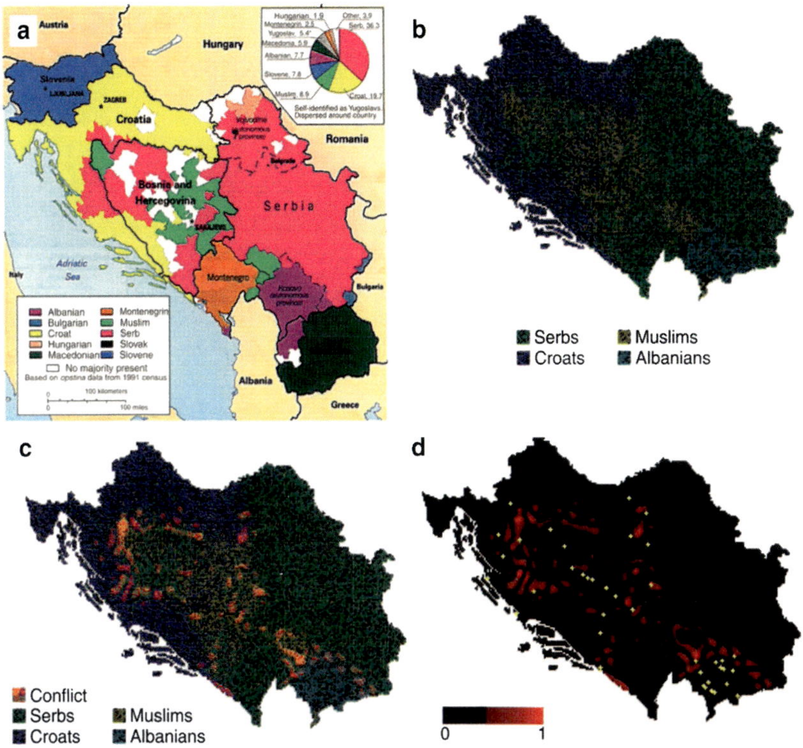


**Fig. 12.1** (**a**) Census data from 1991 shown here in map form was converted into a spatial representation and used in an agent-based simulation shown in (**b**). Our prediction of populations likely to be in conflict with neighboring groups (*red overlay*, **c**, **d**) agrees well with the location of cities reported as sites of major fights and massacres (*yellow dots*, **b**)

variation, with essentially equivalent agreement obtained for filter diameters ranging from 18 to 60 km, a range that is in agreement with intuition about the size of conflict areas. Below or above this range, poorer agreement occurs.

We extended the analysis of Yugoslavia to consider the role of internal boundaries [2]. For this purpose we improved the analysis to include the fractional population of ethnic groups in each pixel. Topographical boundaries reduce slightly the maximum propensity from 0.63 to 0.57, and administrative borders to 0.56. The correlations of predicted and reported violence changes were essentially the same, with correlations of 0.86 and 0.85, respectively. That political boundaries do not have a greater impact on the calculated violence implies that they do not align with the geographical boundaries between groups. We also extended the area to include Macedonia and Slovenia, parts of the Socialist Federal Republic of Yugoslavia before gaining independence (Fig. 12.2). With the political boundaries the correlation is still 0.85; however, when political boundaries are not included, the correlation is reduced considerably to 0.72.

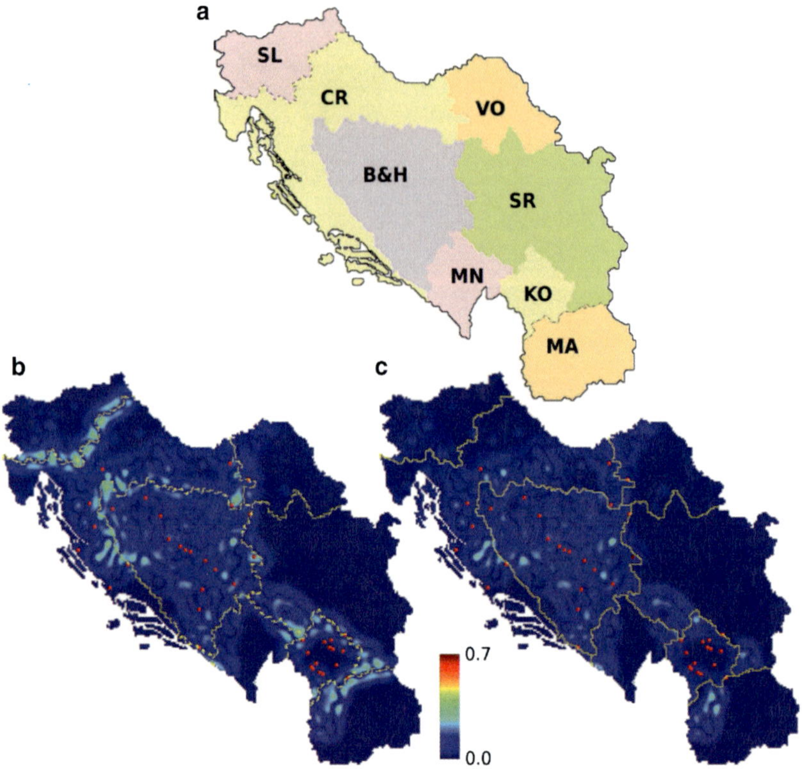


**Fig. 12.2** (**a**) Map of the area of the former Yugoslavia showing administrative provinces. Propensity to violence calculated without (**b**) and with (**c**) administrative boundaries, using a characteristic length of 21 km. Locations of boundaries are shown on both plots as *solid* and *dashed yellow lines*, respectively. Sites of reported violence are shown as *red dots* [23]. Spurious violence is predicted along the borders of Slovenia and Macedonia when boundaries are not included. Province labels are *SL* Slovenia, *CR* Croatia, *VO* Vojvodina*, *B&H* Bosnia & Herzegovina, *SR* Serbia, *MN* Montenegro, *KO* Kosovo*, *MA* Macedonia (*Autonomous administrative provinces of Serbia.)

The lower correlation is specifically due to a high calculated propensity to violence along the borders of Slovenia with Croatia, and of Macedonia with Serbia and Kosovo. These areas, however, were peaceful—consistent with the predictions when boundaries are included. Our results suggest that these political borders were instrumental in reducing ethnic violence, whereas the violence in other areas of Yugoslavia was not prevented because of poor alignment of borders with population groups.

## India

We studied conflict in India as a second case study of the ethnic violence model. We constructed a spatial representation of India on a district level from maps at http://www.censusindia.net, and obtained the distribution of ethno-cultural groups from the

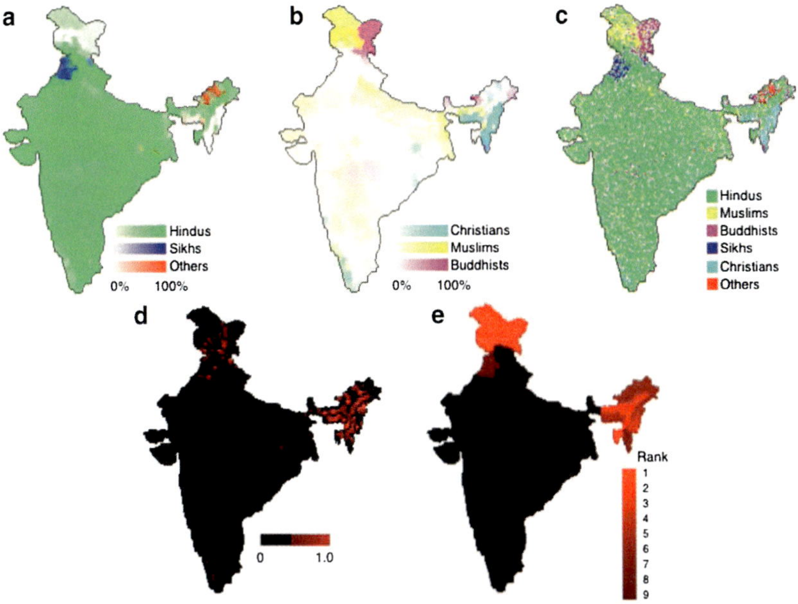


**Fig. 12.3** (**a**), (**b**) Spatial representation of Indian census data from 2001 of six indicated groups was converted into an agent-based simulation shown in (**c**). Our prediction of conflict-prone areas (*red areas* in **d**) agrees with states where major ethnic violence has been reported (*red areas* in **e**) between 1999 and 2002, with the *red shading* intensity corresponding to the rank order of states by number of incidents

2001 Census data at http://www.indiastat.com. The result can be seen in the form of three-color maps in Fig. 12.3a, b representing the relative densities of Hindus, Muslims, Christians, Sikhs, Buddhists, and Others (primarily Jains). The agent model is shown in Fig. 12.3c and the prediction of ethnic violence is indicated in Fig. 12.3d. Predictions correspond very well to the primary locations of extremist violence of government reports as given by indiastat.com (Fig. 12.3e) and confirmed using independent sources, particularly in Kashmir, Punjab, and the states of Northeast India. There are some additional areas of lesser violence that were also predicted by the model, particularly Jharkhandan eastern state created in 2000 which has recently experienced some violence [1, 26]. Consistent with predicted results the violence in this region is not as prevalent as in other violence-prone areas of India. Statistical correlation measures of conflict proximity yield a correlation of 0.998 when the threshold is set above the value of predicted violence in Jharkhand. If the threshold is set lower, so that violence in Jharkhand is included in predicted but not in reported cases, the correlation falls to 0.92. Including reported violence in Jharkhand when comparing at the lower threshold increases the correlation to 0.98. Additional details are provided in the supporting online text. The range of filter diameter values for which good agreement was obtained overlaps that of the former Yugoslavia. However, it is shifted to larger values, up to around 100 km. This may reflect the larger granularity of data, but perhaps also the effect of violence itself on separation. Unlike

Yugoslavia, in India the census was performed during ongoing violence. Since violence accelerates the process of separation, groups in conflict are likely to have separated substantially and reflect the high end of group sizes susceptible to violence.

## Switzerland

A successful model should explain the presence of peace as well as violence. We considered the coexistence of groups in Switzerland, where within-state boundaries are important. Switzerland is known as a country of great stability, without major internal conflict despite multiple languages and religions [27, 28]. Switzerland is not a well-mixed society, it is heterogeneous geographically in both language and religion. The alpine topography and the federal system of strong cantons have been noted as being relevant to coexistence; their importance can be seen in Napoleon's statement, after the failure of his centralized Helvetic Republic, that "nature" had made Switzerland a federation [29–31]. But the existence of both alpine and non-alpine boundaries between groups and the presence of multiple languages and religions within individual cantons suggest partition is not essential for peaceful coexistence in Switzerland. In identifying the causes of peace, the literature has focused on socio-economic and political conditions including a long tradition of mediation and accommodation, social cleavages that "cross-cut" the population rather than coincide with each other, unwritten and written rights of proportionality (fairness) and cultural protectionism, a federal system with strong sub-national units, a civil society that fosters unity, direct democracy through frequent referenda, small size, historical time difference between cleavage in language and religion, neutrality in international warfare, and economic prosperity [27–29, 32–36]. Geography plays an unclear, presumably supporting, role in these frameworks. The analysis of coexistence in Switzerland is also part of a broader debate about whether social and geographical aspects of federalism promote peace or conflict [37–39].

Switzerland's linguistic and geographic heterogeneity is shown in Fig. 12.4. We consider the potential for violence due to linguistic (Fig. 12.5) and religious (Fig. 12.6) groups, each in turn. Initial analyses and the sequence of historical boundary formation suggested considering topographical barriers when discussing language groups, and political barriers when considering religious groups. The geography of languages primarily reflects the extent of invasions prior to the existence of current political boundaries and has remained stable in most areas for over a 1,000 years [28]. The modern state was established afterwards, and religious conflict played a role in establishing the internal political boundaries [28–30]. Census data were obtained for 2,634 municipalities (communes) in Switzerland (bfs.admin.ch), yielding a high spatial resolution.

Physical boundaries such as mountain ranges and lakes or national and subnational political boundaries that establish local autonomy may prevent the violations of cultural norms and enable self-determination, inhibiting the triggers of violence. By creating autonomous domains of activity and authority, the boundaries shield groups of the characteristic size from each other when they correspond with their

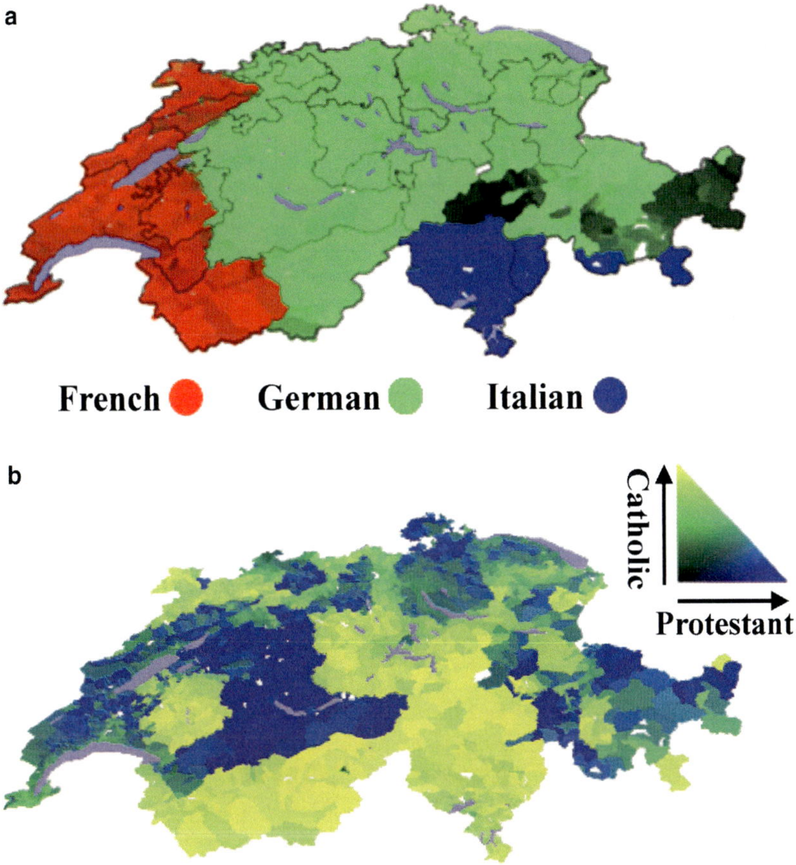


**Fig. 12.4** Maps of Switzerland showing the 2000 census proportion of (**a**) linguistic groups, (**b**) Catholic and Protestant (Mercator projection)

geographical domains. Where explicit boundaries do not exist, such as in mixed cantons where alpine boundaries are absent, violence might be expected, and the results of the model in these areas serve as a particularly stringent test of the theory.

*Language and Topographical Barriers.* We study the three main language groups—German, French and Italian (Fig. 12.6a)—which together comprise 91 % of the total population in the 2000 census (Romansh, the fourth official language, accounts for less than 2 %). We considered only the effect of physical boundaries due to lakes and mountain ranges (Fig. 12.6b, c). We determined the presence of topographical boundaries using an edge detection algorithm on topographical heights (Fig. 12.6d). This process identifies where there is a sharp change in height, i.e., a cliff, or steep incline, that runs for a significant distance forming a natural boundary. Elevation data with a spatial resolution of approximately 91 m [40] was coarsened to pixels of size $9.1 \times 9.1$ km. Edges were identified where there was an increase of more than 1.8 km in height over a distance of 9.1 km (11.5°) using a discretized Laplacian differential

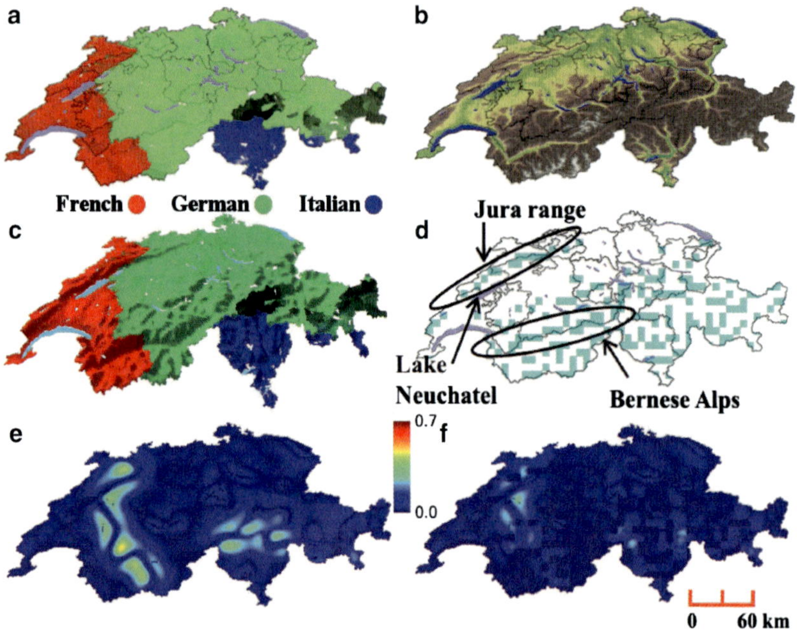


**Fig. 12.5** Maps of Switzerland showing (**a**) proportion of linguistic groups according to the 2000 census, (**b**) elevation within Switzerland, (**c**) overlay of linguistic groups onto a digital elevation model, and (**d**) topographical features including lakes (*blue*) and ridges extracted using edge detection (*cyan*). Comparison of calculated propensity (*color bar*) to violence between linguistic groups without (**e**) and with (**f**) the inclusion of topographical features as boundaries using a characteristic length scale of 24 km. Mercator projection, except C which is the Europe Albers projection. The distance scale is approximate

operator [41] with a mask size of a single pixel. The conclusions are robust to variations in the elevation angle. Calculations of the propensity to violence are reported here (Fig. 12.6e, f) for the characteristic length of 24 km. Without boundaries, the correlation of the wavelet filter yields a maximum propensity to violence value of 0.48. With topographical boundaries the maximum propensity is reduced to 0.30. Between the German and French-speaking areas to the northwest, the Jura mountain range and Lake Neuchatel, and to the south, the Bernese Alps, are mitigating boundaries. The interface between Lake Neuchatel and the Bernese Alps through the canton of Fribourg has no mitigating boundary, but is almost straight—neither side is surrounded by the other, so the propensity is low. Between the Italian and German-speaking areas, the Lepontine Alps dramatically reduce the calculated propensity.

The Jura range is, however, a porous boundary, and the highest residual propensity is adjacent to it in the northwest of the canton of Bern, which, unique in Switzerland, is historically known to be an area of "intense" linguistically based conflict, including arson, bombings, and other terrorist tactics [36, 42]. We obtained a correlation higher than 0.95 between predicted and reported violence, consistent with the hypothesis of the model. Manifesting Swiss willingness to create political boundaries, the conflict led to a referendum, and in 1978 the

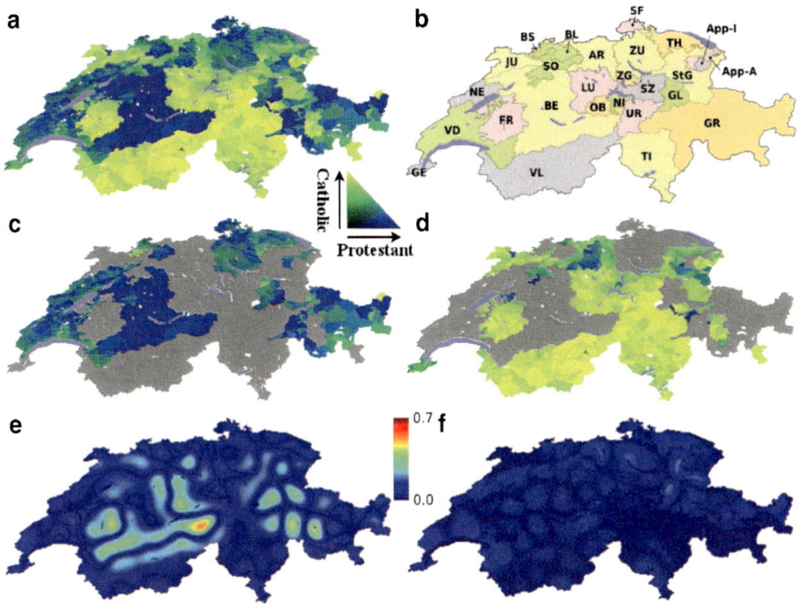


**Fig. 12.6** Maps of Switzerland (Mercator projection) showing (**a**) proportion of Catholic (*yellow*) and Protestant (*blue*) according to the 2000 census, (**b**) cantons, (**c**) and (**d**) cantons (and Graubünden circles) that are majority Protestant and Catholic, respectively, using the same color map as (**a**). Comparison of propensity to violence between religious groups without (**e**) and with (**f**) the inclusion of administrative boundaries using a characteristic length scale of 24 km. Propensity value scale is shown by *color bar*. Canton abbreviations are *GE* Genève, *SO* Solothurn, *ZG* Zug, *VL* Valais, *BS* Basel-Stadt, *GL* Glarus, *VD* Vaud, *BL* Basel-Landschaft, *TI* Ticino, *NE* Neuchatel, *AR* Aargau, *GR* Graubünden, *FR* Fribourg, *LU* Lucerne, *App-A* Appenzell-Ausserhoden, *BE* Bern, *OB* Obwalden, *App-I* Appenzell-Innerrhoden, *JU* Jura, *NI* Nidwalden, *StG* St. Gallen, *UR* Uri, *SF* Schaffhausen, *TH* Thurgau, *SZ* Schwyz, *ZU* Zurich

modern-day canton of Jura was created out of part of the north of what was then the canton of Bern [30]. While the conflict underlying the unrest was linguistic, local votes led to separation by majority religion. However, conflict did not end, and a proposal to shift the French-speaking Protestant areas of Bern to join French-speaking Catholic Jura is currently being considered [43]. Our results suggest that a calculated propensity to violence of 0.3 should be considered just at the threshold for actual violence, even under the social and political conditions prevailing in Switzerland. Remarkably, at this threshold high correlations (above 0.8) also are found in the former Yugoslavia. Thus, similar propensities for violence in different social contexts result in violence.

*Religious Groups and Political Barriers.* The two main religious groups of Switzerland are Protestant and Catholic. The Swiss federal political system separates the country into 26 "cantons" and "half-cantons" considered as semi-autonomous political units (Fig. 12.6). Moreover, this schema is repeated within the largest canton by area, Graubünden, whose sub-cantonal divisions called circles (*kreise*) have a distinctive political autonomy [27, 35]. We obtained canton boundaries from mapping resources (www.gadm.org, www.toposhop.admin.ch). Circles boundaries were identified by district lists (www.gis.gr.ch). In the 2000 census, Roman Catholic and Protestant affiliations account for 77 % of the total population. Less than 8 % subscribe to other religions, and the remainder have no religious affiliation or did not specify one. Without boundaries, the maximum calculated propensity to violence is very high (0.57), and with political borders it is only 0.20. Without Graubünden circles, the propensity increases to a quite high 0.42, still well above the threshold. Because of a 10 % decline in religious affiliation in recent years, we considered also the 1990 census, with similar conclusions.

The separation of religions by canton is apparent geographically and historically. In some cases the area of a canton includes small enclaves embedded in another canton whose majority religion corresponds to the canton to which they belong. Still, there are exceptions to the separation of religions by canton. In each case the geography is sufficient to limit the propensity to violence. For example, there is an area of Protestant majority in the far north of the Catholic canton of Fribourg. It is, however on a long appendage and therefore is not surrounded by Catholic areas, and so has a low propensity to violence according to the analysis. Historical evidence is found in conflict in the 1500s [30]. The Reformation led to cantons adopting a Protestant or retaining a Catholic identity. A brief war resulted in a peace treaty that established religious freedom by canton. The canton Appenzell was split by religious differences into two "half-cantons" Innerrhoden and Ausserrhoden. The political independence of circles (*kreise*) in Graubünden also provided religious autonomy [35]. The intentional formation of political boundaries in regions that would have violence according to the model, and the subsequent model propensity below the threshold associated with a lack of actual violence are consistent with the hypothesis on the role of boundaries in peaceful coexistence.

This work is part of a broader effort to use new methods for quantitative analysis of patterns of violence and their prevention [3, 4, 7–9, 12, 14, 44, 45]. There is also interest in ethnic group interactions across national borders [46–48]. We have shown that groups that are not well mixed but are geographically separated by natural or political boundaries into autonomous domains are peaceful in both Switzerland and the former Yugoslavia. Our model provides good agreement between reported and predicted violence in the former Yugoslavia and India, and it clarifies the role of borders in Switzerland including ambiguities of mixed language and religion Swiss cantons by showing that in most cases the natural geography of the populations conspires to lead to a low level of violence, so that additional boundaries were not necessary; where they were needed, as in Graubünden, they were established. The highest calculated propensity to violence is between linguistic groups in the northern part of the canton of Bern, where historically unresolved real world tensions actually exist. Historical evidence suggests that for religious groups the boundaries in Switzerland were created to provide autonomy to a group with a shared identity and avoid conflict among multiple groups. Ongoing efforts to reduce tensions in Bern include introducing new political boundaries. The many political, social, and economic factors that play roles in reducing violence [27–29, 32–37] build on a strong foundation of geographical borders.

Our analysis indicates that both administrative and natural barriers can play a significant role in mitigating conflict between religious and linguistic groups. Accordingly, when partition within a country is viewed as an acceptable form of conflict mitigation, such partition can give rise to highly stable coexistence and peace.

**Acknowledgments** We thank Stuart Pimm, Irving Epstein, and Lawrence Susskind for helpful comments on the manuscript, Michael Widener, Urbano França, Casey Friedman, and Blake Stacey for help with a figure and formatting. This work was supported in part by AFOSR under Grant FA9550-09-1-0324 and ONR under Grant N000140910516.

## References

1. Lim, M., Metzler, R., & Bar-Yam, Y. (2007). Global pattern formation and ethnic/cultural violence. *Science, 317*, 1540.
2. Rutherford, A., Harmon, D., Werfel, J., Bar-Yam, S., Gard-Murray, A., Gros, A., et al. (2011). Good Fences: The importance of setting boundaries for peaceful coexistence. NECSI Report. arXiv:1110.1409.
3. Horowitz, D. L. (2000). *Ethnic groups in conflict* (2nd ed.). Berkeley/Los Angeles: University of California Press.
4. Harff, B., & Gurr, T. R. (2004). *Ethnic conflict in world politics* (2nd ed.). Boulder: Westview.
5. Huntington, S. (1996). *The clash of civilizations and the remaking of world order*. New York: Simon & Schuster.
6. Chirot, D., & Seligman, M. E. P. (Eds.). (2001). *Ethnopolitical warfare: Causes, consequences, and possible solutions*. Washington, DC: American Psychological Association.
7. Reynal-Querol, M. (2002). Ethnicity, political systems, and civil wars. *Journal of Conflict Resolution, 46*, 29.
8. Gulden, T. R. (2002). Spatial and temporal patterns in civil violence: Guatemala, 1977–1986. *Politics Life Sciences, 21*, 26.
9. Buhaug, H., & Gates, S. (2002). The geography of civil war. *Journal of Peace Research, 39*, 417.
10. Varshney, A. (2003). *Ethnic conflict and civic life: Hindus and Muslims in India*. New Haven: Yale University Press.
11. Toft, M. D. (2003). *The geography of ethnic violence: Identity, interests, and the indivisibility of territory*. Princeton: Princeton University Press.
12. Fox, J. (2004). *Religion, civilization, and civil war: 1945 through the new millennium*. Lanham, MD: Lexington Books.
13. Mann, M. (2004). *The dark side of democracy: Explaining ethnic cleansing*. New York: Cambridge University Press.
14. Lustick, I. S., Miodownik, D., & Eidelson, R. J. (2004). Secessionism in multicultural states: Does sharing power prevent or encourage it? *American Political Science Review, 98*, 209.
15. Chayes, A., & Minow, M. L. (Eds.). (2003). *Imagine coexistence: Restoring humanity after violent ethnic conflict*. San Francisco: Jossey-Bass.
16. Kaufmann, C. (1998). When all else fails: Ethnic population transfers and partitions in the twentieth century. *International Security, 23*, 120.
17. Easterly, W., & Levine, R. (1997). Africa's growth tragedy: Policies and ethnic divisions. *Quarterly Journal of Economics, 112*, 1203.
18. Collier, P., & Hoeffler, A. (1998). On the economic causes of civil war. *Oxford Economic Papers, 50*, 563.
19. Bates, R. (2000). Ethnicity and development in Africa: A reappraisal. *The American Economic Review, 90*(2), 131–134.
20. Fearon, J. D., & Laitin, D. D. (2003). Ethnicity, insurgency, and civil war. *American Political Science Review, 97*, 75–90 (2003).

21. Posner, D. N. (2004). Measuring ethnic fractionalization in Africa. *American Journal of Political Science, 48*(4), 849–863.
22. Ivanov, P. Ch., et al. (1996). *Nature, 383*, 323.
23. Daubechies, I. (1992). *Ten lectures on wavelets*. Philadelphia: SIAM.
24. Map of Yugoslavia. (1991). Courtesy of the University of Texas Libraries. http://www.lib.utexas.edu/maps/europe/yugoslav.jpg.
25. Petrovic, R. (1992). The national composition of Yugoslavia's population, 1991. *Yugoslav Survey, 33*, 3.
26. Chaudhuri, K. (2001). Frontline. http://www.hinduonnet.com/fline/fl1802/18020330.htm. Accessed 20 Jan 2001.
27. Lijphart, A. (1977). *Democracy in plural societies: A comparative exploration*. New Haven: Yale University Press.
28. Schmid, C. L. (1981). *Conflict and consensus in Switzerland*. Berkeley: University of California Press.
29. Martin, W. (1931). A history of Switzerland: An essay on the formation of a confederation of states. London: G. Richards.
30. Hof, U. (1991). *Im, Geschichte der Schweiz. Kohlhammer Urban-Taschenbücher* (vol. 188). Stuttgart: Kohlhammer.
31. Bonaparte, N. (2002). In V. Monnier (Ed.), *Proclamation de St. Cloud, 30 septembre 1802 in Bonaparte et la Suisse: Travaux Preparatoires de l'Acte de Mediation (1803)*. Geneva: Helbing & Lichtenhahn.
32. Steiner, J. (1974). *Amicable agreement versus majority rule: Conflict resolution in Switzerland*. Chapel Hill: University of North Carolina Press.
33. Glass, H. E. (1977). Ethnic diversity, elite accommodation and federalism in Switzerland. *Publius, 7*, 31.
34. Linder, W. (2010). *Swiss democracy: Possible solutions to conflict in multicultural societies* (3rd ed.). New York: Palgrave Macmillan.
35. Head, R. C. (2002). *Early modern democracy in the Grisons: Social order and political language in a Swiss Mountain Canton: 1470–1620*. Cambridge: Cambridge University Press.
36. McRae, K. D. (1983). *Conflict and compromise in multilingual societies: Switzerland*. Waterloo, ON: Wilfrid Laurier University Press.
37. Church, C. H. (2004). *The politics and government of Switzerland*. New York: Palgrave MacMillian.
38. Christin, T., & Hug, S. (2006). Federalism, the geographic location of groups, and conflict, CIS Working Paper No. 23, Center for Comparative and International Studies, ETH Zurich and University of Zurich.
39. Shearer, W. (2008). Determine indicators for conflict avoidance. Global Pattern Formation and Ethnic/Cultural Violence. https://www.sciencemag.org/content/317/5844/1540/reply
40. Jarvis, A., Reuter, A., Nelson, H. I., & Guevara, E. (2008). Hole-filled SRTM Version 4. http://srtm.csi.cgiar.org.
41. Bovik, A. (2009). *Essential guide to image processing*. Burlington: Academic.
42. Keech, W. R. (1972). Linguistic diversity and political conflict: Some observations based on four Swiss cantons. *Comparative Politics, 4*, 387.
43. Citizens to settle territorial Jura conflict, Swissinfo, May 4, 2009. http://www.swissinfo.ch/eng/politics/Citizens_to_settle_territorial_Jura_conflict.html?cid=7377228. Accessed 21.02.11.
44. Cederman, L. E., & Gleditsch, K. S. (2009) Introduction to special issue on "disaggregating civil war". *Journal of Conflict Resolution, 53*, 487.
45. Bohorquez, J. C., Gourley, S., Dixon, A. R., Spagat, M., & Johnson, N. F. (2009). Common ecology quantifies human insurgency. *Nature, 462*, 911.
46. Nugent, P., & Asiwaju, A. I. (Eds.). (1996). *African boundaries: Barriers, conduits and opportunities*. London: Pinter.
47. Mitchell, K. (1997). Transnational discourse: Bringing geography back in. *Antipode, 29*, 101.
48. Silberfein, M., & Conteh, A. (2006). Boundaries and conflict in the Mano River region of West Africa. *Conflict Management and Peace Science, 23*, 343.

# Chapter 13
# Food Security and Political Instability: From Ethanol and Speculation to Riots and Revolutions

**Marco Lagi, Karla Z. Bertrand, and Yaneer Bar-Yam**

**Overview** Social unrest may reflect a variety of factors such as poverty, unemployment, and social injustice. Despite the many possible contributing factors, the timing of violent protests in North Africa and the Middle East in 2011 as well as earlier riots in 2008 coincides with large peaks in global food prices. We identify a specific food price threshold above which protests become likely. These observations suggest that protests may reflect not only long-standing political failings of governments but also the sudden desperate straits of vulnerable populations. If food prices remain high, there is likely to be persistent and increasing global social disruption. Underlying the food price peaks we also find an ongoing trend of increasing prices. We extrapolate these trends and identify a crossing point to the domain of high impacts, even without price peaks, in 2012–2013. This implies that avoiding global food crises and associated social unrest requires rapid and concerted action.

In 2011 protest movements became pervasive in countries of North Africa and the Middle East. These protests were associated with dictatorial regimes and were often considered to be motivated by the failings of the political systems in the human rights arena [1–4]. We have shown that food prices are the precipitating condition for social unrest [5–12] and identified a specific global food price threshold for unrest. As we predicted, the trend of food prices reached the threshold in 2012–2013, even without sharp peaks. This points to a danger of spreading global social disruption.

Historically, there are ample examples of "food riots," with consequent challenges to authority and political change, notably in the food riots and social instability across Europe in 1848, which followed widespread droughts [13]. While many other causes of social unrest have been identified, food scarcity or high prices often underlie riots, unrest, and revolutions [14–20]. Today, many poor countries

M. Lagi • K.Z. Bertrand • Y. Bar-Yam (✉)
New England Complex Systems Institute, Cambridge, MA, USA
e-mail: yaneer@necsi.edu

 
P.V. Fellman et al. (eds.), *Conflict and Complexity*, Understanding Complex Systems,
DOI 10.1007/978-1-4939-1705-1_13

rely on the global food supply system and are thus sensitive to global food prices [21]. This condition is quite different from the historical prevalence of subsistence farming in undeveloped countries, or even a reliance on local food supplies that could provide a buffer against global food supply conditions. It is an example of the increasingly central role that global interdependence is playing in human survival and well-being [22–24]. We can understand the appearance of social unrest in 2011 based upon a hypothesis that widespread unrest does not arise from long-standing political failings of the system, but rather from its sudden perceived failure to provide essential security to the population. In food importing countries with widespread poverty, political organizations may be perceived to have a critical role in food security. Failure to provide security undermines the very reason for existence of the political system. Once this occurs, the resulting protests can reflect the wide range of reasons for dissatisfaction, broadening the scope of the protest, and masking the immediate trigger of the unrest.

Human beings depend on political systems for collective decision making and action and their acquiescence to those systems, if not enthusiasm for them, is necessary for the existence of those political systems. The complexity of addressing security in all its components, from protection against external threats to the supply of food and water, is too high for individuals and families to address themselves in modern societies [25]. Thus, individuals depend on a political system for adequate decision making to guarantee expected standards of survival. This is particularly true for marginal populations, i.e. the poor, whose alternatives are limited and who live near the boundaries of survival even in good times. The dependence of the population on political systems engenders its support of those systems, even when they are authoritarian or cruel, compromising the security of individuals while maintaining the security of the population. Indeed, a certain amount of authority is necessary as part of the maintenance of order against atypical individuals or groups who would disrupt it. When the ability of the political system to provide security for the population breaks down, popular support disappears. Conditions of widespread threat to security are particularly present when food is inaccessible to the population at large. In this case, the underlying reason for support of the system is eliminated, and at the same time there is "nothing to lose," i.e. even the threat of death does not deter actions that are taken in opposition to the political order. Any incident then triggers death-defying protests and other actions that disrupt the existing order. Widespread and extreme actions that jeopardize the leadership of the political system, or the political system itself, take place. All support for the system and allowance for its failings are lost. The loss of support occurs even if the political system is not directly responsible for the food security failure, as is the case if the primary responsibility lies in the global food supply system.

The role of global food prices in social unrest can be identified from news reports of food riots. Figure 13.1 shows a measure of global food prices, the UN Food and Agriculture Organization (FAO) Food Price Index [57] and the timing of reported food riots in recent years. In 2008 more than 60 food riots occurred worldwide [58] in 30 different countries [59], 10 of which resulted in multiple deaths [30–40], as shown in the figure. After an intermediate drop, even higher prices at the end of

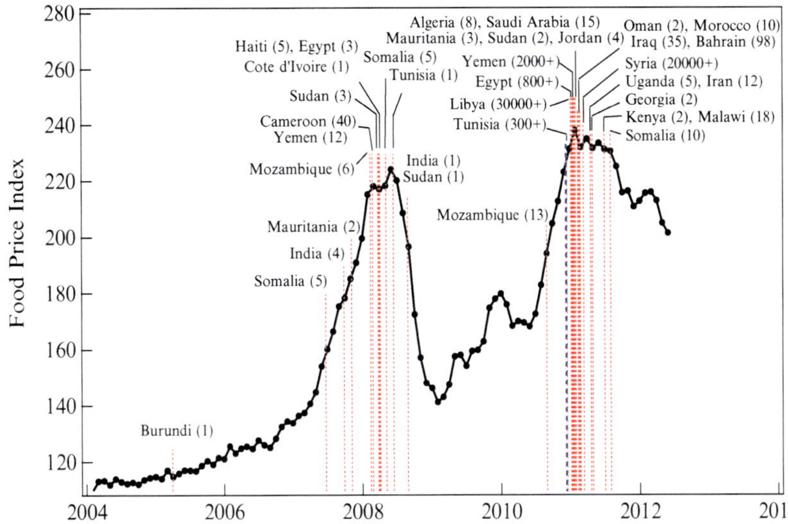


**Fig. 13.1** Time dependence of FAO Food Price Index from January 2004 to June 2012. *Red dashed vertical lines* correspond to beginning dates of "food riots" and protests associated with the major recent unrest in North Africa and the Middle East. The overall death toll is reported in *parentheses* [26–55]. *Blue vertical line* indicates the date, December 13, 2010, on which we submitted a report to the U.S. government, warning of the link between food prices, social unrest, and political instability [56]

2010 and the beginning of 2011 coincided with additional food riots (in Mauritania and Uganda [45, 55]), as well as the larger protests and government changes in North Africa and the Middle East known as the Arab Spring [42–44, 46–54]. There are comparatively fewer food riots when the global food prices are lower. Three of these, at the lowest global food prices, are associated with specific local factors affecting the availability of food: refugee conditions in Burundi in 2005 [26], social and agricultural disruption in Somalia [27], and supply disruptions due to floods in India [28, 39]. The latter two occurred in 2007 as global food prices began to increase but were not directly associated with the global food prices according to news reports. Two additional food riots in 2007 and 2010, in Mauritania [29] and Mozambique [41], occurred when global food prices were high, but not at the level of most riots, and thus appear to be early events associated with increasing global food prices.

These observations are consistent with a hypothesis that high global food prices are a precipitating condition for social unrest. More specifically, food riots occur above a threshold of the FAO price index of 210 ($p < 10^{-7}$, binomial test). The observations also suggest that the events in North Africa and the Middle East were triggered by food prices. Considering the period of time from January 1990 to May 2011 (Fig. 13.1 inset), the probability that the unrest in North Africa and the Middle East occurred by chance at a period of high food prices is $p < 0.06$ (one sample binomial test). This conservative estimate considers unrest across all countries to be a single unique event over this period of just over 20 years. If individual country

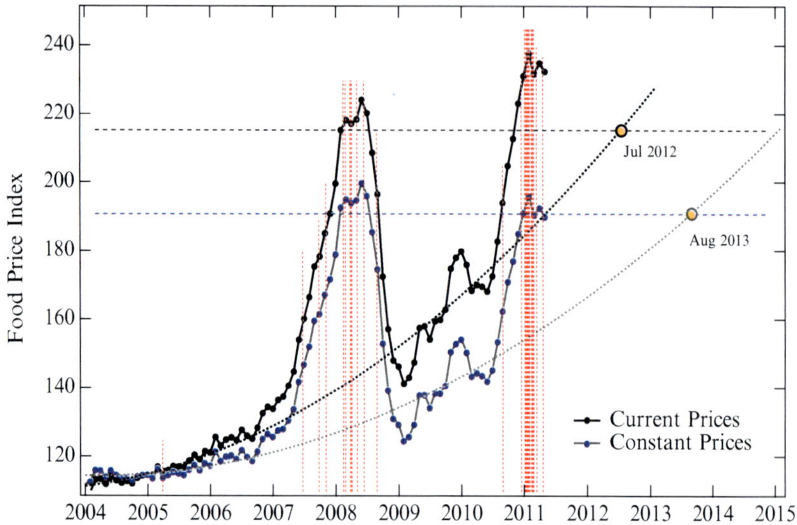


**Fig. 13.2** Time dependence of FAO Price Index at current prices (*upper black curve*) and constant prices (corrected for inflation, *lower blue curve*) from January 2004 to May 2011. *Red dashed vertical lines* correspond to beginning dates of food riots and events associated with the major recent unrest in North Africa and the Middle East. *Black and blue horizontal lines* represent the price threshold above which riots are ignited in current and constant prices, respectively. Index backgrounds are fitted with a third-order polynomial; intersection with the threshold (July 2012 at current prices, August 2013 at prices corrected for world inflation, [66]) represents the point of instability

events are considered to be independent, because the precipitating conditions must be sufficient for mass violence in each, the probability of coincidence is much lower.

A persistence of global food prices above this food price threshold should lead to persistent and increasing global unrest. Underlying the peaks in Fig. 13.1, we see a more gradual, but still rapid, increase of the food prices during the period starting in 2004. It is reasonable to hypothesize that when this underlying trend exceeds the threshold, the security of vulnerable populations will be broadly and persistently compromised. Such a threat to security should be a key concern to policymakers worldwide. Social unrest and political instability of countries can be expected to spread as the impact of loss of security persists and becomes pervasive, even though the underlying causes are global food prices and are not necessarily due to specific governmental policies. While some variation in the form of unrest may occur due to local differences in government, desperate populations are likely to resort to violence even in democratic regimes. A breakdown of social order as a result of loss of food security has been predicted based upon historical events and the expectation that global population increases and resource constraints will lead to catastrophe [60–63]. In 2011 we predicted, as shown in Fig. 13.2, that the underlying trend of increasing prices would reach the threshold of instability in July 2012, considering current prices, and August 2013 if we correct prices for reported inflation. Indeed,

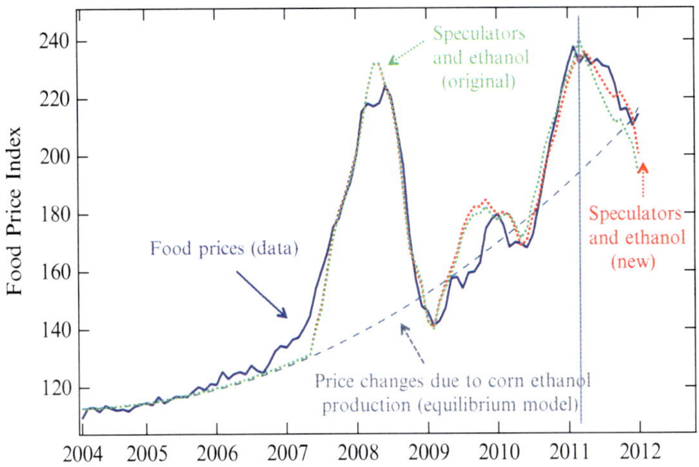


**Fig. 13.3 Food prices and model simulations**. The FAO Food Price Index (*blue solid line*) [57], the ethanol supply and demand model (*blue dashed line*), where dominant supply shocks are due to the conversion of corn to ethanol so that price changes are proportional to ethanol production (see [67], Appendix C) and the results of the speculator and ethanol model (*green* and *red dotted lines*), that adds speculator trend following and switching among investment markets, including commodities, equities and bonds (see [67], Appendices D and E). The *green curve* is the fit extended to the present with the original parameter values, the *red curve* is the fit with new optimized parameters. The *vertical blue bar* marks the end of the original fit in March 2011. Original parameters [67]: $k_{sd}=0.098$, $k_{sp}=1.29$, $\mu_{\text{equity}}\gamma_0=-0.095$, $\mu_{\text{bonds}}\gamma_0=-67.9$. New optimized parameters: $k_{sd}=0.093$, $k_{sp}=1.27$, $\mu_{\text{equity}}\gamma_0=-0.085$, $\mu_{\text{bonds}}\gamma_0=-48.2$

consistent with our analysis, large populations are in distress, as described in a UN report warning of the growing crisis [64]. The FAO food price index has been at or above 210 for much of the relevant period, and we have seen multiple examples of unrest characteristic of threshold food prices [65]. While individual events may not be directly due to food prices, the extent of global social unrest during this period is historically remarkable. As of midyear 2013, prices are still just at the food price threshold [57].

In a paper published in September 2011, we constructed the first dynamic model that quantitatively agreed with food prices [67]. The model was able to fit the FAO Food Price Index time series from January 2004 to March 2011, inclusive, and showed that the dominant causes of price increases during this period were investor speculation and ethanol conversion. It included investor trend following as well as shifting between commodities, equities, and bonds to take advantage of increased expected returns.

In a subsequent paper, we showed that the model was still able to fit food prices up to (then) currently available data (through January 2012) [68]. In Fig. 13.3 we plot both the original fit (green curve), extended to the present using the original parameters, and a new fit obtained by adjusting the parameters to optimize the fit over the entire period (red curve). Without any modification to its assumptions or formulation, by extending its original dynamics, the model proved to be robust, and consistent with the ongoing behavior of food prices. The agreement of the fit with the FAO Food Price Index remained strikingly quantitatively accurate, validating both the descriptive and predictive abilities of the model.

While there have been several suggested origins of the food price increases, we find the dominant ones to be investor speculation and ethanol production. Our analysis shows that the two parts of the dynamics of prices can be directly attributed to the two different causes: the price peaks are due to speculators causing price bubbles, and the background increase shown in Fig. 13.2 is due to corn to ethanol conversion. This intuitive result is made quantitative by the analysis in those papers [67, 68].

Both factors in food prices can be linked directly to recent US governmental actions. Speculator activity has been enhanced by deregulation of the commodities markets that exempted dealers from trading limits [69–71], and subsidies and other policies have been central to the growth of ethanol conversion [72, 73].

The importance of food prices for social stability points to the level of human suffering that may be caused by these price-raising policies. The analysis we presented of the timing of peaks in global food prices and social unrest implies that the 2011 unrest was precipitated by a food crisis that is threatening the security of vulnerable populations. Deterioration in food security led to conditions in which random events trigger widespread violence. The condition of these vulnerable populations could have been much worse except that some countries controlled food prices in 2011 due to the unrest in 2008 [74–85]. Food price controls in the face of high global food prices carry associated costs. Because of the strong cascade of events in the Middle East and North Africa only some countries had to fail to adequately control food prices for events to unfold [86–90]. This understanding suggests that reconsidering biofuel policy as well as commodity market regulations should be an urgent priority for policymakers. Reducing the amount of corn converted to ethanol, and restricting commodity future markets to bona fide risk hedging would reduce global food prices [67, 68]. The current problem transcends the specific national political crises to represent a global concern about vulnerable populations and social order.

Our analysis of the link between global food prices and social unrest supports a growing conclusion that it is possible to build mathematical models of global economic and social crises [91–100]. Identifying a signature of unrest for future events is surely useful. Significantly, prior to the unrest, on December 13, 2010, we submitted a government report [56] analyzing the repercussions of the global financial crises, and directly identifying the risk of social unrest and political instability due to food prices (see Fig. 13.1). This report, submitted 4 days before the initial human trigger event, the action of Mohammed Bouazizi in Tunisia [101, 102], demonstrates that it is possible to identify early warning signs before events occur. Likewise, our reports in 2011 anticipated the 2013 unrest. Prediction is a major challenge for socio-economic analysis. Understanding when and whether prediction is possible is important for science and policy decisions. Our predictions are conditional on the circumstances, and thus allow for policy interventions to change them. Whether policy makers will act depends on the various pressures that are applied to them, including both the public and special interests.

**Acknowledgments** We thank Blake Stacey, Amaç Herdağdelen, Alexander Gard-Murray, Andreas Gros, and Shlomiya Bar-Yam for helpful comments on the manuscript, and Urbano França for editorial assistance. This work was supported in part by AFOSR under grant FA9550-09-1-0324 and ONR under grant N000140910516.

## References

1. Zakaria, F. (2011, February 03). How democracy can work in the Middle East. *TIME Magazine*. http://www.time.com/time/magazine/article/0,9171,2046038,00.html.
2. Mason, P. (2011, February 05). Twenty reasons why it's kicking off everywhere. *BBC News Idle Scrawl*. http://www.bbc.co.uk/blogs/newsnight/paulmason/2011/02/twenty_reasons_why_its_kicking.html.
3. Shah, A. (2011, March 06). Middle East and North African unrest. *Global Issues*. http://www.globalissues.org/article/792/mideast-north-africa-unrest.
4. Beheshtipour, H. (2011, April 21). Analysis of current ME developments. *Press TV*. http://www.presstv.ir/detail/175942.html.
5. Arezki, R., & Bruckner, M. (2011). Food Prices and Political Instability. IMF Working Paper 11/62.
6. MacKenzie, D. (2011, February 03). Can complexity theory explain Egypt's crisis? *New Scientist*. http://www.newscientist.com/article/dn20082-can-complexity-theory-explain-egypts-crisis.html.
7. Lynch, D. (2008, April 30). Tension in Egypt shows potency of food crisis. *USA Today*. http://www.usatoday.com/money/world/2008-04-29-egypt-food-shortage_N.htm.
8. Fraser, E., & Rimasr, A. (2011, January 30). The psychology of food riots. *Foreign Affairs*. http://www.foreignaffairs.com/articles/67338/evan-fraser-and-andrew-rimas/the-psychology-of-food-riots.
9. McCann, S. (2011, January 31). America and the Middle East food riots. *American Thinker*. http://www.americanthinker.com/2011/01/america_and_the_middle_east_fo.html.
10. Kudlow, L. (2011, January 28). Food riots: Is Bernanke partially to blame? *Money and Politics*. http://www.cnbc.com/id/41317486/Food_Riots_Is_Bernanke_Partially_to_Blame.
11. Bellemare, M. F. (2011, June 28). Rising food prices, food price volatility, and political unrest. *SSRN*. http://ssrn.com/abstract=1874101.
12. Brinkman, H.-J., & Hendrix, C. S. (2010, August 02). *Food insecurity and conflict: Applying the WDR framework*. World Development Report 2011. http://wdr2011.worldbank.org/food.
13. Dowe, D., Haupt, H.-G., Langewiesche, D., & Sperber, J. (Eds.). (2001). *Europe in 1848: Revolution and reform*. New York: Berghahn Books.
14. Rudé, G. (1964). *The crowd in history: A study of popular disturbances in France and England, 1730–1848*. Hoboken, NJ. Wiley.
15. Quinault, R., & Stephenson, J. (1974). *Popular protest and public order*. Crows Nest, Australia: Allen & Unwin.
16. Walton, J., & Seddon, D. (1994). *Free markets & food riots: The politics of global adjustment*. Hoboken, NJ: Wiley.
17. Gurr T. R. (1970). *Why men rebel*. Princeton, NJ: Princeton University Press.
18. Tilly, C. (1978). *From mobilization to revolution*. New York, NY: McGraw-Hill.
19. Humphreys, M., & Weinstein, J. M. (2007). Demobilization and reintegration. *Journal of Conflict Resolution, 51*, 531.
20. Blattman, C., & Miguel, E. (2010). Civil war. *Journal of Economic Literature, 48*, 3.
21. Senauer, B., & Venturini, L. (2005). *The globalization of food systems: A conceptual framework and empirical patterns*. The Food Industry Center, University of Minnesota. http://ageconsearch.umn.edu/bitstream/14304/1/tr05-01.pdf.
22. Bar-Yam, Y. (2002). *Encyclopedia of life support systems*. Oxford, UK: EOLSS.

23. MacKenzie, D. (2008, April 5). The collapse of civilization. *New Scientist, 2650*, 28.
24. MacKenzie, D. (2008, October 22). Why the financial system is like an ecosystem. *New Scientist, 2679*, 8–9.
25. Bar-Yam, Y. (1997). *Dynamics of complex systems*. New York, NY: Perseus Press.
26. Refugee killed in Burundi riot. (2005, April 1). *BBC News*. http://news.bbc.co.uk/2/hi/africa/4400811.stm.
27. Somalia: food riot leaves 5 dead. (2007, June 25). *USA Today*. http://www.usatoday.com/news/world/2007-06-25-somalia-unrest_N.htm.
28. Food riots continue in south Bengal. (2007, October 7). *Indian Muslim News and Information*. http://www.indianmuslims.info/news/2007/oct/07/food_riots_continue_south_bengal.html.
29. U.S. Department of State, Mauritania. (2007). http://www.state.gov/e/eeb/ifd/2008/100984.htm.
30. Healy, B., & Munckton, S. (2008, April 19). Global food crisis: Biofuels threaten hunger. *Green Left*. http://www.greenleft.org.au/node/39360.
31. Food riots fear after rice price hits a high. (2008, April 6). *khilafah.com*. http://www.khilafah.com/index.php/news-watch/71-archive/2412-food-riots-fear-after-rice-price-hits-a-high.
32. Worth, R. F. (2008, February 25). Rising inflation creates unease in Middle East. *The New York Times*. http://www.nytimes.com/2008/02/25/world/middleeast/25economy.html.
33. Egyptian court convicts 22 for food riots. (2008, December 15). *The Hindu*. http://www.hindu.com/thehindu/holnus/003200812151875.htm.
34. Kate, D. T. (2008, March 27). Grain prices soar globally. *The Christian Science Monitor*. http://www.csmonitor.com/World/Asia-Pacific/2008/0327/p01s02-woap.html.
35. Delva, J. G. (2008, April 24). World Food Program launches emergency call for Haiti. *Reuters*. http://www.reuters.com/article/2008/04/25/us-haiti-food-idUSN4O35282220080425.
36. Cote d'Ivoire: Country slashes food tax after deadly protests. (2008, April 3). *The Guardian*. http://www.afrika.no/Detailed/16453.html.
37. Five killed in food riots in Somali capital. (May 6, 2008). *The Straits Times*. http://www.straitstimes.com/print/Free/Story/STIStory_234396.html.
38. One protester killed, 18 wounded at Tunisian food price demo. (2008, June 15). *NowPublic*. http://www.nowpublic.com/world/one-protester-killed-18-wounded-tunisian-food-price-demo.
39. Food riots as Indian floods destroy 250,000 homes. (2008, August 27). *Reuters*. http://uk.reuters.com/article/2008/08/27/uk-india-floods-idUKBOM28314020080827.
40. Sanders, E. (2008, September 28). Darfur front lines may be shifting to camps. *Chicago Tribune*.http://articles.chicagotribune.com/2008-09-28/news/0809270321_1_kalma-ali-abdel-khaman-tahir-sudan-analyst-eric-reeves.
41. Mozambique to reverse bread price hikes: Minister. (2010, September 8). *Angola Press*. http://www.portalangop.co.ao/motix/en_us/noticias/africa/2010/8/36/Mozambique-reverse-bread-price-hikes-minister,a8c51400-b677-48fc-98c2-9d49f205ac3e.html.
42. Tunisia death toll was 300: UN. (2011, May 22). *news.com.au*. http://www.news.com.au/world/tunisia-death-toll-was-300-un/story-fn6sb9br-1226060394302.
43. Culprits sought after deadly riots. (2011, January 10). *thetelegraph.com*. http://www.dailytelegraph.com.au/culprits-sought-after-deadly-riots/story-fn6e1m7z-1225984730779.
44. Kouddous, S. A. (2011, January 29) Crackdown in Cairo. *The Pulitzer Center*. http://pulitzercenter.org/articles/cairo-egypt-habib-al-adly-trial-postponed-tahrir-square-revolution?format=print.
45. Mauritania's Bouazizi died today. (2011, January 23). *Dekhnstan*. http://dekhnstan.wordpress.com/2011/01/23/mauritanias-bouazizi-died-today/.
46. McDoom, O. (January 31, 2011). Sudanese student dies after protests - Activists. *Reuters*. http://uk.reuters.com/article/2011/01/31/uk-sudan-protests-idUKTRE70U21620110131.
47. Man dies after setting himself on fire in Saudi Arabia. (2011, January 22). *BBC News*. http://www.bbc.co.uk/news/world-middle-east-12260465.
48. Yemen: Death toll there at 11, Morocco: Five pro-democracy protesters dead. (2011, February 21).*Eagainst.com*.http://eagainst.com/articles/yemen-death-toll-there-at-11-morocco-thousands-march-for-reform/.

49. Deaths in Oman protests. (2011, February 27). *Al Jazeera*. http://english.aljazeera.net/news/middleeast/2011/02/201122711285085 2905.html.
50. Klapper, B. (2011, June 27). Libya death toll. Associated Press. http://www.huffingtonpost.ca/tag/libya-death-toll.
51. Haj, A. A. (2011, May 21). Yemen's president warns of al-Qaida takeover. Associated Press. http://www.deseretnews.com/article/700137365/Yemens-president-warns-of-al-Qaida-takeover.html.
52. McCrummen, S. (2011, February 26). Iraq 'Day of Rage' protests followed by detentions, beatings. *The Washington Post*. http://www.washingtonpost.com/wp-dyn/content/article/2011/02/26/AR2011022601854.html.
53. Bahrain. (2011, April 14). 'Thirty-one protesters killed, 600 arrested', since February, group says. *Adnkronos*. http://www.adnkronos.com/IGN/Aki/English/Security/Bahrain-Thirty-one-protesters-killed-600-arrested-since-February-group-says_311903716211.html.
54. Mroue, B. (2011, May 21). Syria death toll from violence tops 900. Associated Press. http://www.huffingtonpost.com/2011/05/21/syria-death-toll-from-vio_n_865178.html.
55. Kron, J. (April 21, 2011). Protests in Uganda over rising prices grow violent. *The New York Times*. http://www.nytimes.com/2011/04/22/world/africa/22uganda.html?_r=1.
56. Lagi, M., & Bar-Yam, Y. (December 2010). *Socio-economic impact of the financial crisis: Complex systems scoping analysis*. Technical Report. New England Complex Systems Institute.
57. Food and Agriculture Organization of the United Nations. (2014). *FAO Food Price Index*. http://www.fao.org/worldfoodsituation/wfs-home/foodpricesindex/en/.
58. Food crisis less severe than 2008 on rice price, OECD says. (2011 March 02). http://www.businessweek.com/news/2011-03-02/food-crisis-less-severe-than-2008-on-rice-price-oecd-says.html.
59. The food crisis. (2011, February 24). http://www.nytimes.com/2011/02/25/opinion/25fri2.html?_r=1.
60. Malthus, T. R. (1798). *An essay on the principle of population*. London: St. Paul's Church-yard.
61. Ehrlich, P. R. (1968). *The population bomb*. New York: Ballantine Books.
62. Diamond, J. (2005). *Collapse: How societies choose to fail or succeed*. New York: Viking Press.
63. Tainter, J. (1988). *The collapse of complex societies*. Cambridge, UK: Cambridge University Press.
64. Food and Agriculture Organization of the United Nations. (2010). *The state of food insecurity in the world: Addressing food insecurity in protracted crises*. Rome, Italy: FAO.
65. Bar-Yam, Y., Lagi, M., & Bar-Yam, Y. (2013). South African Riots: Repercussion of the Global Food Crisis and US Drought. http://necsi.edu/research/social/southafrica/.
66. Shane, M. (2010, December 22). *Historical consumer price indices for baseline countries/regions (in percent) 1969–2010*. International Financial Statistics, International Monetary Fund.
67. Lagi, M., Bertrand, K. Z., & Bar-Yam, Y. (2011). Food crises: A quantitative model of food prices including speculators and ethanol. arXiv:1108.2455v1 [physics.soc-ph]. http://arxiv.org/pdf/1108.2455v1.
68. Lagi, M., Bar-Yam, Y., Bertrand, K. Z., & Bar-Yam, Y. (2012 ). UPDATE February 2012—The Food Crises: Predictive validation of a quantitative model of food prices including speculators and ethanol conversion. arXiv:1203.1313v1 [physics.soc-ph]. http://arxiv.org/pdf/1203.1313v1.
69. How deregulation fueled the financial crisis. (2009, January 13). http://www.marketoracle.co.uk/Article8210.html.
70. United States. Office of the Federal Register. (2009, March 24). *Federal Register* (*Washington, DC), 74*(55), 12283.
71. Stewart, S., & Waldie, P. (2008, May 31). Who is responsible for the global food crisis? http://www.theglobeandmail.com/report-on-business/who-is-responsible-for-the-global-food-crisis/article688931/.

72. Hahn, R. W., & Cecot, C. (2009). The benefits and costs of ethanol: An evaluation of the government's analysis. *Journal of Regulatory Economics, 35*, 275.
73. Rismiller, C. W., & Tyner, W. E. (2009). *Cellulosic biofuels analysis: Economic analysis of alternative technologies*. Tech. Rep. 09-06. Dept. of Agricultural Economics, Purdue University, West Lafayette, Indiana. See Sect. 3–3.
74. Hanley, C. J. (2011, March 28). It's costly to keep Egypt's daily bread cheap. Associated Press.http://www.msnbc.msn.com/id/42303838/ns/business-world_business/t/its-costly-keep-egypts-daily-bread-cheap/.
75. Bracco II, D. (2011). The bakers of Abbassia. *Prime*. http://www.primecollective.com/the-bakers-of-abbassia/.
76. Cameroon sets up body to control food prices. (2011, February 3). *Reuters*. http://uk.reuters.com/article/2011/02/03/uk-cameroon-food-idUKTRE7122NY20110203.
77. Browne, A., & Frangos, A. (2010, November 18). China takes steps to control prices. *Business with the Wall Street Journal*. http://www.theaustralian.com.au/business-old/news/china-takes-steps-to-control-prices/story-e6frg90x-1225955318934.
78. Ukraine to control food prices. (2011, February 18). *CNC*. http://www.cncworld.tv/news/v_show/12591_Ukraine_to_control_food_prices.shtml.
79. Karim, M. (2011, January 7). Measures taken to control food prices. *bdnews24.com*. http://www.bdnews24.com/details.php?cid=2&id=188224.
80. Vietnam food prices climb amid inflation battle. (2011, July 14). *thanhniennews.com*. http://www.thanhniennews.com/2010/Pages/20110717180620.aspx.
81. Dushanbe limits food price increases. (2011, March 01). *Central Asia Economic Newswire*. http://centralasianewswire.com/Business/Dushanbe-limits-food-price-increases/viewstory.aspx?id=3417.
82. Malik directs ICT admin to control food prices. (2010, October 25). *Daily Times*. http://www.dailytimes.com.pk/default.asp?page=2010%5C10%5C25%5Cstory_25-10-2010_pg7_28.
83. Bharath stresses need to cut food imports. (2010, December 10). *Guardian Trinidad and Tobago*. http://test.guardian.co.tt/?q=news/general/2010/12/10/bharath-stresses-need-cut-food-imports.
84. Nyesiga, D. (2011, May 16). Gov't considers new rules to check high commodity prices. *Rwanda Chamber*. http://www.rwandachamber.org/news/govt+considers+new+rules+to+check+high+commodity+prices.
85. Abdussalam, A. (2011, January 11). News focus: Government taking measures to curb basic necessary price spiral. *ANTARA News*. http://www.embassyofindonesia.org/news/2011/01/news052.htm.
86. First death in Algerian riots. (2011, January 8). *Ahram Online*. http://english.ahram.org.eg/NewsContent/2/8/3387/World/Region/First-death-in-Algerian-riot--ress.aspx.
87. Bouazza, B. B., & Ganley, E. (2011, January 12). Jobless youths in Tunisia riot using Facebook. Associated Press. http://www.keprtv.com/news/tech/113357919.html.
88. Flood, Z. (2011, January 10). At least 20 killed in economic protests in Tunisia and Algeria. Associated Press. http://www.telegraph.co.uk/news/worldnews/africaandindianocean/tunisia/8250477/At-least-20-killed-in-economic-protests-in-Tunisia-and-Algeria.html.
89. Mangalindan, J. P. (2011, January 26). Social networks seen going dark in Egypt. *CNN Money*. http://tech.fortune.cnn.com/2011/01/26/social-networks-seen-going-dark-in-egypt/.
90. Dreibus, T. C. (2011, January 27) Wheat fluctuates as prices at five-month high spur liquidation. *Bloomberg BusinessWeek*. http://www.businessweek.com/news/2011-01-27/wheat-fluctuates-as-prices-at-five-month-high-spur-liquidation.html.
91. Lim, M., Metzler, R., & Bar-Yam, Y. (2007). Global pattern formation and ethnic/cultural violence. *Science, 317*, 1540.
92. MacKenzie, D. (2011, March 3). I predict a riot: Where the next dictator will fall. *New Scientist*. http://www.newscientist.com/article/mg20928024.000-i-predict-a-riot-where-the-next-dictator-will-fall.html.
93. Pozen, R. C., & Bar-Yam, Y. (2008, November 18). There's a better way to prevent "bear raids". *Wall Street Journal*. http://online.wsj.com/article/SB122697410070336091.html.

94. Harmon, D., & Bar-Yam, Y. (2008). *Technical report on the sec uptick repeal pilot.* Tech. Rep. 2008–11. New England Complex Systems Institute.
95. Vedantam, S. (February 16, 2009). The computer as a road map to unknowable territory, *Washington Post.* http://www.washingtonpost.com/wp-dyn/content/article/2009/02/15/AR2009021501794.html.
96. Harmon, D., Stacey, B., Bar-Yam, Y., & Bar-Yam, Y. (2010, November). Networks of economic market interdependence and systemic risk. arXiv:1011.3707v2.
97. Keim, B. (2010, November 18). Analysis of 2008 collapse shows economy networked for failure. *Wired.* http://www.wired.com/wiredscience/2010/11/economics-of-network-collapse.
98. Harmon, D., et al. (2010, February). Predicting economic market crises using measures of collective panic. arXiv:1102.2620v1.
99. Keim, B. (2011, March 18). Possible early warning sign for market crashes. *Wired.* http://www.wired.com/wiredscience/2011/03/market-panic-signs.
100. Kelland, K. (2011, March 14). Analysis: Scientists who predict change enjoy fertile times. *Reuters.* http://www.reuters.com/article/2011/03/14/us-science-complex-change-idUSTRE72D4Z720110314.
101. Abouzeid, R. (2011, January 21). Bouazizi: The man who set himself and Tunisia on fire. *TIME Magazine.* http://www.time.com/time/magazine/article/0,9171,2044723,00.html.
102. Watson, I., & Karadsheh, J. (2011, March 22). The Tunisian fruit seller who kickstarted Arab uprising. *CNN.* http://articles.cnn.com/2011-03-22/world/tunisia.bouazizi.arab.unrest_1_fruit-vendor-sidi-bouzid-abedine-ben-ali?_s=PM:WORLD.

# Chapter 14
# South African Riots: Repercussion of the Global Food Crisis and US Drought

**Yavni Bar-Yam, Marco Lagi, and Yaneer Bar-Yam**

**Overview** High and volatile global food prices have led to food riots and played a critical role in triggering the Arab Spring revolutions in recent years. The severe drought in the USA in the summer of 2012 led to a new increase in food prices. Through the fall, they remained at a threshold above which the riots and revolutions had predominantly occurred. Global prices at this level create conditions where an exacerbating local circumstance can trigger unrest. Global corn (maize) prices reached new highs, and countries that depend mostly on maize are more likely to experience high local food prices and associated pressures toward social unrest. Here we analyze the conditions in South Africa, which is a heavily maize-dependent country, and we find that consumer food indices have increased dramatically. Coinciding with the food price increases this summer, massive labor strikes in mining and agriculture have led to the greatest single incident of social violence since the fall of apartheid in 1994. Worker demands for dramatic pay increases reflect that their wages have not kept up with drastic increases in the prices of necessities, especially food. Without attention to the global food price situation, more incidents of food-based social instability are likely to arise. Other countries that have manifested food-related protests and riots in 2012 include Haiti (prior to the impact of Hurricane Sandy) and Argentina. Moreover, these cases of unrest are just the most visible symptom of widespread suffering of poor populations worldwide due to elevated food prices. Our analysis has shown that policy decisions that would directly impact food prices are decreasing the conversion of maize to ethanol in the USA, and reimposing regulations on commodity futures markets to prevent excessive speculation, which we have shown causes bubbles and crashes in these markets. Absent such policy actions, governments and companies should track and mitigate the impact of high and volatile food prices on citizens and employees.

---

Y. Bar-Yam • M. Lagi • Y. Bar-Yam (✉)
New England Complex Systems Institute, Cambridge, MA, USA
e-mail: yaneer@necsi.edu

 
P.V. Fellman et al. (eds.), *Conflict and Complexity*, Understanding Complex Systems,
DOI 10.1007/978-1-4939-1705-1_14

On August 2, 2012, a violent riot at a platinum mine in South Africa resulted in three deaths [8]. Subsequent events throughout the platinum mining region included a particularly violent incident on August 16 at the Marikana mine, which resulted in 34 strikers killed and about 80 more injured—the most violent such incident since the end of apartheid in 1994 [18, 20, 25]. The protests were taken up by gold miners [12] and agricultural workers, resulting in the destruction of over 120 acres of crops [21]. Observers considered these incidents to be so severe that they may undermine the country's peaceful reputation [20, 25]. The only event of comparable magnitude since the end of apartheid was a wave of xenophobic violence in May of 2008.

The conditions that give rise to social violence are often poorly understood. Here, however, we provide evidence for a link between the violence in South Africa and the rapidly rising food prices that have affected many parts of the world, including triggering widespread food riots in 2007–2008 and the uprisings in North Africa and the Middle East often called the Arab Spring in 2010–2011 [16]. At the beginning of August 2012, when the labor riots started, corn (maize) prices rose to record highs driven not only by a drought in the US Midwest but also by other underlying causes that have increased food prices rapidly in recent years. The earlier xenophobic violence [3] coincided with the 2007–2008 global food price peak that is linked to food riots in 30 countries.

In earlier papers [1, 2, 13–16] we have shown that high and spiking global food prices have triggered riots and revolutions, and warned that predicted future spikes in global food prices would lead to additional social unrest [16]. We constructed a quantitative model that identified the causes of the rises in global food prices to be biofuels, particularly the conversion of corn to ethanol in the USA, and increased financial speculation in the agricultural commodities futures market [14]. In addition to the global analyses, we have analyzed the origins of violence in Yemen [9] and the increase in cost of Mexico's maize imports [17].

This summer, a severe drought in the USA interrupted a declining food price trend and contributed to a new price peak, and the effects are being felt worldwide [13]. While the drought's impact on supply is important, our analysis reveals high levels of speculation that can amplify its impact on food prices well beyond the natural consequences of the supply shock. Maize and wheat prices rose rapidly through July. At the beginning of August, requests for a repeal three of the US ethanol mandate, which would have led to a significant increase in the food supply had it been granted, may have limited the price increases on the futures market [4, 13, 30]. Prices remained relatively constant from August through November. However, they remained at the threshold above which violence was found in 2007–2008 and 2010–2011, as seen in Fig. 14.1.

When prices are significantly higher than the threshold, as they were in 2007–2008 and 2010–2011, widespread violence can be expected. When the prices are proximate to the threshold, incidents of violence should be more sensitive to the specifics of local conditions. Due to cultural differences, different countries rely primarily upon different grains, mostly maize, wheat or rice (Fig. 14.2).

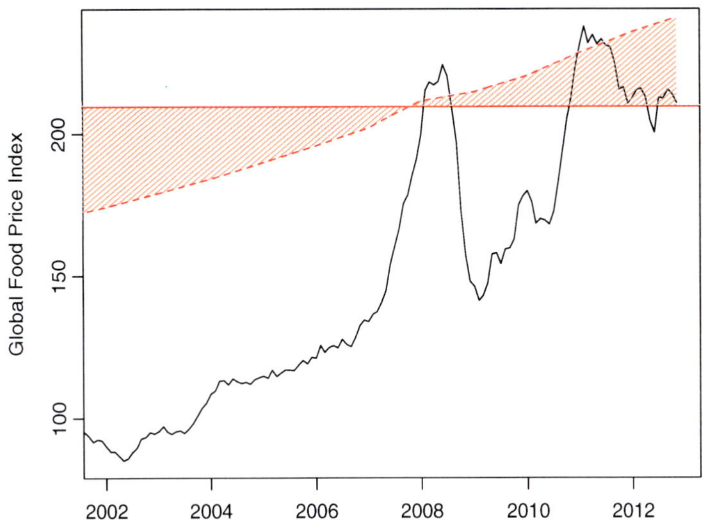


**Fig. 14.1** Global food price index since 2002 [11]. The threshold above which widespread food riots and revolutions occurred in 2008–2009 and 2010–2011 [16] is shown both without (*solid red*) and with (*dashed red*) inflation. Whether incomes of poor populations increase with inflation depends on local conditions and national policies. Since mid-2011 the food price index has hovered around the threshold value

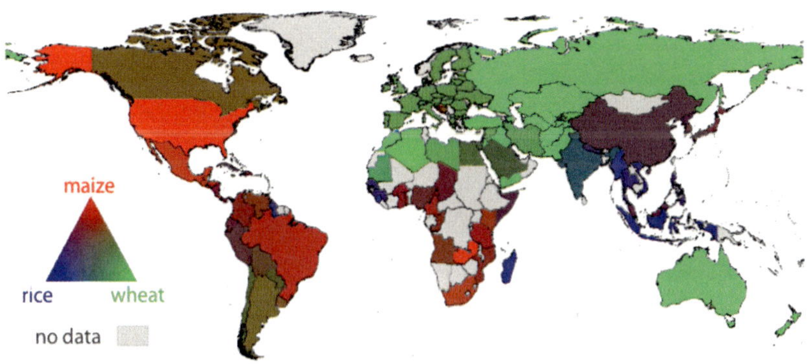


**Fig. 14.2** Map showing different levels of consumption for maize (*red*), wheat (*green*), and rice (*blue*), by country. Map is from [10], based on data from [31]

Also, local prices are not always fully coupled to global prices. Finally, national policies play an important role due to food subsidies and whether salaries increase with inflation, which may be pegged more or less directly to the food prices. These considerations suggest that local conditions and the distinct impacts of different grain prices can play a pivotal role in determining the societal response to price increases.

The prices of different grains (Fig. 14.3) are partially but not completely linked, with rice having a more distinct behavior than maize and wheat [29]. In particular,

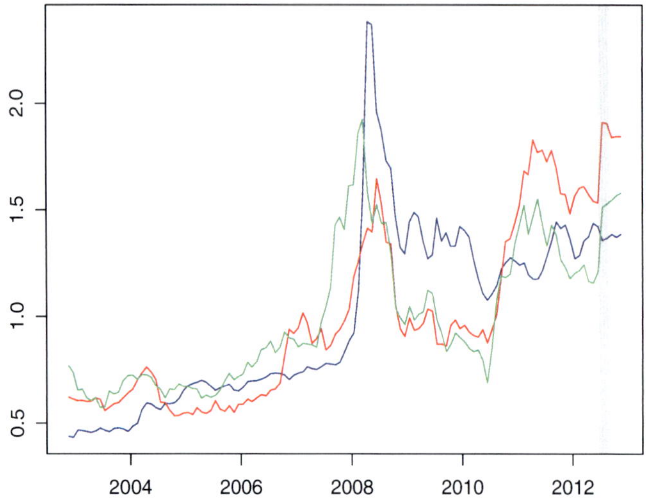


**Fig. 14.3** Futures prices for maize (*red*), wheat (*green*), and rice (*blue*) since 2002, normalized by the average price of each grain for that time period. *Shaded bar* is the period of severe drought in the USA in the summer of 2012

maize prices in the summer of 2012 reached and persisted at historic highs, and wheat prices were also high. The price of rice has been relatively stable since 2009 and well below the peak reached in 2008.

In South Africa, maize is the staple grain. In 2009, the most recent year of data available from the UN Food and Agricultural Organization [7], more than half of the per capita calorie intake was from grains, of which 57 % was from maize and 32 % from wheat. Poorer consumers are more dependent on maize and wheat and their ability to afford them is more sensitive to their prices than higher income consumers. The unusually violent and deadly worker riots at platinum mines starting in August of 2012 [8, 18] coincided both with record global maize prices and record high prices for basic food items in South Africa. Figure 14.4 shows the Consumer Price Index for bread and cereals for South Africa since 2002 [28] and the date of the first deadly riot in the mines.

The riots have been attributed to labor issues and corruption in both the government and union organizations [18]. However, worker demands include significant wage increases amid claims that they are being paid "hunger wages" that do not cover basic necessities for their families [5, 6], and strikes have stopped where wage increases were granted [27]. Moreover, while labor strikes are not uncommon in South Africa, these events have been widely considered to be the most violent in the country since the end of apartheid in 1994 [18, 20]. This is consistent with the view that when people are unable to feed themselves and their families, desperation leads to social unrest [16]. Previously, xenophobic riots in May of 2008 stood out as the bloodiest violence since apartheid [3]. These riots coincided with food riots around the world during a previous peak of global food prices. The riots have been attributed to anger about foreigners competing for limited resources—an anger which would be exacerbated by high food prices.

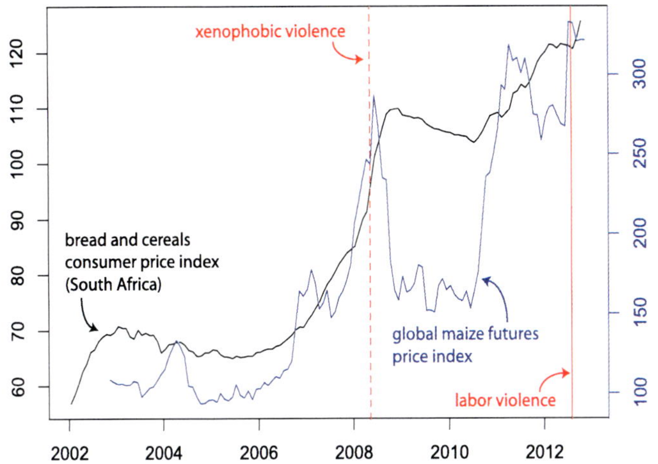


**Fig. 14.4** Consumer price index for bread and cereals in South Africa since 2002 (*black, left axis*) [28] and global maize prices (*blue, right axis*). *Red solid vertical line* indicates beginning of deadly riots in platinum mines, and *red dashed line* indicates period of severe xenophobic riots. Local prices have increased with global prices but have not correspondingly decreased when prices declined

Food prices in South Africa have doubled since 2006 and the increases are strongly associated with observed periods of extreme violence. Our current results, as well as recent news of food-related protests in Haiti [22, 23, 26] and Argentina [19, 24], combined with our previous analysis of the role of food prices in food riots and the Arab Spring point to the importance of food prices in social unrest worldwide and the suffering of poor populations. The relatively constant prices during the fall of 2012, and the comparatively low prices of rice should not keep policymakers from recognizing that food prices are historically high. The deregulation of commodity futures markets and the diversion of almost 50 % of the US maize crop to ethanol are ill-advised policies that should be changed. Awareness of the impact of global food prices should also influence countries around the world to mitigate the impact on poor populations, including workers at the low end of the wage scale.

**Acknowledgments** We thank Peter Timmer, Dominic Albino, Karla Bertrand, Urbano França, and Clare Fitzgerald for helpful comments on the manuscript.

## References

1. Albino, D. K., Bertrand, K. Z., & Bar-Yam, Y. (2012). *Food for fuel: The price of ethanol*. arXiv:1210.6080 [q-fin.GN]. http://necsi.edu/research/social/foodprices/foodforfuel/.
2. Albino, D. K., Bertrand, K. Z., & Bar-Yam, Y. (2012, September 28). *Food FAQ*. Tech. Rep. no. 2012-09-01. New England Complex Systems Institute. http://necsi.edu/research/social/foodprices/foodfaq/.
3. Bearak, B., & Dugger, C. W. (2008, May 20). South africans take out rage on immigrants. *NY Times*. http://www.nytimes.com/2008/05/20/world/africa/20safrica.html.

4. Bertrand, K. Z., Lindsay, G., & Bar-Yam, Y. (2012, September 28). Necsi food briefing. *New England Complex Systems Institute*. http://necsi.edu/research/social/foodprices/briefing/.
5. Corcoran, B. (2012, November 16). S African farm workers clash with police in wages protest. *The Irish Times*. http://www.irishtimes.com/newspaper/world/2012/1116/1224326666923.html.
6. de Waal, M. (2012, November 08). South africa: Striking workers burn vineyards in protest over 'hunger wages'. *The Guardian*. http://www.guardian.co.uk/world/2012/nov/08/south-africa-strike-vineyards.
7. Food and Agriculture Organization. (2009). Food balance sheets 2009. *FAOSTAT*. http://faostat3.fao.org/home/.
8. Freed, J. (2012 August 02). Riot leaves 3 dead at aquarius platinum's mine. *Financial Review*. http://beta.afr.com/p/business/companies/riot_leaves_dead_at_aquarius_platinum_RhaRRngij0h0npnYXGnOQK.
9. Gros, A., Gard-Murray, A., & Bar-Yam, Y. (2012). Conflict in yemen: From ethnic fighting to food riots. arXiv:1207.5778 [physics.soc-ph]. http://necsi.edu/research/social/yemen/.
10. Herdadelen, A., Gros, A., & Bar-Yam, Y. (2011, August 23). *Global grain consumption map*. Tech. Rep. no. 2011-08-03. New England Complex Systems Institute. http://necsi.edu/research/economics/grainconsumption.html.
11. Index Mundi. http://www.indexmundi.com/commodities/ Retrieved 13.12.12.
12. Jamasmie, C. (2012 December 20). Labour unrest forces harmony gold to close south african mine. *Mining.com*. http://www.mining.com/labour-unrest-forces-harmony-gold-to-close-south-%african-mine-67655/.
13. Lagi, M., Bar-Yam, Y., & Bar-Yam, Y. (2012, July 23). Update July 2012 the food crises: The us drought. arXiv: 1209.6376 [physics.soc-ph]. http://necsi.edu/research/social/foodprices/updatejuly2012/.
14. Lagi, M., Bar-Yam, Y., Bertrand, K., & Bar-Yam, Y. (2011). The food crises: A quantitative model of food prices including speculators and ethanol conversion. arXiv:1109.4859 [q-fin.GN]. http://necsi.edu/research/social/foodprices.html.
15. Lagi, M., Bar-Yam, Y. Bertrand, K., & Bar-Yam, Y. (2012, March 6). Update february 2012 the food crises: Predictive validation of a quantitative model of food prices including speculators and ethanol conversion. arXiv:1203.1313 [physics.soc-ph]. http://necsi.edu/research/social/foodprices/update/.
16. Lagi, M. Bertrand, K., & Bar-Yam, Y. (2011). The food crises and political instability in North Africa and the Middle East. arXiv:1108.2455v1 [physics.soc-ph]. http://necsi.edu/research/social/foodcrises.html.
17. Lagi, M., Gard-Murray, A., & Bar-Yam, Y. (2012). *Impact of ethanol conversion and speculation on mexico corn imports*. Tech. Rep. no. 2012-05-01. New England Complex Systems Institute.http://necsi.edu/research/social/foodprices/mexico/.
18. Lakmidas, S. (2012, November 14). Analysis: South African mining grapples with decline of union num. *Reuters*. http://www.reuters.com/article/2012/11/14/us-safrica-mining-idUSBRE8AD16S20121114.
19. Lifschitz, A., Popper, H., & Allen, V. (2012, December 21). Two killed in supermarkets looting in Argentina. *Reuters*. http://www.reuters.com/article/2012/12/21/us-argentina-violence-idUSBRE8BK0TZ20121221.
20. McClenaghan, M. (2012, October 18). South African massacre was the tip of an iceberg. *Bureau of Investigative Journalism*. http://www.thebureauinvestigates.com/2012/10/18/south-african-massacre-was-the-tip-of-an-iceberg/.
21. Mokhema, T. (2012, November 06). South Africa vineyards burned for second day protests grow. *Bloomberg Businessweek*. http://www.businessweek.com/news/2012-11-06/south-africa-vineyards-burn%ed-for-second-day-protests-grow.
22. Morrison, J. (2012, October 07). Embassy row: Unrest in haiti. *Washington Times*. http://www.washingtontimes.com/news/2012/oct/7/embassy-row-unrest-in-haiti/.
23. Placide, R. P. (2012, September 19). Soaring prices of basic commodities in haitian street markets. *Haiti Chery*. http://www.dadychery.org/2012/08/14/soaring-food-prices-in-haiti/.

24. President resigns after riots leave 22 dead in Argentina. (2012, December 28). *DailyMail*. http://www.dailymail.co.uk/news/article-91169/President-resigns-riots-l%eave-22-dead-Argentina.html.
25. Riot police open fire on south african miners. (2012, August 17). *ABC News/Reuters*. http://www.abc.net.au/news/2012-08-17/south-africa-miners-killed/4204582.
26. Several thousand people protest haiti gov't. (2012, September 30). *AP Newswire*. http://www.huffingtonpost.com/huff-wires/20120930/cb-haiti-protest/.
27. South Africa police fire at Marikana mine protesters. (2012, September 19). *BBC*. http://www.bbc.co.uk/news/world-africa-19646264.
28. Statistics South Africa. http://www.statssa.gov.za/timeseriesdata/timeseriesdata.asp. Retrieved 13.12.12.
29. Timmer, C. P. (October, 2008). Causes of high food prices. *Asian Development Bank, 128.*
30. Timmer, P. (August, 2012). Personal correspondence.
31. U.S. Department of Agriculture. http://fas.usda.gov. Retrieved 23.07.11.

# Chapter 15
# Conflict in Yemen: From Ethnic Fighting to Food Riots

**Andreas Gros, Alexander S. Gard-Murray, and Yaneer Bar-Yam**

**Overview** Yemen is considered a global terrorist base for Al-Qaeda and in recent years rampant violence is threatening social order. Here we show that the socio-economic origins of violence recently changed. Prior to 2008, violence can be attributed to inter-group conflict between ethnically and religiously distinct groups. Starting in 2008, increasing global food prices triggered a new wave of violence that spread to the endemically poor southern region with demands for government change and economic concessions. This violence shares its origins with many other food riots and the more recent Arab Spring. The loss of social order and the opportunities for terror organizations can be best addressed by directly eliminating the causes of violence. Inter-group violence can be addressed by delineating within-country provinces for local autonomy of ethnic and religious groups. The impact of food prices can be alleviated by direct food price interventions, or by addressing the root causes of global food price increases in US policies that have promoted conversion of corn to ethanol and commodity speculation. Addressing the food prices is the most urgent concern as a new bubble in food prices has been projected to begin before the end of 2012.

Violence has been common in Yemen since the founding of a modern state in the southern Arabian peninsula nearly 100 years ago [1, 2], but recent attacks and social disruption are particularly severe [3–5]. Yemen is one of the global bases of Al-Qaeda [6–10], with the attacks on the USS Cole in 2000 and American Embassy in 2008 as the most well-known local incidents. Yemeni Al-Qaeda has also been involved in global terror activities including an alleged attempt to bomb a

A. Gros • A.S. Gard-Murray • Y. Bar-Yam (✉)
New England Complex Systems Institute, Cambridge, MA, USA
e-mail: yaneer@necsi.edu



P.V. Fellman et al. (eds.), *Conflict and Complexity*, Understanding Complex Systems,
DOI 10.1007/978-1-4939-1705-1_15

Detroit-bound plane on Christmas Day of 2009. At the beginning of that year, Yemeni Al-Queda joined with the smaller Saudi Al-Queda to form Al Queda of the Arabian Peninsula (AQAP). The current social disruption increases concerns that Yemen may become an even stronger terrorist base, threatening security worldwide.

Here we show that the nature of violence in Yemen has changed between 2005 and 2011 from being ethnically and religiously based to being dominated by the effects of increases in food prices on an impoverished population. During the period of ethnic and religious inter-group violence, geographical locations of incidents are consistent with a theory that predicts areas of violence based upon the geographical composition of the population [11, 12], building on a tradition of geographic analysis in social science [13–15]. In contrast, the later period of violence begins at the time of globally increasing food prices in 2007, and spread from areas of ethnic conflict in the north to the endemically poor southern part of Yemen.

Our results have immediate implications for strategies to reduce violence and limit the growth of terrorist influence. Rather than direct military and security operations, effective interventions may require eliminating the primary economic and social drivers of violence. First, the immediate economic drivers can be relieved by addressing the problems of the global food market, which has been implicated more broadly in the revolutions in North Africa and the Middle East [16–18]. Second, by providing partial internal autonomy to ethnic and religious groups, the origins of the longer term inter-group violence can also be alleviated [12].

The interventions we identify would not only address the growing security risks but also improve the living conditions of millions of people, reducing severe poverty, social disruption, and endemic violence. Indeed, these two goals are directly linked as the social and economic conditions are the origins of social disorder, in whose shadow terrorist activities can grow.

Our analysis of violence in Yemen begins from an understanding of the role of geography in conflict between distinct self-identifying groups defined by properties like ancestry, culture, language, and religion [11, 12]. In this paper we use the term ethnic violence to describe this kind of inter-group conflict. Such violence is typically though not exclusively directed against or by civilians. When self-identifying groups are either sufficiently separated or sufficiently well-mixed, violence is unlikely. Separation limits inter-group friction, while integration inhibits inter-group alienation. Ethnic violence occurs most frequently in areas that have a certain intermediate degree of population separation, but in which control of the area is not separated accordingly, i.e. neither political nor physical boundaries exist to allow for local autonomy. In places where self-identifying groups separate into geographical patches of a critical size, in the range of 20–60 km, a group is able to impose its cultural norms, religious values, language differences, and in-group social signaling within public spaces. These spaces may include public squares, markets, restaurants, places of worship, and schools. However, when social expectations are violated because of the proximity of other ethnic domains, the resulting friction is likely to cause radicalization of some members of the population. Even a small radicalized minority is enough to lead to endemic conflict, and the propensity for violence becomes high. The violence may engage political and military components. Still, the origin of the conflict in the self-identity of the groups is likely to be manifest in violence directed against those who are not politically or militarily powerful. For patches larger than the critical geographical size individuals remain largely within their own domains and de facto local sovereignty exists. If patches are smaller than the critical size, ethnic groups cannot impose their own norms and expectations about behavior in public spaces, allowing for the peaceful coexistence of the multiple ethnic groups that are present. Natural and political boundaries can increase autonomy to allow for separation that can prevent violence in areas where it would otherwise occur. Tests of ethnic violence in various parts of the world have indeed shown that ethnic violence occurs in the vicinity of patches of a critical size without well-defined boundaries [11, 12].

In contrast to ethnic violence, social unrest reflecting socio-economic despair is often directed against authorities that fail to satisfy the most basic needs of the population, especially available or affordable food. Indeed, the relationship between food prices and social unrest has been demonstrated [16, 19–22]. Food riots around the world in 2007–2008 and 2010–2011 were triggered by steep increases in food prices. Since Yemen is one of the poorest countries in the Arab World [23], increases in food prices severely impact a large portion of the population [24]. According to the World Bank's 2007 Poverty Assessment Report, 35 % of the country's population is classified as poor [25].

In order to perform a more detailed and quantitative analysis we start by considering the ethnic geography of Yemen. There are four commonly described self-identifying ethnic and religious groups in Yemen: Zaydi Shiites, Ismaili Shiites, mainstream (Shafi'i) Sunnis, and Salafi (Wahhabi) Sunnis. Together these groups are estimated to represent 99 % of the population (55 % Sunni and about 44 % Shiite) [26]. Yemen's societal structure has a strong tribal aspect, especially in rural areas [1, 27]. While neither tribal allegiances nor political attitudes necessarily align with their members' religious denomination [6], it is nevertheless reasonable to assume as a first approximation that conflict arises between self-identified ethnic and religious groups. Obtaining data about the geographical distribution of these groups is difficult as there is no direct census and the distribution has changed in recent times, especially due to the spread of Salafism [28]. Moreover, since political and religious affiliations may be linked, various movements including the Moslem Brotherhood may have both political and religious connotations. For our analysis of ethnicity and violence we use spatial demographic data from 2004 [29] to identify the populated areas and an approximate map of the spatial distribution of the four major groups in 2000 obtained from a compilation of sources [30] to identify ethnic compositions, as shown in Fig. 15.1. The approximate nature of the available data limits the precision of the calculations we perform. Demographic dynamics, specifically the spread of Salafism in recent years changed the sectarian associations across Yemen. Our conclusions only depend on very general features of spatial geography, specifically the presence of groups of a given geographic size in a region of the country. The conclusions are therefore robust to all but very specific localized changes relative to surrounding areas. This is a strength of our method, especially in

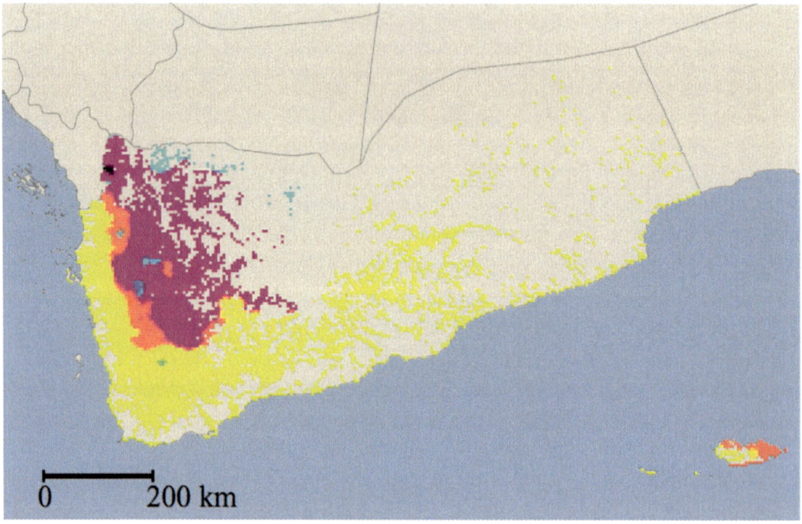


**Fig. 15.1** Estimated spatial distribution of major self-identifying groups in Yemen in 2000 [30]; Zaydi Shiite: *magenta*, mainstream Sunni: *yellow*, Ismaili: *cyan*, Salafi: *black*

application to areas where data is poor and changes are ongoing. Data on violent incidents was obtained from the Worldwide Incident Tracking System (WITS) [31] from which we selected the incidents that involved civilian casualties.

We calculated the propensity for violence in any given populated area by identifying patches of ethnic groups of a critical size of 56 km. This size is consistent with the value that provides predictive success in other countries [11, 12]. Mathematically we use a wavelet filter [11] that weighs the presence of ethnic types in a circular area around a focal point against the presence of ethnic types in the surrounding area. If ethnic types are well mixed or the whole area is populated uniformly by only one type, the output of the filter is small. However, if the inner area is populated by a different type than the surrounding area, forming an ethnic island or peninsula, the output will be high. We perform this analysis for focal points on a fine regular mesh throughout Yemen with results shown in Fig. 15.2. Actual incidents of violence involving civilians are indicated for each year from 2005 through 2011.

We quantify the level of agreement between our prediction and the data on violent incidents by correlating maps of shortest distances to locations of violent incidents and locations of predicted violence. We calculate the distance to the closest violent incident and the closest location of predicted violence at every point on the spatial mesh for a given year. We consider a location of predicted violence to be any point where the violence potential is above a threshold of 0.48 (the average propensity to violence plus two standard deviations).

We performed the analysis for the western part of Yemen, in which most of Yemen's population resides (see vertical dashed line in Fig. 15.2). Figure 15.3 shows

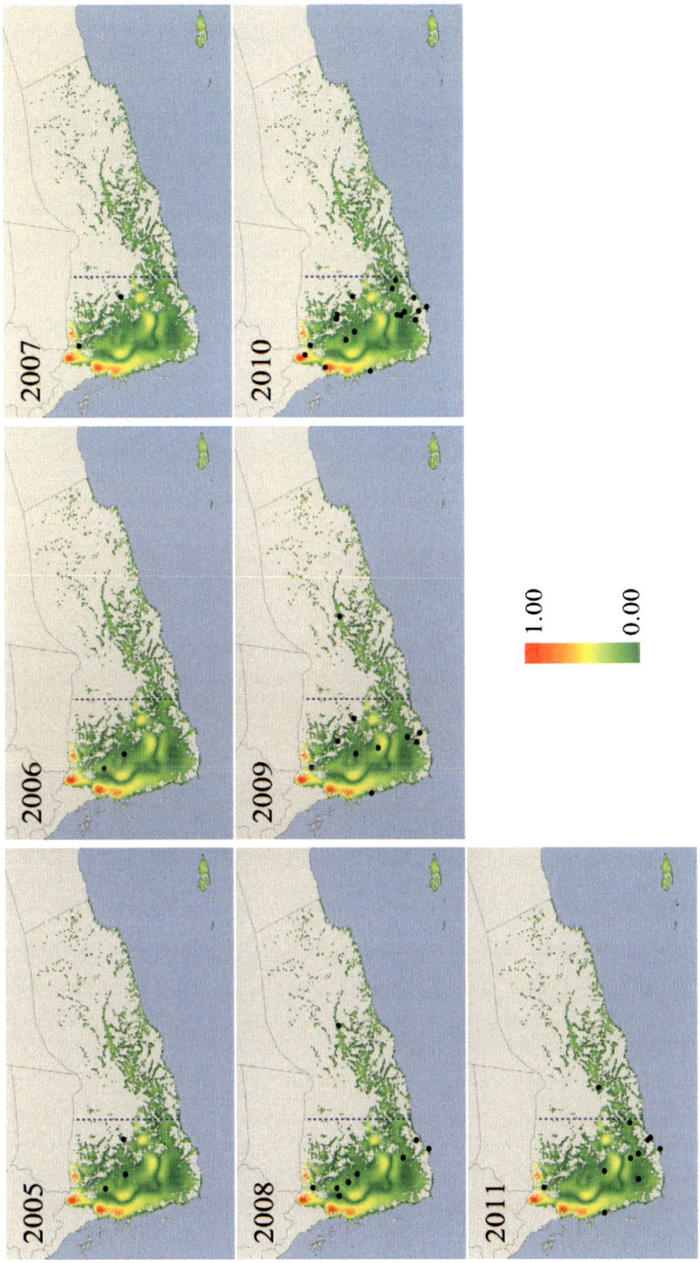


**Fig. 15.2** Propensity for violence (*color bar*) and incidents of violence (*black dots*) in populated areas of Yemen. *Dashed vertical line* delineates the western part of Yemen that we consider in the correlation analysis (Fig. 15.3). Much of the area to the east has a population density of less than 1 person per square km

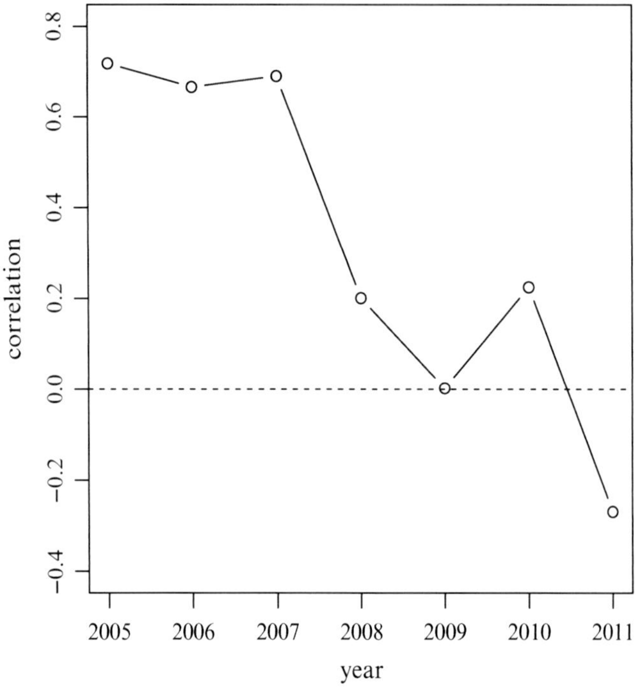


**Fig. 15.3** Correlations between shortest distances to locations of predicted and actual violence; the confidence values for the years 2005, 2006, and 2007 are 98.00 %, 96.58 %, and 99.23 %, respectively

the correlations between shortest distances to actual and predicted locations of ethnic violence in the west over time. The correlation values are approximately 0.70 for 2005, 2006, and 2007, and drop to between $-0.2$ and 0.20 for 2008 through 2011, showing a distinct shift away from ethnically motivated conflict. The confidence values for the correlations in years 2005, 2006, and 2007 are 98 %, 97 %, and 99 %, respectively. We calculate confidence intervals using 100,000 trials with random placement of the same number of predicted locations of violence within the western part of Yemen and compare the correlations between the corresponding maps of shortest distances. The correlation values for the years 2005, 2006, and 2007 are lower than reported in previous studies [11, 12], perhaps due to the limitations of the geographic ethnic data and reporting of incidents in the WITS for these years. However, the confidence values for 2005, 2006, and 2007 are still well above 95 %. Our results are consistent with reports that ethnic violence plays a significant role in Yemen resulting in the deaths of more than 2,000 people annually [32]. Some violence is politicized in the form of the Houthi rebellion, but it also has sectarian roots and manifests in violence against civilians in a manner characteristic of ethnic conflict [33–35].

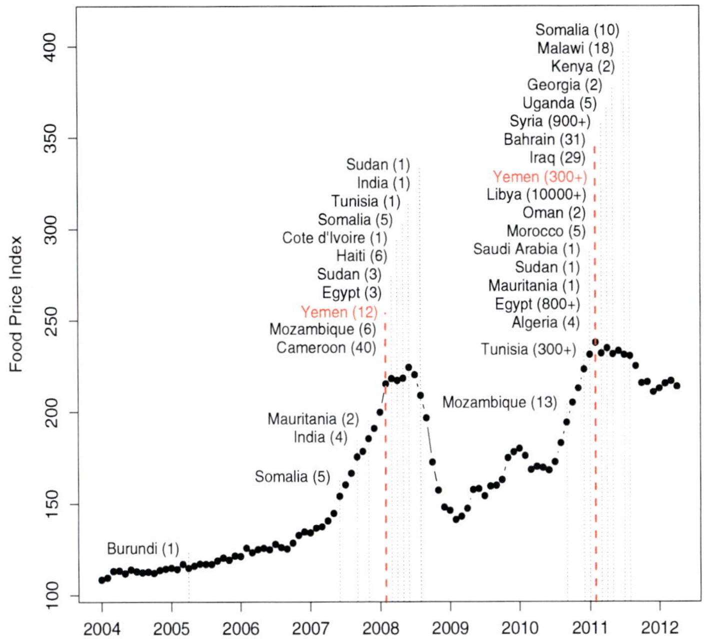


**Fig. 15.4** Global food price index and the occurrence of food riots (number of casualties in *brackets*); food riots in Yemen are marked *red*; reproduced with permission from [16]

The marked drop in correlations after 2007 indicates that the nature of the conflict changed and was no longer solely ethnically motivated. In order to identify the origins of violence after 2007 we turn to an understanding of social unrest in which food prices are a key component [16]. Figure 15.4 shows the global Food Price Index over time and the occurrence of food riots and revolutions associated with the Arab Spring [16]. The dates of the food riots in Yemen in early 2008 and 2011 are marked in red and coincide with similar events in many other countries. The co-occurrence of global food riots with large spikes in food prices is consistent with a causal role of food prices in social unrest. (An alternative hypothesis positing that the spread of Salafism caused the violence in the south is not supported by direct analysis, indicating they were not particularly involved [28].) Figure 15.4 shows that food riots in 2007–2008 and 2010–2011 were not a local phenomenon, but affected a broad spectrum of regions in Africa and the Middle East as well as Haiti and India. We can therefore understand the appearance of social unrest at these times based upon a hypothesis that widespread unrest is not necessarily related to governmental activities, or, in the case of Yemen, to terrorist actions. Instead, social unrest is induced by the government's perceived failure to provide food security to the population [16]. The poverty prevalent in Yemen [3], and southern Yemen's dependency on imported wheat [6, 24, 36], similar to Egypt and Tunisia [37], in combination with rising food prices, are very likely to have been the underlying trigger for violent incidents in 2008 and later.

Geographically, the violence in 2008 expanded from the north to the south. The southern violence can be understood from the recent political and economic history of Yemen. From 1967 through 1990 South Yemen existed as an independent state. The separation between North and South Yemen is partially, but not completely, according to ethnic regions. After unification in 1990, the north dominated and the south was economically marginalized. Ownership of resources was transferred to northern individuals and organizations [38–40]. The corresponding political disaffection manifested in a brief civil war in 1994. In 2007, during the first food price peak, political discontent coalesced into the Southern Movement, which was reenergized by food riots in 2010 to demand a wide range of economic and social concessions [41, 42]. While the expansion of violence to the south is a key change, poverty is also widespread in the north. An increase in violent activity in 2008 and 2010 in the north can be attributed to food based riots overlaid upon the preexisting ethnic conflict. Similar to other countries associated with recent revolutions in North Africa and the Middle East, the unrest based on food riots developed into a broader revolutionary process based upon persistent economic and social conditions, with implications for both local political instability and global terror.

Our work has identified two major sources of violence in Yemen: partial ethnic separation with poorly defined boundaries and unreliable food security for a vulnerable population. These conclusions have direct implications for policy.

The most urgent socioeconomic problem driving violence is high food prices. Recent work has shown that there is likely to be another food price bubble by the end of 2012 [43]. Based on this prediction conditions in Yemen will deteriorate if no mediating policy changes take effect to lower food prices. Current political efforts to broaden the governmental basis through assembling a National Dialogue Conference aim to tackle political grievances but do not address the problems of food prices. The most direct method to achieve food price stability is to provide subsidies as have been implemented in many countries in the face of the inability of the population to afford available food. Such subsidies are, however, difficult to afford for impoverished countries and would require external financing. More fundamentally, while many different factors have been considered for causing the rise of global food prices, a quantitative analysis has shown that the drivers of food price increases originate in US agricultural policies that are affecting food prices globally. These include two distinct domains of domestic policy. The first is subsidies for corn-to-ethanol conversion, which resulted in growth over less than a decade from negligible rates to 40 % of the US corn crop being converted to ethanol [44]. More recently, concerns about their impact has led to the elimination of these subsidies as of December, 2011 [45]. However, regulations that specify the amount of ethanol to be produced continue [46] The second is the elimination of constraints on commodity speculation in 2000 [47], which led to rapid growth of speculative activity through commodity index funds that do not follow supply and demand, and result in speculative bubbles [48, 49]. The Commodity Future Trading Commission is in the process of reimposing constraints on commodity trading to avoid speculative bubbles [50, 51]. However, the market participants are seeking to dilute the impact of these new regulations [52–56] These examples show that increased attention to the impact of food prices and their role in global social unrest necessarily links global security planning and domestic agricultural policy.

The violence that is ethnic in nature could be dramatically reduced by increasing political independence by establishing internal country boundaries between ethnic groups [12]. The paradigmatic example of the use of internal political boundaries to successfully promote peace is that of the Cantons in Switzerland which were established to separate Catholic and Protestant populations at a time when conflict was prevalent. The success of this approach of internal autonomous regions can be considered a model for other areas of the world. The value of a federal system of governance to reduce the propensity for violence in Yemen has been recently suggested [57]. More political self-determination has been demanded by the Southern Movement [42] and would most likely be easier to implement than separate nations. One form of potential boundary is the implementation of road blocks, which are currently used by the government as well as by tribes [58], Al-Qaeda, and Ansar al-Sharia [59]. However, access control is met with hostility where the authority over group territories is not legitimized or established historically. Legitimizing partial autonomy in a context of central government power, in regions determined by the geographical distribution of the main ethnic groups, would be effective according to our analysis.

We have shown that science can directly analyze social disruption and violence and identify their causes, as well as provide recommendations about policy changes to mitigate them. Our framework enables us to consider violence within its socioeconomic context. Terrorist organizations proliferate in the power vacuum in countries in which the government is inherently unable to provide order. For the specific case of Yemen, insurgents benefit from and amplify existing social disruption as the government and military are caught up in conflicts stemming from food insecurity and ethnic differences. Food prices and ethnic conflict can be seen to play a direct role [60]. We recommend the implementation of jointly defined internal political boundaries, within which the different groups can enjoy a degree of self-determination, in addition to lowering food prices through within-country subsidies and global food policy actions, as these measures have the greatest chance of stabilizing Yemen.

**Acknowledgments** We thank Jeb Boone and Charles Schmitz for helpful comments on the manuscript, and Urbano França for editorial assistance. This work was supported in part by AFOSR under grant FA9550-09-1-0324.

## References

1. Dresch, P. (1989). *Tribes, government, and history in Yemen*. Oxford: Clarendon Press.
2. Dresch, P. (2000). *A history of modern Yemen*. Cambridge: Cambridge University Press.
3. Fattah, K. (2011). Yemen: A social intifada in a republic of sheikhs. *Middle East Policy*, *18*, 79.
4. Bernin, P. (2009). Yemen's hidden war. *Le Monde Diplomatique*. Translated by Robert Waterhouse.

5. Boucek, C. (2009). Yemen: Avoiding a downward spiral. *The Carnegie Papers.*
6. Katz, M. N. (2003). Breaking the Yemen - Al Qaeda connection. *Current History, 102*, 40.
7. Al Qaeda in the Arabian Peninsula. (2012). *New York Times.* http://topics.nytimes.com/top/reference/timestopics/organizations/a/al_qaeda_in_the_arabian_peninsula/.
8. Masters, J. (2012). *Al Qaeda in the Arabian Peninsula (Aqap).* New York: Council on Foreign Relations.
9. Phillips, S. (2007). *What comes next in Yemen? Al-Qaeda, the tribes, and state-building.* Washington, DC: Carnegie Endowment for International Peace.
10. Prados, A. B., & Sharp, J. M. (2007). *Yemen: Current conditions and U.S. relations.* Washington: Congressional Research Service.
11. Lim, M., Metzler, R., Bar-Yam, Y. (2007). Global pattern formation and ethnic/cultural violence. *Science, 317*, 1540.
12. Rutherford, A., Harmon, D., Werfel, J., Bar-Yam, S., Gard-Murray, A., Gros, A., et al. (2011). PLoS ONE 9(5): e95660. doi:10.1371/journal.pone.0095660 (May 21, 2014). Good fences: The importance of setting boundaries for peaceful coexistence. arXiv:1110.1409v1 [physics.soc-ph]. http://arxiv.org/abs/1110.1409.
13. Berry, B., & Marble, D. (1968). *Spatial analysis: A reader in statistical geography.* Englewood Cliffs, NJ: Prentice-Hall.
14. Chorley, R., & Haggett, P. (Eds.). (1967). *Models in geography.* London: Methuen.
15. Harvey, D. (1969). *Explanation in geography.* London: Edward Arnold.
16. Lagi, M., Bar-Yam, Y., Bertrand, K. Z., Bar-Yam, Y. (2011). The food crises: A quantitative model of food prices including speculators and ethanol conversion. arXiv:1109.4859v1 [q-fin. GN]. http://arxiv.org/abs/1109.4859.
17. Ciezadlo, A. (2011). *The New Arab Revolt: What happened, what it means, and what comes next.* New York: Council on Foreign Relations.
18. Brown, L. R. (2011). The new geopolitics of food. *Foreign Policy 186*, 54–63.
19. Arezki, R., & Brükner, M. (2011). Food prices and political instability. International Monetary Fund Working Paper.
20. Bellemare, M. F. (2011, June 28). Rising food prices, food price volatility, and political unrest. *Social Science Research Network.* http://dx.doi.org/10.2139/ssrn.1874101.
21. Bush, R. (2011). Food riots: Poverty, power, and protest. *Journal of Agrarian Change, 10*, 119.
22. Walton, J., & Seddon, D. (1994). *Free markets and food riots.* Oxford: Blackwell.
23. Central Intelligence Agency. (2011). *The world factbook.* Washington, DC: Central Intelligence Agency.
24. Yemen: Soaring food prices force more people below poverty line — WFP. (2008). Integrated Regional Information Networks.
25. *Yemen Poverty Assessment, Vol. I: Main Report.* (2007). The Government of Yemen, the World Bank, and the United Nations Development Program.
26. *Yemen.* (2007). Institut Européen de Recherche sur la Coopération Euro-Arabe (MEDEA). http://www.medea.be/en/countries/yemen/yemen/.
27. Manea, E. M. (1998). La tribu et l'Etat au Yémen. In *Islam et changement social* (pp. 205–218). Lausanne: Editions Payot. English translation at http://www.al-bab.com/yemen/soc/manea1.htm.
28. Bonnefoy, L. (2012). *Salafism in Yemen: Transnationalism and Religious Identity.* London: C. Hurst & Co. Publishers Ltd.
29. *Villages.* (2004). Yemen Ministry of Public Health & Population. http://www.mophp-ye.org/docs/Data/shp_villages_2004.zip.
30. Izady, M. (2011). *Yemen: Religious and ethnic composition.* http://gulf2000.columbia.edu/images/maps/Yemen_Ethno_Religious_lg.jpg.
31. *Worldwide Incidents Tracking System.* (2012). U.S. National Counterterrorism Center. https://wits.nctc.gov/FederalDiscoverWITS.
32. In-depth: Guns out of control: The continuing threat of small arms. (2006, 21 May). *IRIN* (21 May 2006). http://www.irinnews.org/InDepthMain.aspx?InDepthId=8&ReportId=41050.

33. Yemen: Sectarian tensions rise. (2010, November 30). *IRIN*. http://www.irinnews.org/Report/91245/YEMEN-Sectarian-tensions-rise.
34. Yemen: Over 40 killed in sectarian clashes. (2012, January 25). *IRIN*. http://www.irinnews.org/Report/94724/YEMEN-Over-40-killed-in-sectarian-clashes.
35. France-Presse, A. (2001, November, 27). Yemen: Over 40 killed in sectarian clashes. *Pakistan Today*. http://www.pakistantoday.com.pk/2011/11/27/news/foreign/20-killed-70-injur%ed-in-yemen-ethnic-violence/.
36. United States Department of Agriculture Foreign Agricultural Service. (2012). *Production, supply and distribution online*. http://www.fas.usda.gov/psdonline/.
37. North Africa Brief. (2011). *FAO/GIEWS Global Watch*.
38. *Yemeni Civil War (1990–1994)*. (2000). GlobalSecurity.org. http://www.globalsecurity.org/military/world/war/yemen1.htm.
39. Whitaker, B. (2009). *The birth of modern Yemen*. http://www.al-bab.com/yemen/birthofmodernyemen.
40. Fault lines: Tracking armed violence in Yemen. (2010). *Small arms survey issue brief*.
41. Is South Yemen preparing to declare independence? (2011). *TIME Magazine*http://www.time.com/time/world/article/0,8599,2081756,00.html.
42. Day, S. (2010). Yemen on the brink – The political challenge of Yemen's Southern Movement. *Middle East Program* (Vol. 108). Washington, DC: Carnegie Endowment for International Peace.
43. Lagi, M., Bar-Yam, Y., Bertrand, K. Z., & Bar-Yam, Y. (2012). Update 2012 – The food crises: Predictive validation of a quantitative model of food prices including speculators and ethanol conversion. arXiv:1203.1313v2 [physics.soc-ph]. http://arxiv.org/abs/1203.1313.
44. Production, Supply and Distribution. (2012). United States Department of Agriculture, Foreign Agricultural Service. http://www.fas.usda.gov/psdonline.
45. Pear, R. (2012). After three decades, tax credit for ethanol expires. *New York Times*. http://www.nytimes.com/2012/01/02/business/energy-environment/after-three-d%ecades-federal-tax-credit-for-ethanol-expires.html.
46. Federal and State Ethanol and Biodiesel Requirements. (2007). U.S. Energy Information Administration. http://205.254.135.7/oiaf/aeo/otheranalysis/ethanol.html.
47. Commodity Futures Modernization Act of 2000. (2000). (Pub. L. 106–554, Sec. 1(a)(5) [H.R. 5660], Dec. 21, 2000, 114 Stat. 2763, 2763A-365). U.S. Federal Government.
48. Kaufman, F. (2010). The food bubble: How Wall Street starved millions and got away with it. *Harper's Magazine, 321*, 27–34.
49. Ghosh, J. (2011). The unnatural coupling: Food and finance. *Journal of Agrarian Change, 10*, 72.
50. Position limits for futures and swaps. (2011). *Federal Register, 76*, 71626.
51. Patterson, S., & Trindle, J. (2011). CFTC raises bar on betting. *Wall Street Journal*.
52. Trindle, J. (2012). CFTC may change application of speculative limits. *Market Watch – The Wall Street Journal*. http://www.marketwatch.com/story/cftc-may-change-application-of-speculative-limits-2012-05-17.
53. Brush, S. (2012). CFTC proposes easing of Dodd-Frank speculation limit rules. *Bloomberg*. http://www.bloomberg.com/news/2012-05-18/cftc-proposes-easing-of-dodd-frank-speculation-limit-rules.html.
54. Speculative Position Limits. (2011). *Futures Industry Association*.
55. Donahue, C. S. (2010, January 26). Re: CFTC Proposed Rulemaking on "Federal Speculative Position Limits for Referenced Energy Contracts and Associated Regulations," 75 Fed. Reg. 4144. *CME Group*.
56. Damgard, J. M. (2011). Re: Position Limits for Derivatives (RIN 3038-AD15 and 3038-AD16). *Futures Industry Association*.
57. The roots of Yemen's uprising: Interview with Khaled Fattah. (2011, July 31). *Studies in Ethnicity and Nationalism*. http://senjournal.wordpress.com/2011/07/31/the-roots-of-yemen%E2%80%99s-uprising-interview-with-khaled-fattah/.

58. Dawood, A. (2012). Bani Hushaish tribesment block Sana'a–Mareb road. *Yemen Times*. http://www.yementimes.com/en/1561/news/673/Bani-Hushaish-tribesmen-block-Sana%E2%80%99a-Mareb-Road.htm.
59. Qaed, S., & Anees, A. (2012). Different armed groups govern Aden. *Yemen Times*. http://yementimes.com/en/1564/report/718/Different-armed-groups-govern-Aden.htm.
60. Arc of Convergence: AQAP, Ansar al-Sharia and the Struggle for Yemen. (2012, June 21). *CTC Sentinel*. http://christopher-swift.com/publications/arc-of-convergence-aqap-ansar-al-sharia-and-the-struggle-for-yemen.

# Chapter 16
# Complexity and the Limits of Revolution: What Will Happen to the Arab Spring?

**Alexander S. Gard-Murray and Yaneer Bar-Yam**

**Overview** The recent social unrest across the Middle East and North Africa has deposed dictators who had ruled for decades. While the events have been hailed as an "Arab Spring" by those who hope that repressive autocracies will be replaced by democracies, what sort of regimes will eventually emerge from the crisis remains far from certain. Here we provide a complex systems framework, validated by historical precedent, to help answer this question. We describe the dynamics of governmental change as an evolutionary process similar to biological evolution, in which complex organizations gradually arise by replication, variation, and competitive selection. Different kinds of governments, however, have differing levels of complexity. Democracies must be more systemically complex than autocracies because of their need to incorporate large numbers of people in decision-making. This difference has important implications for the relative robustness of democratic and autocratic governments after revolutions. Revolutions may disrupt existing evolved complexity, limiting the potential for building more complex structures quickly. Insofar as systemic complexity is reduced by revolution, democracy is harder to create in the wake of unrest than autocracy. Applying this analysis to the Middle East and North Africa, we infer that in the absence of stable institutions or external assistance, new governments are in danger of facing increasingly insurmountable challenges and reverting to autocracy.

Revolutions can greatly alter societies. Learning about their potential outcomes is important for both participants and policymakers. If we can identify patterns across all revolutions, past unrest may inform our understanding of present crises. Doing so is especially critical as the upheaval across the Middle East and North Africa continues to unfold. For a revolution to be successful it must do more than depose or alter the current government; it must also create a new government in line with

A.S. Gard-Murray • Y. Bar-Yam (✉)
New England Complex Systems Institute, Cambridge, MA, USA
e-mail: yaneer@necsi.edu



P.V. Fellman et al. (eds.), *Conflict and Complexity*, Understanding Complex Systems,
DOI 10.1007/978-1-4939-1705-1_16

the intentions of the revolutionaries. A framework that can provide insight into outcomes is therefore necessary in order to evaluate whether a given revolution is likely to be successful.

The literature on revolutions has dealt with a wide variety of revolutionary events and outcomes. The conditions cited to explain the outcomes of revolution are themselves very diverse and include, among others, the effect of violence on leader selection [1, 2], the dynamics of large organizations [3], the slow nature of state-building [4], the rational decisions of participants [5], elite incentives [6], and leader characteristics [7]. The definition of revolution ranges widely, from great "social" revolutions that reshape society [8] to lesser "political" ones that only change leadership [9]. The diversity of definitions and the range of conditions and factors affecting revolutionary outcomes make it difficult to imagine a general theory that could identify the consequences of present unrest.

As broad and deep as this literature is, it has been criticized by the sociologist Charles Tilly for failing to treat revolutions as "complex but lawful phenomena" such as floods or traffic jams. Rather than sui generis events, he argues that they are part of a spectrum of social change, with no natural boundaries isolating them from the course of human events. Thus, any definition considering revolutions to be ontologically distinctive is inherently limited. In this vein, the recent literature is concerned with the specifics of power of groups and institutions that are able to achieve long-range goals [10].

Here we construct a theory of governmental change from the perspective of complex systems, which can be used to explain and perhaps anticipate the outcomes of revolutions. By embedding social systems in the much broader context of complex systems, we are not limited by the available data on states to formulate hypotheses and frame relevant insights. The available societal data can then serve to validate the mapping of general mathematical principles onto social processes, rather than testing the principles themselves—just as the more commonly applied statistical methods are applicable to physical, biological, and social systems and must be used correctly in each case. The focus of our analysis is on the complex challenge of forming a functional government, and the implications of the difficulties in doing so in a revolutionary context. In this way, complexity theory may provide an additional perspective that may synthesize and extend existing theories about the role that violence, institutions, and time play in revolutions. Our theory's applications are not limited to any particular type or scale of revolution. Because it characterizes governmental change generally, it can be applied broadly, with special relevance for periods of rapid and disruptive change such as revolutions. This is helpful when assessing an unfolding event such as the Arab Spring, when it is still unclear whether an event will meet a particular definition of a revolution [11]. In this work we do not consider the causes of revolution themselves [12, 13].

The crux of our theory is that revolutions disrupt the complex web of dependencies within governments and between governments and other parts of society, making it more likely that simpler systems, such as autocracy, rather than more complex ones, like democracy (for the current discussion taken to mean representative democracy) will result. We provide empirical support for this theory using data on the outcomes of unrest and governmental changes during the period 1945–2000. In these events higher levels of disruptive violence result in greater incidence of autocratic outcomes. We also show how the sequence of events of widely studied violent revolutions are consistent with our theory, as they revert to autocracy after failing to achieve stable democracy. In evaluating the consistency of objectives of the revolutionaries or observers with likely outcomes, we conclude that whether the Arab Spring will produce democracies hinges on whether existing institutional assets are sufficiently robust to survive the revolutions intact and support the subsequent transition to democracy.

Modern governments are highly complex systems. Complex systems are formed out of many independently acting yet simultaneously interdependent parts. Governments are made up of many people interacting within institutional frameworks (e.g., bureaucrats, judges, and soldiers) equipped with resources such as money, buildings, and weapons, and regulated by laws, norms, procedures, and precedents. It is not merely the quantity of parts, but their interaction and interdependence which give rise to the complexity of the whole.

Highly complex functional systems do not arise spontaneously. In biology, it has long been recognized that complex organisms must evolve gradually over many generations from simpler ones; it is probabilistically impossible for random association to produce the necessary relationships between molecules. This constraint generally arises in large-scale complex systems such as ecologies, economies, and governments. Societal forms do not coincide with biological ones, and the dynamics of social change are not exactly the same as biological evolution. Nevertheless, the constraint that highly complex systems do not arise spontaneously applies to complex social organizations. They, too, accumulate structural and functional complexity incrementally as described by Darwin's theory of evolution [14].

The general conditions for an evolutionary process are replication with heredity and variation, and selection with competition [15]. These conditions can be found in the historical change of governments. Properties of governments are replicated, as new regimes inherit elements of the old and political systems are copied from one country to another; sometimes, they are imposed by force, other times they are adopted voluntarily. Innovations and composites of existing governmental forms lead to variation over time. The innovations are subject to selection through competition between countries. Whether this competition is military, economic, or cultural, national success makes a country's form of government more attractive as a model, as well as better able to impose itself on others via conquest or influence. On the other hand, military or economic failure makes a particular government vulnerable to replacement. This process corresponds to the way genes or traits are selected through the competition between lifeforms: Less fit governmental forms are more likely to fail, while more fit ones are more likely to succeed and spread. While competitive success may be described in economic and social terms, it is ultimately characterized by the ability to survive and replicate.

While evolution does not have an inherent objective, the process can result in the progressive development of higher levels of complexity developed from simpler forms [16]. We overcome some conceptual difficulties by formalizing complexity in terms of the description of systems at multiple scales [17]. Collective behaviors of components give rise to complex behaviors at larger scales, clarifying the connection between interdependence and complexity. Thus, new and more complex structures are necessarily dependent on the existence of prior, less complex ones [18]. To the extent that a more complex structure provides a survival advantage, it will be selected for.

Governments, like organisms, may be under evolutionary pressure that results in them becoming more complex. Governmental complexity must increase as the complexity of its environment—the society it governs and other states—increases, in order for Ashby's Law of Requisite Variety to be satisfied [19]. This principle states that in order for a system to survive in a complex environment, its control mechanism must be correspondingly complex. The demands and needs of defense, international relations, infrastructure, services, law and order, industrial oversight, and priorities of subpopulations challenge governments. The large variety of necessary actions must be matched by the ability of the government to act. Formally, the complexity of the task as measured by the variety of actions needed to be effective must be matched by the complexity of the organization as measured by the variety of actions that it can make in responding to changing conditions. A complex organization may not be effective, but an effective organization must be sufficiently complex.

A few notes of clarification are necessary. First, the process is not always monotonically increasing in complexity. Complexity can be lost as well as gained, a principle most dramatically demonstrated by historical cases of societal "collapse" [20]. That higher levels of complexity arise based upon previously existing lower levels of complexity does not mean that they always prevail in individual or species competitions. Single celled organisms are necessary for the creation of multicellular organisms, even though bacterial diseases can kill mammals. The fall of Rome might also be considered as a death of a large and more complex organism due to invasion by simpler ones, illustrating how the fate of a single organism, or type, is not ensured by evolutionary process. Second, while states are sometimes portrayed as being formally planned, the successful implementation of such plans must depend on an incremental accumulation of complexity. Constitutions can be written quickly, but if they call for institutions which are more complex than existing ones, they will take significant time to implement. Third, the process by which complexity increases does not necessarily occur at a steady rate. Complexity may increase more rapidly in one period than previously, as described by the theory of punctuated equilibrium in biological evolution [4, 21]. But even in cases of rapid increase, higher levels of complexity are still dependent on previously existing structures. Fourth, the analysis does not itself place a value on more or less complex organisms or governments, but does provide guidance about the conditions under which certain structures are likely to be successful, and thus the likelihood of achieving the objective if one or another is desired. Fifth, the coexistence of many different types of biological organisms suggests we similarly should not expect a single ultimate or ideal governmental form, and newer forms should continue to emerge.

The assertion that governmental complexity has increased via incremental evolution is well supported by history. Early political entities, dominated by personalistic hierarchies led by chiefs, kings, and emperors, were far simpler organizations than today's rational-bureaucratic states [22, 23]. Governments gradually increased their complexity by raising armies, instituting taxes and systems to collect them, building bureaucracies, and so on [4]. In recent times, the established geography of nations often constrains the evolution of governments in that new options cannot grow organically from smaller successful versions. Instead, variations must happen in situ. This is different from corporations and other social organizations. Both historically and structurally, the complexity of modern states is dependent on the institutions and institutional frameworks of earlier ones. This simplified picture, while necessarily concealing the chaotic and nonlinear nature of history, nevertheless captures the general trend towards building greater complexity on earlier foundations.

Revolutions inherently involve rapid change that breaks with this cumulative process [7]. Officials including soldiers, police, judges, and bureaucrats may be removed from office, and both their immediate contributions to government function and their experience may be lost. Laws may be abrogated. Different groups of individuals may become engaged in or empowered by new political arrangements. The more the public sees radical change as necessary, the more likely such disruption of the previous system will be seen as an improvement.

These changes may be considered positive in many regards. The overthrown regime may be corrupt, repressive, or incompetent. The rapid and violent overthrow of even the worst government, however, may still have negative consequences for the development of a new government. By overturning existing structures which could serve as the basis for future structures, revolutions can endanger the very transitions they aim to effect. To the extent that the accumulated complexity of the previous regime is disrupted in the revolution, any new government is forced to build itself without a foundation.

The ongoing needs of society require a functioning government. Moreover, if a new regime is to succeed it must be not merely as effective as the previous one, but given that the old regime's failings prompted a revolution, better. To that end, the post-revolutionary government must restore the lost complexity of essential structures quickly. Revolutions are often prompted by failures of the preexisting regime to meet the basic needs of its people—failures which are often manifest in social crises caused by high food prices, unemployment, or war. The social crisis must be addressed promptly in order for any new government to succeed and ward off further unrest. The time available for the government to organize itself and act is therefore short.

These conditions—the loss of complexity and the brief time available to restore it—constrain all post-revolutionary governments. But different forms of government are differentially complex, and less complex ones will fare better under these conditions.

Forms of government may be distinguished in many ways. Among the most important is to whom political authority is assigned. Autocracy and democracy represent two very distinct modes of assigning political authority, and the difference between them entails a difference in complexity. This difference in turn affects their robustness when established during a revolution.

An autocratic government is by its nature less complex than a democratic one. Autocracies do not require the extensive decision-making structures that are necessary to gather popular opinion, negotiate with other branches of government, or hold elections or votes in representative bodies. Incorporating many independently acting and interacting interests requires special institutions and procedures: legislatures and representatives, competitive elections and debates. An autocrat may choose to take these actions but they are not required by the form of government. Many authoritarian governments do not conform to the pure autocratic type. Rulers may share power with a group like a family, party, or military [24]. Arrangements like these should entail more political complexity than pure autocracy but need not have the level of complexity necessary for a representative democracy. Modern democracies also involve a multitude of non-governmental institutions such as civic associations and an independent media. Whatever form these institutions take, the need to distribute political authority among many individuals and the diverse groups of which they are a part constitutes an extra layer of complexity within democracy.

In applying our analysis to revolutions that aim to create democratic government we immediately identify the associated difficulties. Having gained power, a provisional government must address many questions in order to build its own institutions. There is no single democratic template. The many kinds of democracy—direct and representative, federal and centralized, presidential and parliamentary—reflect its complexity. While autocracies may choose embellishments, democracies must determine which of a large number of alternatives to implement. Every constitutional choice can become a subject of contention, since each decision might benefit one segment of the population over another.

Quite aside from the problem of structural transformation, the "complexity gap" between the institutions of the old regime, minus whatever was lost in the revolution, and those required for the new government to function, are liable to overwhelm attempts to create democracies. Even though more complex structures may be capable of responding to higher complexity challenges, the prospective benefits may be stymied by the challenge of creating those structures.

Given these theoretical constraints, we can hypothesize that disruptive revolutionary events will favor the development of autocracies over democracies, even when the impetus of the revolution itself is to create democracy. The degree to which autocracies are favored should increase when there is a greater disruption of the pre-existing governmental form. The level of disruption may be identified as a first approximation by the level of violence that takes place. The historical record of revolutionary outcomes supports these hypotheses.

In Fig. 16.1 we summarize data from 1945 to 2000. We track the changes in country regime type in the 10 years following revolutionary events. Events are separated into partially overlapping sets based on the particular definition and source used to gather the events. The different sets of events are clustered in three groups. Events that can be characterized as violent revolutions were more likely to produce

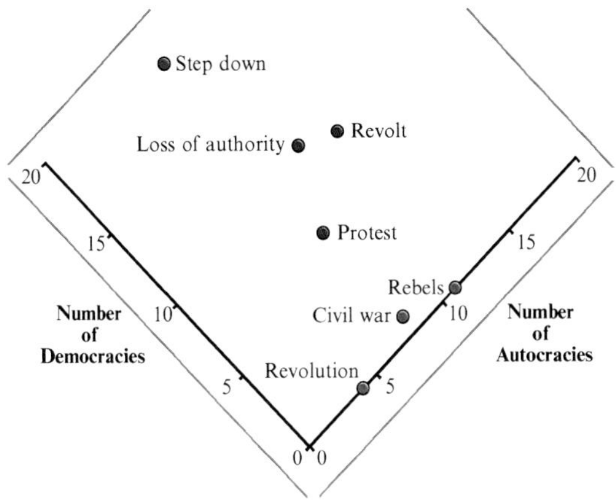


**Fig. 16.1** Historical outcomes of governmental changes during the period 1945–2000. Each *dot* indicates the number of democratic regimes and the number of autocratic regimes from different event collections: Step down, Revolt and Civil war [25]; Protests and Rebels [26]; Loss of authority [27]; and Revolution [28]. Collections of events with higher levels of organized violence and social disruption are indicated by *red dots*. Events characterized by mass protest or political turmoil without significant levels of violence are indicated by *purple dots*. Those in which a leader voluntarily stepped down are indicated by a *blue dot*. The more violent and disruptive events are more likely to result in autocracies. Regime type is obtained from the Polity project [28, 29], which characterizes historical governments with daily resolution using multiple coders [30]

autocracies. Mass protests and political turmoil were about equally likely to produce democracy as autocracy. More orderly transitions in which leaders voluntarily resigned were the most likely to produce democracy. This is consistent with the hypothesis that violence is more likely to produce autocracy than more peaceful transitions.

While the theory points to the role of disruptive violence itself as a culprit in the instability of democracy after revolution, the data may be consistent with initial conditions playing a causal role in both the extent of violence and the resultant form of governance. Still, the primary conclusion is the same—where violence takes place, the likelihood of autocratic over democratic outcomes is increased.

Though data prior to 1945 is less extensive, we can examine several earlier violent revolutions [9] in greater detail to show how a pattern of revolution, attempted democracy, failed government, and resurgent autocracy repeats itself. The pattern illustrates the difficulty disruptive transitions have in creating stable democratic institutions, even when such institutions are the desired outcome of the revolutionary activity. Through this discussion we also anchor the concepts of the theory in the process of revolution captured in historical narrative, enabling easier application to current events.

The French Revolution of 1789 spawned a series of attempts at representative governments: the National Constituent Assembly, the Legislative Assembly, the National Convention, the Directory. None of these deliberative bodies was able to achieve stability. Instead they were impeded by factional infighting and coups.

Dominant groups or leaders drifted towards authoritarianism and repression. Violence and upheaval continued until a military coup replaced the revolutionary governments with autocracy.

It would be easy to blame the failure of the post-revolutionary democratic assemblies and the triumph of autocracy on the ambition of a few autocratically inclined men like Maximilien Robespierre or Napoleon Bonaparte. But to do so would mask the underlying weakness of the system they subverted. The problem was not that there were no democrats—there was frequent opposition to the autocrats—but that democratic techniques failed to produce governments of sufficient effectiveness to stabilize the country.

The revolution had purged civil institutions of many experienced members, along with the army's aristocratic officer corps [2]. The new democratic governments battled civil strife and invaders as well as inflation and hunger, but lacked the apparatus their deposed predecessors possessed. Although elections were held, those elected were never able to hold power for long enough to build the kind of legitimacy that would have discouraged coups. Napoleon, having seized power, could rely on the army as a functioning organization to enable his rule. Whether his policies were good for France can be debated elsewhere; he was nonetheless able to establish an effective and stable government through which to pursue his aims.

Half a century later, Europe was engulfed by the Revolutions of 1848, whose spread mirrored today's Arab Spring in many respects. In France, the monarchy was again overthrown and replaced by a provisional government, which was again unable to resolve the divisions in French politics. Eventually the elected president, Louis-Napoleon Bonaparte (heir to the Emperor), returned the country to autocracy, which brought stability with relatively little popular resistance.

In the Russian Revolution of 1917, the Tsarist autocracy was replaced by a Provisional Government which proclaimed a republic. Democratic processes spread throughout the country, as worker's councils and professional and labor unions formed. But these institutions, while in line with the ideology of the new regime, also diluted its ability to control the state. The Provisional Government suffered from a series of internal crises and cabinet reshuffles, and was unable to prevent the emergence of a competing power center in the form of the Petrograd worker's council. The result was an uneasy system of "dual power," as the Provisional Government's hold on state institutions was gradually weakened in favor of the worker's council (or Soviet). Soon after, the Petrograd Soviet launched a second revolution, starting a civil war that ended 4 years later with a return to autocracy under the Communist Party.

From these examples a common pattern emerges. First, autocratic regimes are brought down by revolution and new democratic governments are proclaimed. But new legislatures and presidents struggle to meet the challenges before they and their constituents become increasingly dissatisfied. Factionalism emerges as different groups, reacting to and manifesting the weakness of emerging institutions, try to impose their will. Violence erupts, either between the government and the populace or between factions, as immature institutions fail to contain the conflict. Within a few years the new democracies are replaced by regimes generally just as autocratic as the ones recently overthrown, if not more so [2, 5, 31].

This process can be translated into the language of complexity: a revolution brings rapid change, displacing or destroying accumulated complexity. Subsequently, governments try to create democracy, but find the gap between their present and intended complexity is too great and are replaced by autocracy. The problem is not that democracy is flawed; the stability and prosperity of mature democratic states testifies to that. But the greater complexity of democracy means that it is harder to build when time is short and the foundations for more complex structures have been eroded or did not exist in the first place.

Not all revolutions end in autocracy. The extent of disruption plays a pivotal role. Particularly in the last several decades, several revolutions—in the Philippines in 1986, in South Korea in 1987, and across Eastern Europe in 1989—have managed to oust autocratic regimes and replace them with more or less stable democracies. These are not counter-examples, but rather demonstrations of the constructive side of our conclusions. In all of these nations, the revolutions were largely non-violent and forced out autocrats without overturning the apparatus of government. These revolutions also managed to take advantage of the existing store of complexity represented by previously powerless democratic elements, such as parliaments with only symbolic roles and tightly managed elections. Put in place by autocratic governments to provide the semblance of representational government, these powerless institutions later provided a basis for a democratic government without significant changes to their structure. Furthermore, these regimes all received external support during their critical early years. The retention of complexity in the form of previous government structures, combined with external support, allowed democracy to grow incrementally and stabilize.

Even in violent revolutions, maintaining existing complexity can help support post-revolutionary democracy. In the American Revolution, the groundwork for democracy was laid prior to 1776 in elected colonial assemblies. When the Articles of Confederation outlined a federal government for the new nation, they were able to build on functioning democratic foundations that had existed for decades. The responsibilities of the federal government could then grow over time, while the state governments continued to address the needs of their populations as they had done previously.

For today's revolutionaries in the Middle East, the complexity of governmental formation does not mean democracy is impossible, but it does mean that it will be very hard to create. Our analysis does not incorporate distinct cultural, historical, economic, and other conditions, but the fundamental principles should provide helpful insight. With regard to Egypt and Tunisia, many observers are concerned that the pace of democratization has been slow and the old regime remains essentially in place. But this path, while it risks the persistence of the old guard, may offer the best hope for a smooth transition. Preserving and gradually changing some of the existing structures of government may ultimately be more likely to succeed. Efforts to dramatically accelerate change may cause a reversion to the prior form of autocratic government, or extended social disorder.

Contrast this with Libya and Syria, where the revolutions have been far more violent, threatening the prospects for a smooth transition. Libya already faces emerging fragmentation within the National Transition Council. The conflict created many militias which may be competitors for power in the absence of a strong and accepted government. The same weapons which overthrew Muammar Qaddafi may now fuel violent conflict between revolutionary and ethnic factions. In Syria, severe repression means much of the old regime's army and bureaucracy will be too discredited to participate in a new government. Bashir Assad's determination to maintain power regardless of the level of civilian casualties, and the apparent loyalty of his security apparatus, means that if the revolution does succeed it will have to destroy much of the Assad regime's institutions, leaving little foundation on which to build a complex, democratic post-Assad government.

The preceding analysis has focused on the internal dynamics of new governments. External interventions, whether through aid or military action, have the potential to alter these dynamics. In Libya, NATO airstrikes contributed to the rebels' victory, though how external interventions might promote the formation of a stable government is less straightforward. Saudi Arabian military intervention in Bahrain stopped the incipient revolution there. Still, the potential of foreign involvement should not be overstated. The relevance of diverse local imperatives in shaping the creation of complex democratic institutions cautions against the external imposition of such institutions.

The likelihood of a persistent democracy can be increased if new internal disruptions are inhibited by external intervention. First, the political structure must be stabilized and economic stresses reduced, allowing governments time to develop. All such external interventions must, however, be carefully considered. Evolution is always shaped by the environment: national institutions that develop in the presence of external aid may become reliant on it. Thus, the form of assistance should be carefully designed to be like scaffolding—to be removed rather than to become integral to the functioning of the system [14].

A separate category of constructive intervention is the prevention of global conditions that increase stress on vulnerable countries. In this regard, high and volatile global food prices are a key ongoing culprit for social unrest and political instability that can and should be addressed [13].

Aside from such interventions, the fundamental dynamics of building complex systems like governments constrain the outcomes of revolutions. Even if external interventions do occur their success is far from assured. Governments and societies are evolving systems which develop over long periods of time. Stability can neither be assumed nor instantly restored after revolution. Recognizing the difficulties that revolutions create for post-revolutionary governments and societies may help guide our response to social unrest. Building a successful government begins, rather than ends, with the revolution.

**Acknowledgments** We thank Greg Lindsay, Karla Z. Bertrand, Dominic Albino, Urbano França, and Yavni Bar-Yam for editorial assistance, Lawrence E Susskind, Robert H. Bates, and Dietrich Rueschemeyer for helpful comments on the manuscript. This work was supported in part by AFOSR under grant FA9550-09-1-0324, ONR under grant N000140910516.

## References

1. Gurr, T. R. (1988). War, revolution, and the growth of the coercive state. *Comparative Political Studies*, *21*, 45.
2. Becker, J., & Goldstone, J. A. (2005). State development after revolutions - Rapid state building or transforming existing structures under pressure? In M. Lange & D. Rueschemeyer (Eds.), *States and development: Historical antecedents of stagnation and advance* (pp. 183–210). New York: Palgrave Macmillan.
3. Michels, R. (2001). *Political parties: A sociological study of the oligarchical tendencies of modern democracy*. Ontario: Batoche Books.
4. Rueschemeyer, D. (2005) Building states: Inherently a long-term process? An argument from theory. In M. Lange & D. Rueschemeyer (Eds.), *States and development: Historical antecedents of stagnation and advance* (pp. 143–164). New York: Palgrave Macmillan.
5. Weede, E., & Muller, E. N. (1997) Consequences of revolutions. *Rationality and Society*, *9*, 327 (1997).
6. Acemoglu, D., & Robinson, J. A. (2006). *Economic origins of dictatorship and democracy*. Cambridge: Cambridge University Press.
7. Foran, J., & Goodwin, J. (1993). Revolutionary outcomes in Iran and Nicaragua: Coalition fragmentation, war, and the limits of social transformation. *Theory and Society*, *22*, 209 (1993).
8. Skocpol, T. (1979). *States and social revolutions: A comparative analysis of France, Russia, and China*. New York: Cambridge University Press.
9. Goldstone, J. A. (1998). *The encyclopedia of political revolutions*. Washington, DC: Congressional Quarterly.
10. Rueschemeyer, D. (2010, October). On the state and prospects of comparative democratization research. *Comparative Democratization Section of the American Political Science Association (APSA-CD) Newsletter*, *8*, 1.
11. Stinchcombe, A. L. (1999). Ending revolutions and building new governments. *Annual Review of Political Science*, *2*, 49.
12. Goldstone, J. A., Bates, R. H., Epstein, D. L., Gurr, T. R., Lustik, M. B., Marshall, M. G., et al. (2010). A global model for forecasting political stability. *American Journal of Political Science*, *54*, 190.
13. Lagi, M., Bertrand, K. Z., & Bar-Yam, Y. (2011, August 10). *The food crises and political instability in North Africa and the Middle East*. arXiv:1108.2455.
14. Bar-Yam, Y. (2005). *Making things work: Solving complex problems in a complex world*. Cambridge, MA: NECSI Knowledge Press.
15. Lewontin, R. C. (1970) The units of selection. *Annual Review of Ecology and Systematics*, *1*, 1.
16. Bonner, J. T. (1988). *The evolution of complexity by means of natural selection*. Princeton, NJ: Princeton University Press.
17. Bar-Yam, Y. (1997). *Dynamics of complex systems*. Boulder, CO: Westview Press.
18. Bar Yam, Y. (2002). Complexity rising: From human beings to human civilization, a complexity profile. In *Encyclopedia of life support systems*. Oxford: UNESCO/EOLSS.
19. Ashby, W. R. (1956). *An introduction to cybernetics* (Chap. 11). London: Chapman & Hall.
20. Diamond, J. (2005). *Collapse: How societies choose to fail or succeed*. New York: Viking Penguin.
21. Goldstone, J. A. (2001). Toward a fourth generation of revolutionary theory. *Annual Review of Political Science*, *4*, 139.
22. Flannery, K. V. (1972). The cultural evolution of civilizations. *Annual Review of Ecology and Systematics*, *3*, 399.
23. Richerson, P. J., & Boyd, R. (1972). Complex societies: The evolutionary origins of a crude superorganism. *Human Nature*, *10*, 253.
24. Boix, C., & Svolik, M. (2011). The foundations of limited authoritarian government: Institutions and power-sharing in dictatorships, Working paper.

25. Svolik, M. (2009). Power-sharing and leadership dynamics in authoritarian regimes. *American Journal of Political Science, 53*, 477.
26. Goemans, H. E., Gleditsch, K. S., & Chiozza, G. (2009). Introducing Archigos: A data set of political leaders. *Journal of Peace Research, 46*, 269.
27. Marshall, M. G., & Marshall, D. R. (2011). Coup d'Etat events, 1946–2010: Codebook. Center for Systemic Peace.
28. Marshall, M. G., Jaggers, K., & Gurr, T. R. (2012). Polity IV project: Political regime characteristics and transitions, 1800–2010. Center for Systemic Peace.
29. Gleditsch, K. S. (2008). Modified polity P4 and P4D data, version 3.0. http://privatewww.essex.ac.uk/~ksg/polity.html.
30. Munck, G. L., & Verkuilen, J. (2002). Conceptualizing and measuring democracy: Evaluating alternative indices. *Comparative Political Studies, 35*, 5.
31. Huntington, S. P. (1968). *Political order in changing societies*. London: Yale University Press.

Made in the USA
Columbia, SC
29 January 2026

43ba0263-e407-436b-8d16-9988cbd9ebc4R02